S0-EJH-965

Convergent Evolution

Vienna Series in Theoretical Biology
Gerd B. Müller, Günter P. Wagner, and Werner Callebaut, editors

The Evolution of Cognition, edited by Cecilia Heyes and Ludwig Huber, 2000

Origination of Organismal Form: Beyond the Gene in Development and Evolutionary Biology, edited by Gerd B. Müller and Stuart A. Newman, 2003

Environment, Development, and Evolution: Toward a Synthesis, edited by Brian K. Hall, Roy D. Pearson, and Gerd B. Müller, 2004

Evolution of Communication Systems: A Comparative Approach, edited by D. Kimbrough Oller and Ulrike Greibel, 2004

Modularity: Understanding the Development and Evolution of Natural Complex Systems, edited by Werner Callebaut and Diego Rasskin-Gutman, 2005

Compositional Evolution: The Impact of Sex, Symbiosis, and Modularity on the Gradualist Framework of Evolution, by Richard A. Watson, 2006

Biological Emergences: Evolution by Natural Experiment, by Robert G. B. Reid, 2007

Modeling Biology: Structure, Behaviors, Evolution, edited by Manfred D. Laubichler and Gerd B. Müller, 2007

Evolution of Communicative Flexibility: Complexity, Creativity, and Adaptability in Human and Animal Communication, edited by Kimbrough D. Oller and Ulrike Greibel, 2008

Functions in Biological and Artificial Worlds: Comparative Philosophical Perspectives, edited by Ulrich Krohs and Peter Kroes, 2009

Innovation in Cultural Systems: Contributions from Evolutionary Anthropology, edited by Michael J. O'Brien and Stephen J. Shennan, 2009

The Major Transitions in Evolution Revisited, edited by Brett Calcott and Kim Sterelny, 2011

Transformations of Lamarckism: From Subtle Fluids to Molecular Biology, edited by Snait B. Gissis and Eva Jablonka, 2011

Convergent Evolution: Limited Forms Most Beautiful, by George R. McGhee Jr., 2011

Convergent Evolution

Limited Forms Most Beautiful

George R. McGhee Jr.

The MIT Press
Cambridge, Massachusetts
London, England

For information about special quantity discounts, please email special_sales@mitpress.mit.edu

This book was set in Syntax and Times Roman by Toppan Best-set Premedia Limited. Printed and bound in the United States of America.

Library of Congress Cataloging-in-Publication Data

McGhee, George R.
Convergent evolution : limited forms most beautiful / George R. McGhee Jr.
p. cm.
Includes bibliographical references and index.
ISBN 978-0-262-01642-1 (hardcover : alk. paper)
1. Convergence (Biology) I. Title.
QH373.M34 2011
576.8—dc22

2011007840

10 9 8 7 6 5 4 3 2 1

For Marae
’S òg a thug mi gràdh dhuit.

Contents

Series Foreword

Biology is becoming the leading science in this century. As in all other sciences, progress in biology depends on interactions between empirical research, theory building, and modeling. But whereas the techniques and methods of descriptive and experimental biology have evolved dramatically in recent years, generating a flood of highly detailed empirical data, the integration of these results into useful theoretical frameworks has lagged behind. Driven largely by pragmatic and technical considerations, research in biology continues to be less guided by theory than seems indicated. By promoting the formulation and discussion of new theoretical concepts in the bio-sciences, this series intends to help fill the gaps in our understanding of some of the major open questions of biology, such as the origin and organization of organismal form, the relationship between development and evolution, and the biological bases of cognition and mind.

Theoretical biology has important roots in the experimental biology movement of early-twentieth-century Vienna. Paul Weiss and Ludwig von Bertalanffy were among the first to use the term *theoretical biology* in a modern scientific context. In their understanding the subject was not limited to mathematical formalization, as is often the case today, but extended to the conceptual problems and foundations of biology. It is this commitment to a comprehensive, cross-disciplinary integration of theoretical concepts that the present series intends to emphasize. Today theoretical biology has genetic, developmental, and evolutionary components, the central connective themes in modern biology, but also includes relevant aspects of computational biology, semiotics, and cognition research, and extends to the naturalistic philosophy of sciences.

The "Vienna Series" grew out of theory-oriented workshops organized by the Konrad Lorenz Institute for Evolution and Cognition Research (KLI), an international center for advanced study closely associated with the University of Vienna. The KLI fosters research projects, workshops, archives, book projects, and the journal *Biological Theory*, all devoted to aspects of theoretical biology, with an emphasis on integrating the developmental, evolutionary, and cognitive sciences. The series editors welcome suggestions for book projects in these fields.

Gerd B. Müller, University of Vienna and KLI
Günter P. Wagner, Yale University and KLI
Werner Callebaut, Hasselt University and KLI

Preface: Limited Forms Most Beautiful

In 1859, Charles Darwin concluded his momentous book *On the Origin of Species* by musing on the grandeur of the evolutionary view of life: "from so simple a beginning endless forms most beautiful . . . have been, and are being, evolved" (Darwin 1859, 490). A century and a half after Darwin published those words, many evolutionary biologists are questioning whether the evolution of life actually does produce "endless forms." In fact, in many cases we see that evolution has produced the same form—or a very similar one—over and over again in many independent species lineages, repeatedly, on timescales of hundreds of millions of years.

The phenomenon of convergent evolution has led many of us to suggest that that famous sentence should perhaps be rewritten to read, "from so simple a beginning *limited forms* most beautiful . . . have been, and are being, evolved." The purpose of this book is to reveal to the reader how ubiquitous the phenomenon of convergent evolution is in life and that it occurs on all levels of evolution, from tiny organic molecules to entire ecosystems of species, and even in the way in which we think.

Why does evolution produce limited forms most beautiful? The phenomenon of convergent evolution leads directly to the concept of evolutionary constraint—that is, the number of evolutionary pathways available to life is in fact not endless, but is quite restricted. In the final two chapters of the book, the causes of convergent evolution and the implications of evolutionary constraint will be examined in detail. Those implications are not only of scientific but also of philosophical interest. And those implications all concern, in one way or another, an essential question: How predictable is the evolutionary process?

My fascination with convergent evolution dates back to my first visit, as a small boy, to a natural history museum and zoological garden. I was astonished to see that the eyes of the deadly pit viper, gazing coldly at me through the glass enclosure of its pen, were identical to the eyes of my beloved black cat back at home. Both animals had eyes with very odd-looking vertical-slit pupils and irises of a beautiful gold color. How could it be that this deadly snake, a cold-blooded scaly reptile so vastly different from a warm, furry mammal, had eyes like a cat? In my later university studies I began to realize that the number of evolutionary possibilities was not endless, and that there was a very good reason why the eyes of a predatory snake and a predatory mammal are so similar. By the end of this book, I hope the reader will also share my fascination with an evolutionary process that produces limited forms most beautiful.

I thank the Konrad Lorenz Institute for Evolution and Cognition Research, Altenberg, Austria, for the fellowship that enabled me to work at the institute in 2010 and to bring this book to completion. I thank Gerd Müller, director of the KLI, for all of his help in preparing this book for the Vienna Series in Theoretical Biology. I thank all my colleagues (in particular Karl Niklas and Doug Erwin) who have discussed the scientific implications of convergent evolution with me over the years. I thank Simon Conway Morris, Celia Deane-Drummond, and Michael Ruse for our discussions of the philosophical implications of convergent evolution while we attended a conference on the subject at the Vatican Observatory, Castel Gandolfo, Italy (and I thank Simon for inviting me to the conference). Needless to say, philosophical opinions expressed in this book are my own. Finally, I thank my wife, Marae, for her patient love.

1 What Is Convergent Evolution?

The question "How common is convergence?" remains unanswered and may be unanswerable. Our examples indicate that even the minimum detectable levels of convergence are often high, and we conclude that at all levels convergence has been greatly underestimated.
—Moore and Willmer (1997, 1)

Recognizing Convergent Evolution

A porpoise looks like a fish. It has a fusiform, streamlined body like that of a swordfish or a tuna. It has four fins on the ventral side of its body, instead of four legs. It has a large fin at its posterior end, instead of a tail. And it even has a vertical fin centered on its back, so it looks very much like a shark when it is swimming through the water toward you.

Astonishingly, all appearances to the contrary, a porpoise is not a fish; it is a mammal. It possesses all the distinctive combination of mammalian traits: a porpoise is placental, gives live birth, nurses its young with specialized mammary glands, has an endothermic metabolism, has three bones in its inner ears, has mammalian milk teeth, reproduces via internal fertilization, and so on. But it has lost other traits that are found in most mammals: it has no legs, no tail, no fur, and has instead evolved fins like those of a fish in place of legs and a tail.

The porpoise inherits its mammalian traits from its mammalian ancestors, and the possession of these traits indicates to us that the porpoise belongs to the evolutionary lineage, or clade, of the Mammalia. These traits are synapomorphies, derived traits that are inherited from a common ancestor and that define membership within a particular clade.

The fins of the porpoise are not directly inherited from fish ancestors; they are independently derived convergent traits. That is, porpoises have evolved fins that have converged on the morphology of the fins that are seen in the fish, and, even though the fins of the porpoise look very much like those of a fish, they are in fact not fish fins but rather are mammal fins.

Distinguishing synapomorphic traits from convergent traits is critical in the recognition of convergent evolution. In the literature (particularly the older literature), synapomorphic traits are sometimes called secondary homologous traits, and convergent traits are called homoplastic traits (Lecointre and Le Guyader 2006). Thus, the critical determination, for the purpose of recognizing convergent evolution, is sometimes referred to as the determination of secondary homology versus homoplasy (a term that I have always found to be most uneuphonious—it sounds like some type of disease).

There are actually three ways in which homoplastic traits—traits that look similar but that are not inherited from a common ancestor—might arise in evolution: convergence, parallelism, and reversion. Let us consider a hypothetical example in which we know the evolutionary relationships between six species, illustrated by the cladogram given in figure 1.1. Species 1, 2, and 3 all possess synapomorphy S, and thus belong to the monophyletic clade S. Species 4, 5, and 6 all possess synapomorphy T, and thus belong to the monophyletic clade T. Although clade S and clade T have diverged in their evolution, they nevertheless evolved from a common ancestor that evolved the derived trait R, a trait that all six species still possess by inheritance; thus, trait R is a synapomorphy for the larger monophyletic clade R, which contains both clade S and clade T.

So far, so good. Now let us suppose that in clade S, the new trait Z arises by evolutionary modification of the preexisting trait A, as seen in species 3 (figure 1.2). However, in clade T a new trait very similar to trait

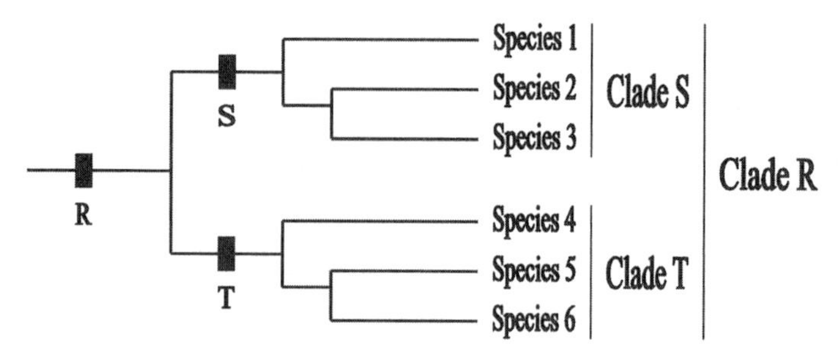

Figure 1.1
Cladogram of evolutionary relationships between six hypothetical species.

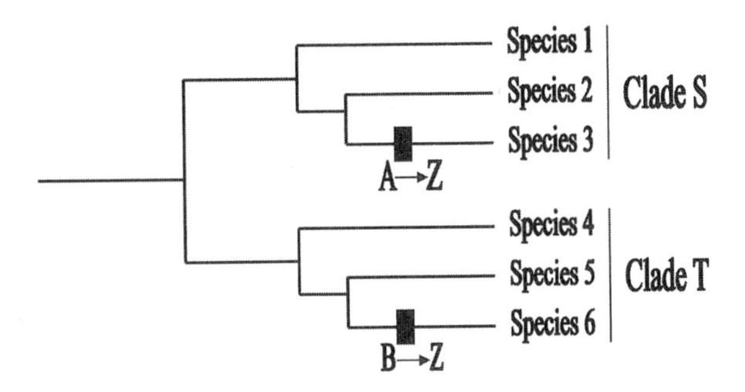

Figure 1.2
Convergent evolution of trait Z in species 3 and 6.

Z also arises, but by evolutionary modification of the preexisting trait B, as seen in species 6 (figure 1.2). Trait Z is thus a convergent trait, a homoplastic trait, having evolved independently in species 3 and species 6. If we were to mistakenly think that trait Z was a synapomorphic trait, we would mistakenly think that species 3 and species 6 were closely related sister species, since they both possess trait Z, and we would thus mistakenly include species 3 and species 6 in the new clade Z. The erroneous clade Z would thus be a polyphyletic clade. If, for example, we were to include swordfish and porpoises together in the clade "fins" because we mistakenly believed that fins were a synapomorphic trait, we would have created an erroneous polyphyletic clade because the fins present in these two species are convergent traits, not synapomorphies.

Parallel evolution is a particular type of convergent evolution, where parallelism is "a similarity that has appeared independently in different closely related taxa," closely related in that the same trait has independently evolved "from the same ancestral character in different taxa. A parallelism is a special case of convergence" (Lecointre and Le Guyader 2006, 541). A hypothetical case of parallel evolution is illustrated in figure 1.3. All six species possess the same ancestral character R in the original cladogram given in figure 1.1. In figure 1.3, however, trait R has been modified into trait Z independently, in parallel, in both species 3 and species 6.

The conceptual difference between the evolutionary scenarios illustrated in figures 1.2 and 1.3 lies in the traits that have been modified in producing the convergent trait Z. In figure 1.2, two separate traits, A and B, have been modified in two separate lineages to produce the same convergent trait Z. In figure 1.3, only one trait, the ancestral trait R, has

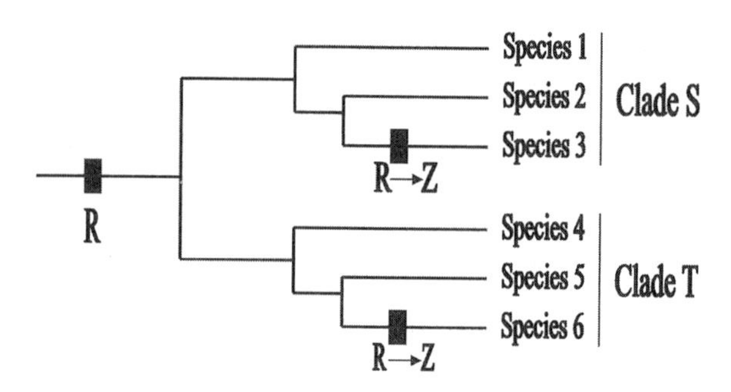

Figure 1.3
Parallel evolution of trait Z in species 3 and 6.

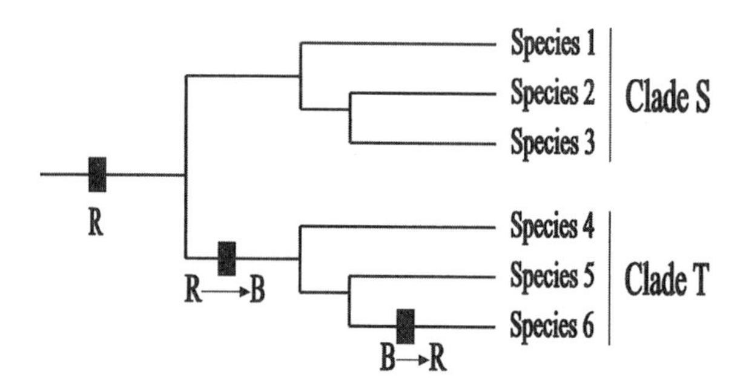

Figure 1.4
Reverse evolution of trait R in species 6 within clade T, whereas all the species of clade S simply inherit trait R from a common ancestor.

been modified in two separate lineages to produce the parallel convergent trait Z.

Last, a homoplastic trait can also be produced by the process of reverse evolution. In figure 1.4, trait R has been modified into trait B, a new derived trait possessed by species 4 and species 5. In species 6, however, trait B has been modified back into trait R—an evolutionary reversion. Species 1, 2, and 3 all possess trait R by inheritance from a common ancestor. Since species 6 did not inherit trait R directly from that ancestor, trait R is a homoplastic reversion in this species. Consequently, it would be erroneous to conclude that species 6 was more closely related to species 1, 2, and 3 than it is to species 4 and 5 simply because it also possesses trait R. We would create an erroneous polyphyletic clade if we were to mistakenly include species 6 in clade S along with species 1, 2, and 3. Instead, species 6 possesses synapomorphy T (figure 1.1) and belongs in clade T along with species 4 and 5, even though it does not possess trait B as species 4 and 5 do (figure 1.4). At present, the importance and amount of homoplasy in evolution that is produced by reversion is uncertain (Spinney 2007).

It is easy for us to see the conceptual differences between convergent, parallel, and reverse evolution in figures 1.2–1.4 because we already knew the true evolutionary relationships of the six hypothetical species (figure 1.1). In real life, however, evolutionary relationships are not known beforehand, and we must carefully analyze all traits that we see are shared by different species. Do these species share these traits simply because they all inherit them from a common ancestor? Or are these traits convergent—have they independently arisen in species that belong to separate evolutionary lineages, in separate clades? A porpoise is not a fish, even though it looks like one.

Convergent Evolution versus Parallel Evolution?

In modern phylogenetic systematics, we see that parallel evolution is simply a special case of convergent evolution (Lecointre and Le Guyader 2006). In the literature (particularly the older literature), not only is parallel evolution sometimes erroneously considered to be different from convergent evolution, but in some cases it is considered to be the actual disproof of convergent evolution. For example, in the next chapter we will see that eyes have convergently evolved in animals in some 49 independent lineages (table 2.15). It was long thought that each separate animal lineage convergently evolved its type of eye from separate original traits, in a case of classic convergent evolution as illustrated in figure 1.2. However, we now know that all animal eyes are produced by modifying the same conserved regulatory gene present in the animal genome (the *Pax-6* gene), and thus the evolution of eyes is a case of parallel convergent evolution, as illustrated in figure 1.3. The recent molecular discovery that the same ancestral character was independently, in parallel, modified in the evolution of eyes was mistakenly taken by Rey et al. (1998) to be a disproof of convergent evolution: "An oft invoked example of convergent evolution has been compound eyes of insects versus singular eyes of vertebrates. This proved to be a wrong example, however, because development of all metazoan eyes recently has been shown to be under control of the same regulatory gene that encodes Pax-6 protein" (Rey et al. 1998, 6212).

All life on Earth evolved from a single common ancestor. We know this because all life on Earth uses the exact same coding molecule, DNA, and the exact same molecular construction: DNA codes for RNA, and RNA codes for the same 20 amino acids that are assembled to produce proteins, the building blocks of life. This fact leads us to the concept of "deep homology" and parallelism in evolution: "One of the most important, and entirely unanticipated, insights in the past 15 years was the recognition of an ancient similarity in patterning mechanisms in diverse organisms, often among structures not thought to be homologous on morphological or phylogenetic grounds. In 1997, prompted by the remarkable extent of similarities in genetic regulation between organs as different as fly wings and tetrapod limbs, we suggested the term 'deep homology' to describe the sharing of the genetic regulatory apparatus that is used to build morphologically and phylogenetically disparate animal features" (Shubin et al. 2009, 818). However, rather than viewing deep homology as a disproof of convergent evolution, Shubin et al. (2009) recognize that it is proof of parallel convergent evolution: "The deep homology of generative processes and cell-type specification mechanisms in animal development has provided the foundation for the *independent evolution* of a great variety of structures . . . common genetic mechanisms are used to generate diverse adaptations and can lead to the *parallel evolution* of novelties" (Shubin et al. 2009, 818; italics mine).

In actual practice it is sometimes quite difficult to differentiate convergent and parallel evolution. Rokas and Carroll (2008) give unambiguous examples of both types of evolution at the molecular level: both dolphins and rhinos have independently evolved the amino acid valine at the same amino acid site. The dolphins, however, modified alanine to valine at that site, whereas the rhinos modified tyrosine to alanine. The same amino acid, valine, was independently evolved from two different amino acid precursors in a process of convergent evolution (the process illustrated in figure 1.2). In contrast, the filamentous ascomycetes *Aspergillus clavatus* and *Aspergillus oryzae* have independently evolved glutamic acid at the same amino acid site in parallel, in that both modified their ancestral amino acid asparagine to glutamic acid at that site (the process illustrated in figure 1.3).

In other cases the determination is not so easy. The wings of birds and bats are overwhelmingly used in the literature as examples of convergent evolution. And indeed these are two very distantly related animals—the bats belong to the clade Mammalia and the birds to the clade Dinosauria. We have to go back in time to the Carboniferous, some 340 million years ago, to find a common ancestor for these two animals. Both have wings, but the birds modified dinosaurian forelimbs to wings whereas bats modified mammalian forelimbs to wings in what was a process of convergent evolution (the process illustrated in figure 1.2). Or perhaps it was not? Even though dinosaurian forelimbs are very different from mammalian forelimbs, they are both forelimbs. And both dinosaurs and mammals have inherited (and modified) their forelimbs from their ancient amphibian ancestors. One could thus just as well argue that the evolution of wings in birds and bats is a case of parallel evolution: the same ancestral character (forelimbs) was modified in parallel to produce wings in birds and in bats (the process illustrated in figure 1.3). Arguments of deep homology would push that parallelism even further back in time, in that the genetic regulatory mechanisms used to produce forelimbs were evolved before the Cambrian diversification of animal life over 540 million years ago.

In the chapters that follow, I will not draw a sharp distinction between convergent evolution and parallel evolution. Again, from the perspective of modern phylogenetic systematics, parallel evolution is simply a type of convergent evolution (Lecointre and Le Guyader 2006). Where I think the distinction is useful, and reasonably well demonstrated, I will point out examples of these two evolutionary processes in operation.

Evolutionary Constraint and Convergent Evolution

Convergent evolution was once thought to be almost exclusively produced by functional constraint, the old maxim that "form follows function." Given the same function, evolution via natural selection will produce the same form to serve that function. However, modern studies in evolutionary development ("evo-devo" to its practitioners) have revealed that a great deal of convergent evolution of life on Earth is in fact due to developmental constraint: the phenomenon of "deep homology" that we considered in the previous section of the chapter is just one example. Ancient homologous regulatory genes have been independently switched on and off many times in the evolution of life to produce convergent forms, such as the many types of eyes seen in animals (which we will consider in the next chapter). The same forms have been produced by the repeated channeling of evolution along the same developmental trajectory. Natural selection has a limited repertoire of potential forms from which to choose, and convergent evolution is the result.

In order to understand the phenomenon of convergent evolution, we now need to carefully consider the action of both functional and developmental constraints in the evolutionary process: "Patterns of parallel evolution can provide even stronger illustrations of the need to distinguish explanations based on the similarity of natural selection from those involving developmental bias or genetic channeling" (Brakefield 2006, 364). Throughout this book I will point out examples of functional and developmental constraint in operation. The theoretical analysis of evolutionary constraint in producing convergent evolution will be considered in detail in chapter 7.

Mimicry, Camouflage, and Convergent Evolution

Mimicry is a form of convergent evolution in which one species independently evolves a morphology very similar to that of another species simply in order to fool a third species. In Batesian mimicry, a harmless mimic converges on the morphology of a harmful model species in a case of false warning—potential predators will avoid the harmless mimic, thinking it is the harmful model. An example is the harmless New Mexico milk snake (*Lampropeltis triangulum celaenops*), which looks almost identical to the deadly Arizona coral snake (*Micruroides euryxanthus*). In Müllerian mimicry, two harmful species converge on a common morphology in a case of real warning—potential predators easily recognize the harmful species because the warning morphology is twice as abundant in nature as it would be if it was characteristic of only a single species. An example is the yellow-and-black striping seen in both bees and wasps.

Mimicry is similar to camouflage, where species evolve morphologies that converge on the form of either a living or a nonliving model in order to blend into the surroundings so that the camouflaged species cannot be seen. An example of the former is the giant walkingstick (*Megaphasma dentricus*), an insect that so closely resembles the leafless twig of a plant that it is unnoticed by predators. An example of the latter is the white fur color that is convergently evolved by numerous animal species that live in snowy habitats, such as the Asian snow leopard (*Uncia uncia*) and the Arctic polar bear (*Thalarctos maritimus*).

All the myriad forms of mimicry and camouflage convergence serve a single function: deception. Predator species repeatedly and independently evolve forms that deceive prey animals into not noticing them (until it is too late), and prey animals independently evolve forms that deceive predators so that the prey either are not noticed or are noticed but avoided. Although the convergence in form that is seen in mimicry and camouflage in nature is sometimes spectacular, the fundamental function of that convergence is rather simple, in that it is all deception. Entire books have been written on the evolution of mimicry and camouflage, and I will not consider these forms of convergence in more detail in this book, with the exception of the carrion-mimic flowers, discussed in chapter 3.

On the Organization of This Book

This book is not intended to be an "Encyclopedia of Convergent Evolution" (an effort that would run to many volumes). Instead, I have concentrated on listing a single species example from as many separate and disparate phylogenetic lineages that I am aware of in which a trait has originated independently by convergent evolution.

A phylogenetic classification of life is absolutely essential in determining whether a similar trait found in separate species is shared by those species simply because they inherited it from a common ancestor—that is, a synapomorphic trait—or whether that trait has arisen independently in each separate species lineage—and is thus a convergent trait. The phylogenetic classification that I use throughout this book is given in abbreviated form in table 1.1 and in detail in the appendix. This classification is modified chiefly from the phylogenetic classification of living life forms by Lecointre and Le Guyader (2006) and the APG II (2003); additional information on extinct life forms is taken from Benton (2005), Donoghue (2005), and Niklas (1997).

For each species example given in the chapters that follow, I have provided an abbreviated list of the major nodes in its phylogenetic classification, which will allow the reader to quickly see the close or distant relationship of a group of species that have independently evolved a similar trait by convergent evolution. I have not repeated the entire sequence of nodes, as that would be redundant with the appendix and would burden the lists of species with information overload. The major clades I have used in the text of this book are given in table 1.1 and are set in boldface type in the appendix, allowing the reader to quickly assemble the entire evolutionary sequence of nodes for any species lineage in which he or she is particularly interested.

For each individual convergent structure or trait that is discussed in the chapters that follow, I will list the phylogenetic lineages of species that have independently evolved that structure or trait in a "step down" fashion from least derived to most derived; that is, I will list lineages that

Table 1.1
The major lineages of life on Earth

Eubacteria
Archaea
Eukarya
– Bikonta
– – Chlorobiota
– – – Embryophyta
– – – – Tracheophyta
– – – – – Lycophyta
– – – – – Euphyllophyta
– – – – – – Moniliformopses
– – – – – – Lignophyta
– – – – – – – Spermatophyta
– – – – – – – – Angiospermae
– – – – – – – – – Euangiosperms
– Unikonta
– – Metazoa
– – – Cnidaria
– – – Bilateria
– – – – Protostomia
– – – – – Lophotrochozoa
– – – – – Ecdysozoa
– – – – Deuterostomia
– – – – – Echinodermata
– – – – – Chordata
– – – – – – Vertebrata
– – – – – – – Chondrichthyes
– – – – – – – Osteichthyes
– – – – – – – – Actinopterygii
– – – – – – – – Sarcopterygii
– – – – – – – – – Tetrapoda
– – – – – – – – – – Batrachomorpha
– – – – – – – – – – Reptiliomorpha
– – – – – – – – – – – Amniota
– – – – – – – – – – – – Sauropsida
– – – – – – – – – – – – – Diapsida
– – – – – – – – – – – – – – Lepidosauromorpha
– – – – – – – – – – – – – – Archosauromorpha
– – – – – – – – – – – – Synapsida
– – – – – – – – – – – – – Therapsida
– – – – – – – – – – – – – – Mammalia

Note: The complete phylogenetic classification of life used in this book is given in the appendix.

occur at the top of table 1.1 first and arrange subsequent lineages in their sequence of evolution as outlined in table 1.1. Thus, in the discussion of the convergent evolution of eyespots in animals given in chapter 2, the nonbilaterian jellyfish species *Leuckartiara octona* (lineage Metazoa: Cnidaria in table 1.1) is listed before the more derived bilaterian rotifer species *Asplanchna brightwelli* (lineage Metazoa: Bilateria: Protostomia: Lophotrochozoa in table 1.1), which itself is listed before the deuterostomous sea cucumber species *Synaptula lamperti* (lineage Metazoa: Bilateria: Deuterostomia: Echinodermata in table 1.1).

I will take the list of nodes for each pair of convergent lineages back to the point where the lineages diverged in evolution so the reader can see how distantly related species are that have independently acquired a convergent trait. Thus, in the previous example, the living sea cucumber *Synaptula lamperti* and the rotifer *Asplanchna brightwelli* both have eyespots, yet you have to go all the way back to the evolutionary split between the protostomous and deuterostomous lineages of bilaterian animals to find a common ancestor for the two species (Bilateria: Protostomia versus Bilateria: Deuterostomia). In order to find a common ancestor for these bilaterian animals and the cnidarian jellyfish *Leuckartiara octona*, you have to go all the way back to the evolution of the metazoans themselves, over 600 million years ago (i.e., Metazoa: Cnidaria versus Metazoa: Bilateria).

Published sources for all the convergent lineages discussed in this book are given in the references. These have partially come from my own lists of convergences, amassed over many years, and all the many textbooks and encyclopedia articles on evolution I have read, all of which contain examples of convergent evolution (in many cases, the same ones). In addition, three compendia that are exclusively devoted to the subject of convergent evolution, and that have been particularly helpful in my analyses, are Conway Morris's *Life's Solution* (2003), Barlow's *Let There Be Sight! A Celebration of Convergent Evolution* (2005), and Berg's *Nomogenesis, or Evolution Determined by Law* (1922), and I recommend these to the reader as they also contain many additional examples not included in this volume.

The reader can quickly grasp the structure of my analysis of the phenomenon of convergent evolution by reading the section headings within each of the individual chapters given in the table of contents of the book. Throughout the book I will discuss the functional and developmental constraints that have resulted in convergent evolution at every level of life, from the external forms of living organisms down to the very molecules from which they are constructed, from their ecological roles in nature to the way in which their minds function. I will also use the analytical techniques of theoretical morphology in that discussion (McGhee 1999, 2001, 2007). In theoretical morphology, the consideration of nonexistent biological form is just as important as that of existent biological form, and can give us valuable insights into the phenomenon of convergent evolution.

2 Convergent Animals

In all cases of two very distinct species furnished with apparently the same anomalous organ, it should be observed that, although the general appearance and function of the organ may be the same, yet some fundamental difference can generally be detected. I am inclined to believe that in nearly the same way as two men have sometimes independently hit on the very same invention, so natural selection, working for the good of each being and taking advantage of analogous variations, has sometimes modified in very nearly the same manner two parts in two organic beings, which owe but little of their structure in common to inheritance from the same ancestor.
—Darwin (1859, 193–194)

Locomotion

Some of the most spectacular examples of convergent evolution are clearly due to the functional constraints of locomotion. Consider one of the most frequently cited cases of convergent evolution: the astonishing morphological similarity between the extinct Mesozoic marine reptile *Ichthyosaurus platyodon* and the living marine mammal *Phocaena phocaena*, the harbor porpoise, or *Tursiops truncatus*, the bottlenose dolphin. Not only do they look amazingly similar to one another, but they all look amazingly similar to large, fast-swimming fishes like *Xiphias gladius*, the swordfish, or *Carcharodon carcharias*, the great white shark. The cartilaginous fishes and the bony fishes both solved the physics of swimming in the dense medium of water back in the Silurian by evolving streamlined, fusiform morphologies (table 2.1). Some 230 million years later, a group of land-dwelling reptiles rediscovered this same morphology in their evolutionary return to the sea. And around 175 million years after that, a group of land-dwelling mammals also rediscovered this same morphology in their own evolutionary return to the sea.

Table 2.1
Convergent evolution of animal swimming morphologies

1 Convergent structure and function: FUSIFORM BODY (missile-shaped form for drag reduction)

Convergent lineages:

1.1 Great white shark (Vertebrata: Chondrichthyes: Elasmobranchii: Lamnidae; *Carcharodon carcharias*)

1.2 Swordfish (Vertebrata: Osteichthyes: Actinopterygii: Teleostei: Xiphiidae; *Xiphias gladius*)

1.3 Ichthyosaur (Osteichthyes: Saurcopterygii: Reptiliomorpha: Amniota: Sauropsida: Diapsida: Ichthyosauria: Ichthyosauridae; *Icthyosaurus platyodon* †Jurassic)

1.4 Harbor porpoise (Amniota: Synapsida: Therapsida: Mammalia: Eutheria: Laurasiatheria: Cetartiodactyla: Cetacea: Odontoceti: Phocaenidae; *Phocaena phocaena*)

2 Convergent structure and function: EEL-SHAPED BODY (elongated, cylindrical form for drag reduction)

Convergent lineages:

2.1 Reed fish (Osteichthyes: Actinopterygii: Cladista: Polypteridae; *Erpetoichthyes calabaricus*)

2.2 European eel (Actinopterygii: Teleostei: Elopomorpha: Anguillidae; *Anguilla anguilla*)

2.3 South American lungfish (Osteichthyes: Sarcopterygii: Dipnoi: Lepidosirenidae; *Lepidosiren paradoxa*)

2.4 Mosasaur (Sarcopterygii: Reptiliomopha: Amniota: Sauropsida: Diapsida: Lepidosauromorpha: Squamata: Scleroglossa: Autarchoglossa: Anguimorpha: Mosasauridae; *Platecarpus ictericus* †Cretaceous)

2.5 Geosaur (Diapsida: Archosauromorpha: Crurotarsi: Crocodilia: Metriorhynchidae; *Geosaurus giganteus* †Jurassic)

3 Convergent structure and function: PADDLE-FORM APPENDAGES (walking appendages modified to paddle shapes for paddling/rowing through water)

Convergent lineages:

3.1 Sea scorpions (Bilateria: Protostomia: Ecdysozoa: Arthropoda: Cheliceriformes: Merostomata: Eurypterida: Eurypteridae; *Eurypterus remipes* †Silurian)

3.2 Marbled diving beetle (Arthropoda: Mandibulata: Hexapoda: Dytiscidae; *Thermonectes marmoratus*)

3.3 Green sea turtle (Bilateria: Deuterostomia: Chordata: Osteichthyes: Sarcopterygii: Reptiliomorpha: Amniota: Sauropsida: Anapsida: Testudines: Chelonioidae; *Chelonia mydas*)

3.4 Plesiosaur (Sauropsida: Diapsida: Lepidosauromorpha: Sauropterygia: Elasmosauridae; *Muraenosaurus leedsii* †Jurassic)

3.5 King penguin (Diapsida: Archosauromorpha: Ornithodira: Dinosauria: Saurischia: Theropoda: Maniraptora: Aves: Sphenisciformes: Spheniscidae; *Aptenodytes patagonica*)

3.6 African manatee (Amniota: Synapsida: Therapsida: Mammalia: Eutheria: Afrotheria: Sirenia: Trichechidae; *Trichechus senegalensis*)

3.7 California sea lion (Eutheria: Laurasiatheria: Carnivora: Caniformia: Otariidae; *Zalophus californianus*)

Note: The geological age of extinct species is marked with a †.

The evolution of an ichthyosaur or a porpoise morphology is not trivial. It can be correctly described as nothing less than astonishing that a group of land-dwelling tetrapods, complete with four legs and a tail, could devolve their appendages and tails back into fins like those of a fish, while also reevolving a dorsal fin. Highly unlikely, if not impossible? Yet it has happened twice, convergently in the reptiles (sauropsid amniotes) and the mammals (synapsid amniotes), two groups of animals that are not closely related. We have to go back in time as far as the Carboniferous to find a common ancestor for the mammals and the reptiles; thus, their genetic legacies are very different. Nonetheless, the ichthyosaur and the porpoise both independently reevolved fins.

An extremely ancient form of swimming morphology can be seen in some of the first chordate animals to evolve back in the Cambrian, not long after the evolution of animals themselves some 600 million years ago. These animals have elongated cylindrical bodies, usually laterally compressed, with caudal fins located along the dorsal and ventral surfaces of the posterior one-third or so of the body. Superficially, these animals look somewhat like arrows, but rather than moving rigidly like an arrow, these animals undulate through the water, with a sinusoidal motion of the body, propelled by the caudal fins at the rear. Such eel-like morphologies can be seen in the extinct conodonts, such as *Eoconodontus notchpeakensis* (the "dawn conodont" from Notchpeak) from the Late Cambrian, which were very ancient craniate chordates possessing a skull but no jaws or vertebrae. The same plesiomorphic, eel-like morphology can be seen in the modern myxinoid hagfish *Myxine glutinosa*, which also have no jaws or vertebrae, and in the modern sea lamprey *Petromyzon marinus*, which does possess vertebrae but still has no jaws.

Surprisingly advanced, derived animals have repeatedly converged on this ancient swimming morphology—including the modern eels themselves, such as the European eel *Anguilla anguilla* (table 2.1). Modern eels are highly modified teleost fish, the Elopomorpha, and the teleost fishes themselves are highly derived actinopterygian fish. Yet a modern eel looks very much like an ancient Cambrian craniate, and swims in the same undulatory fashion, propelled by caudal fins. The teleost fishes were not the first to evolve eel-forms among the ray-fin fishes, however, as this same form was convergently evolved within the plesiomorphic cladistian fishes, a basal group of actinopterygian fishes, as can be seen in the modern reed fish.

This same morphology has been convergently evolved in the second major extant branch of the bony fishes, the lobe-finned sarcopterygians, as can be seen in the South American lungfish *Lepidosiren paradoxa*, which is virtually identical to an eel. And, as one might expect from the previous example of the convergent evolution of fusiform swimming morphologies, these convergences are not confined to disparate types of fish (table 2.1). The Cretaceous mosasaurs were eel-like marine reptiles, closely resembling the mythical sea serpent, and evolved from land-dwelling lizards. The Jurassic geosaurs were eel-like marine archosaurs, a more derived form of reptile than the lizards, yet they too evolved from land-dwelling tetrapods that devolved the tetrapods back into fins and returned to the sea. Both the mosasaurs and geosaurs developed caudal fins by dorsoventrally elongating, and laterally compressing, the caudal vertebrae themselves.

In analyzing evolution from the perspective of theoretical morphology, it is important to consider both nonexistent convergent form and existent convergent form (McGhee 2001, 2007). There is one major group of land animals that has not yet convergently evolved an eel-like swimming morphology—the mammals (see table 2.1). Mammalian "fish-forms" have evolved in the Cenozoic (porpoises, etc.), but no mammalian "eel-forms" have evolved as yet. Why not? At present, the sea otters (*Enhydra lutris*), with their webbed feet, elongated bodies, and long flattened tails used in swimming, might be the best example of a precursor to the convergent evolution of the eel-form in mammals. Careful consideration of nonexistent convergent form may give us valuable insights into the phenomenon of convergent evolution, and we shall return to this topic throughout this book.

Last, extensive convergence in form is seen in animals that have secondarily evolved the capability to paddle or row through water with their former walking appendages (table 2.1). Among the arthropods, the extinct sea scorpions (ancestors of modern land scorpions) modified a pair of their fore appendages into two paddles back in the Silurian, and 420 million years later the modern diving beetles have modified a pair of their hind appendages into paddles. All four of the tetrapods of modern sea turtles have been modified into paddles, and appear very similar in form to the four paddles of the ancient plesiosaurs, reptiles that underwent this same modification some 170 million years before. Among the advanced dinosaurian reptiles, the modern-day penguins have modified their wings—forelimb structures that were themselves modified from walking appendages back in the Jurassic—into paddles for use in swimming rather than in flight. Two separate groups of modern mammals have convergently modified their forelimbs into paddles: the manatees and their kin, and the sea lions and their kin (table 2.1). Although a manatee and a walrus look very similar, they are very different types of mammals. The afrotherian manatee is more closely related to an elephant, and the laurasiatherian walrus is more closely related to a wolf, than they are to each other. Their apparent similarity in form is entirely due to convergent evolution.

The second most commonly cited example of amazing feats in convergent evolution is the evolution of flying in land animals. No fewer than three separate groups of tetrapods have modified their forelimbs into wings: the extinct pterosaurs and the living birds and bats (table 2.2). At first glance, it would seem to be extremely unlikely that an originally quadrupedal animal, using all four legs for walking, would somehow be able to evolve into an animal in which the forelimb is now used as a flying structure. Yet not only has this occurred in the evolution of life, it has occurred three times independently! The wings of birds, bats, and pterosaurs have an amazingly similar appearance, which is clearly due to the functional constraints of locomotion via flying in the thin gaseous medium of the Earth's atmosphere. This similarity is deceptive, however, for closer examination reveals that the wings of the three groups of animals are constructed differently. The wing of a bat largely consists of its hand, in which the fingers are vastly elongated and between which is stretched a webbed membrane of skin. The wing of a pterosaur is similar, but consists of a single finger (the fourth digit) from which a membrane of skin is stretched all the way to the side of the animal. The wing of a bird is even more different, in that the surface of the wing is not constructed of skin membranes, but consists of elongated flight feathers attached to its arm. Still, wings in all three groups are convergent modifications of an original walking structure, the forelimb.

The wings of insects, the fourth major group of animals to evolve powered flight (table 2.2), are radically different from those of the tetrapods. The insects did not modify their forelimbs in the evolution of wings. Rather, insect wings consist of modified gill branches originally present only in the larval stages. Originally present on multiple segments in the abdomens of larvae, these gill structures were expanded in some body segments while being developmentally suppressed in others, leading to the four-wing condition seen in primitive adult insects such as mayflies (Carroll 2006).

In examining the convergent evolution of wings (table 2.2), we begin to see the effect of a different constraint on the evolution of form in nature, in addition to that of functional constraint. All three tetrapod

Table 2.2
Convergent evolution of animal flying morphologies

1 Convergent structure and function: WINGS (flat to curved planar structures for generating lift in active flying)

Convergent lineages:

1.1 Green darner dragonfly (Bilateria: Protostomia: Ecdysozoa: Panarthropoda: Arthropoda: Mandibulata: Hexapoda: Odonata: Aeschnidae; *Anax junius*)

1.2 Hairy devil pterosaur (Bilateria: Deuterostomia: Chordata: Osteichthyes: Sarcopterygii: Reptiliomorpha: Amniota: Sauropsida: Diapsida: Archosauromorpha: Ornithodira: Pterosauria: Rhamphorhynchidae; *Sordes pilosus* †Jurassic)

1.3 Great blue heron (Archosauromorpha: Ornithodira: Dinosauria: Saurischia: Theropoda: Maniraptora: Aves: Neognathae: Neoaves: Ciconiiformes: Ardeidae; *Ardea herodias*)

1.4 Mouse-eared bat (Amniota: Synapsida: Therapsida: Mammalia: Eutheria: Laurasiatheria: Chiroptera: Microchiroptera: Vespertilionidae; *Myotis myotis*)

2 Convergent structure and function: GLIDER MEMBRANES (expanded skin membranes stretched between fore and hind limbs for retarding falling speed in passive flying)

Convergent lineages:

2.1 Flying Mesozoic mammal (Synapsida: Therapsida: Mammalia: Volaticotheria: Volaticotheridae; *Volaticotherium antiquum* †Cretaceous)

2.2 Flying possum or sugar glider (Mammalia: Marsupialia: Diprotodontia: Petauridae; *Petaurus breviceps*)

2.3 Feathertail glider (Marsupialia: Diprotodontia: Acrobatidae; *Acrobatus pygmaeus*)

2.4 Greater glider (Marsupialia: Diprotodontia: Pseudocheiridae; *Petauroides volans*)

2.5 True flying squirrel (Mammalia: Eutheria: Euarchontoglires: Rodentia: Sciuridae; *Glaucomys volans*)

2.6 Scaly-tailed flying squirrel (Eutheria: Euarchontoglires: Rodentia: Anomaluridae; *Anomalurus derbianus*)

2.7 Flying lemur or colugo (Eutheria: Euarchontoglires: Dermoptera: Cynocephalidae; *Cynocephalus volans*)

3 Convergent structure and function: GLIDER BODIES (dorsoventrally flattened, laterally expanded body for retarding falling speed in passive flying)

Convergent lineages:

3.1 Flying diapsid reptile (Sauropsida: Diapsida: Weigeltisauridae; *Coelurosauravus jaekeli* †Permian)

3.2 European flying Mesozoic lizard (Diapsida: Lepidosauromorpha: Kuehneosauridae; *Kuehneosaurus latus* †Triassic)

3.3 North American flying Mesozoic lizard (Lepidosauromorpha: Kuehneosauridae; *Icarosaurus siefkeri* †Triassic)

3.4 Black-bearded flying lizard (Lepidosauromorpha: Lepidosauria: Squamata: Iguania: Agamidae; *Draco melanopogon*)

3.5 Flying paradise snake (Lepidosauria: Squamata: Scleroglossa: Autarchoglossa: Anguimorpha: Serpentes: Colubridae; *Chrysopelea paradisi*)

4 Convergent structure and function: GLIDER FEET (expanded skin membranes stretched between elongated toes in feet for retarding falling speed in passive flying)

Convergent lineages:

4.1 Wallace's flying tree frog (Tetrapoda: Batrachomorpha: Lissamphibia: Batrachia: Anura: Rhacophoridae; *Rhacophorus nigropalmatus*)

4.2 Costa Rican flying tree frog (Lissamphibia: Batrachia: Anura: Hylidae; *Agalychnis spurrelli*)

Table 2.2
(continued)

4.3 Kuhl's flying gecko (Tetrapoda: Reptiliomorpha: Amniota: Sauropsida: Diapsida: Lepidosauromorpha: Squamata: Scleroglossa: Gekkonta: Gekkonidae; *Ptychozoon kuhli*)

5 Convergent structure and function: GLIDER FINS (greatly elongated, horizontally oriented pectoral fins for retarding falling speed in passive flying)

Convergent lineages:

5.1 Flying fish (Osteichthyes: Actinopterygii: Exocoetidae; *Exocetus volitans*)

5.2 Flying gurnard fish (Actinopterygii: Dactylopteridae; *Dactylopterus volitans*)

6 Convergent structure and function: GLIDER APPENDAGES (elongated, flattened appendages and wide, flanged heads for retarding falling speed in passive flying)

Convergent lineages:

6.1 Giant gliding ant (Arthropoda: Mandibulata: Hexapoda: Hymenoptera: Myrmicinae: Cephalotini; *Cephalotes atratus*)

6.2 Mexican gliding ant (Arthropoda: Mandibulata: Hexapoda: Hymenoptera: Pseudomyrmecinae; *Pseudomymex gracilis*)

6.3 South American gliding ant (Arthropoda: Mandibulata: Hexapoda: Hymenoptera: Formicinae; *Camponotus heathi*)

Note: The geological age of extinct species is marked with a †. For data sources, see text.

groups modified their forelimbs in the evolution of wings because only those structures were present for potential modification: unlike the ancient insects, the tetrapods had no gill branches to modify. The triple modification of forelimbs to wings in the tetrapods is an example of developmental constraint, a constraint that did not apply to the insects in this instance. Hypothetically, a group of insects could have modified their forelimbs to wings, like the tetrapods, but they were not confined to that developmental pathway and, as the geological record demonstrates, they evolved along another developmental pathway not open to the tetrapods.

One of the analytical techniques of theoretical morphology involves the generation of nonexistent biological form, usually by mathematical modeling and computer simulation (McGhee 2007). As a thought experiment in theoretical morphology, we can consider the nonexistent mythical animal the dragon, a reptile-like animal with functional fore and hind limbs and functional wings as well. From our previous discussion of the convergent evolution of wings in tetrapods and insects, we can easily see why dragons are nonexistent. First, a group of tetrapods would have to develop an additional pair of forelimbs (the serial replication of a pair of limbs is not developmentally impossible, but extremely rare in the vertebrates) and, secondly, it would have to develop wings from the new pair of forelimbs. For the insects, such a morphological transition would have been relatively simple, as they already possessed three pairs of walking appendages, and it is interesting to note that they never took this potential developmental pathway. Instead, they modified a different structure, their larval gill branches, into flying structures. Later in this chapter we shall consider another group of insects who in fact have convergently modified their first pair of walking appendages into structures no longer used in locomotion.

Considerably more animals have convergently evolved gliding morphologies than powered-flight morphologies (table 2.2). At least seven separate groups of mammals have evolved the capability to glide by stretching skin membranes between their fore and hind limbs, beginning with the volaticoterian mammals in the Cretaceous. These ancient gliders looked amazingly similar to modern-day southern flying squirrels, *Glaucomys volans*, in North America. The true flying squirrels are widely distributed around the planet, except in Africa. Thus, it is fascinating to note that in Africa a separate group of rodents, the scaly-tailed flying "squirrels" (which are not squirrels at all), have convergently evolved the same morphology. In Southeast Asia, another group of mammals, the flying lemurs or colugos, have converged on this morphology, and these animals are not any type of rodent at all but dermopterans, more closely related to primates than to rodents. Last, this same convergent morphology was evolved by very distantly related marsupial mammals, and is found in three different groups of flying possums and marsupial gliders in Australia and parts of Southeast Asia (table 2.2).

Glider morphologies have repeatedly been evolved in the diapsid reptiles, but in a form very different from that found in the mammals. Reptilian gliders have bodies that are dorsoventrally flattened and laterally widened. Within their bodies, the ribs are flattened and greatly elongated laterally, away from the anteroposterior axis of the vertebral column. Species of the modern flying lizard genus *Draco* have bodies that resemble a discus in flight—with a pair of forelimbs and a head attached at one side and a pair of hind limbs and a tail attached to the opposite side. This identical morphology was evolved more than a quarter of a billion years ago by ancient diapsid weigeltisaurid reptiles in the Permian (table 2.2). A very interesting example of parallel evolution of this morphology occurred in the Triassic, where the lepidosauromorph *Kuehneosaurus latus* evolved in Europe and a second group of lepidosauromorphs separately evolved the same form in North America, as seen in *Icarosaurus siefkeri* (Kuhn-Schnyder and Rieber 1986). Last, a modification of the same glider morphology has been evolved by the flying paradise snakes. Rather than flattening and flaring the ribs in the midsection of the body, the entire snake is flattened like a ribbon, and it flares its ribs laterally outward as it glides through the air.

Both groups of animals—the mammals and the reptiles—have convergently evolved glider morphologies with expanded surface-area structures. The construction of those glider structures is very different in the two groups of animals, however: the mammals repeatedly evolve skin membranes stretched between their fore and hind limbs, and the reptiles repeatedly evolve flattened bodies with laterally flared rib cages (table 2.2). Apparently no reptiles ever evolved skin-membrane gliding morphologies, and no mammals ever evolved glider bodies with flattened, flared rib cages. The two groups of animals appear to be developmentally constrained to a particular convergent form for each group.

Skin-membrane gliding structures are also found in another group of animals, but with quite a different construction from those produced by the mammals. These are the flying tree frogs, which possess skin membranes stretched between the elongated toes of their feet. The flying tree frogs also present a modern example of parallel evolution. The Old World frogs, such as Wallace's flying tree frog, belong to an entirely different family from the New World frogs, such as the Costa Rican flying tree frog; we can see that the same glider morphology has independently been evolved by the frogs in two separate regions of the Earth. Lizards have also convergently evolved webbed feet for gliding, as in Kuhl's flying gecko and related species (table 2.2). Interestingly, these geckos also have added a narrow strip of skin running along the sides of their head, body, and tail to provide extra gliding surface area, as their webbed feet are not nearly as large in proportion to their bodies as those found in the flying tree frogs. If any of the lizards are to converge in the future on the glider-membrane morphology found in the mammals, these gecko lizards are perhaps the most likely candidates.

Convergent evolution of gliding morphologies is not confined to land animals. At least two separate groups of fishes have evolved gliding morphologies by expanding and elongating their pectoral fins, which are held horizontally as wing surfaces while the fish leap and then glide through the air (table 2.2). Again, this convergent morphology appears to be confined to actinopterygian fishes, but it is interestingly similar to the modification of forelimbs to wings in the descendants of another group of fish, the sarcopterygians (these descendants being the pterosaurs, birds, and bats; see table 2.2).

Last, convergent gliding morphologies have evolved in at least three separate groups of ants (table 2.2). Curiously, members of two of these gliding ant groups (the cephalotini and pseudomyrmecinae) glide backward, abdomen first (Yanoviak et al., 2005). What is perhaps more odd is the evolution of gliding morphology at all in the ants, given that so many species of these animals have wings and powered flight.

Most all discussions of convergent evolution begin with the spectacular examples of convergence of form in fast swimming animals (fusiform bodies, table 2.1) and flying animals (wings, table 2.2). They often end there as well. Convergence of form due to locomotory functional constraints does not stop with swimming and flying morphologies, however, so let us consider further convergent consequences of these constraints. The act of walking in the gravitational field of the Earth requires that an animal not only resist the pull of gravity in standing up, but also remain standing and balanced in the progressive falling-forward motion used in walking locomotion. Walking requires legs, which are lever assemblages composed of rigid structural elements with connecting tissues and powered by muscle contractions.

Two major groups of animals have convergently evolved legs: the arthropods and the tetrapods (table 2.3). As in the convergent evolution of wings (table 2.2), close examination of the legs of arthropods and tetrapods reveals interesting differences to solving the same functional problem, that of walking via lever assemblages. In tetrapod legs, the rigid structural elements (bones) are located inside the leg, and the connecting tendons and muscles are located on the outside of the leg (and covered with a thin layer of skin tissue). Arthropod legs are exactly the reverse: the rigid structural elements (composed of chitin) are located on the outside of the leg, and the connecting tendons and muscles are located inside the leg. A disadvantage of tetrapod endoskeletal legs is that the soft-tissue muscles and tendons are located on the outside of the leg, and thus are easily damaged (as we all painfully discover at one time or another in our lives). On the other hand, the soft-tissue muscles and tendons of the arthropod exoskeletal leg are protected by the enclosing chitin shell of the leg. A disadvantage of arthropod legs, however, is growth of the leg itself. Arthropods must periodically split and shed the outside rigid covering of the exoskeletal leg (molting), a process not necessary in growth of the endoskeletal legs of the tetrapod.

Walking locomotion has been hugely successful in the evolution of animal life on Earth: there are more species of arthropods on Earth than all other animals species combined, and tetrapod species of vertebrates

Table 2.3
Convergent evolution of animal walking morphologies

1 Convergent structure and function: LEGS (articulated, muscle-controlled lever assemblages for standing and walking in a gravitational field)

Convergent lineages:

1.1 Giant walkingstick arthropod (Bilateria: Protostomia: Ecdysozoa: Panarthropoda: Arthropoda: Mandibulata: Hexapoda: Phasmatodea: Phasmidae; *Megaphasma dentricus*)

1.2 Giraffe tetrapod (Bilateria: Deuterostomia: Chordata: Osteichthyes: Sarcopterygii: Tetrapoda:Reptiliomorpha:Synapsida:Therapsida:Mammalia:Eutheria:Laurasiatheria: Cetartiodactyla: Giraffidae; *Giraffa camelopardalis*)

2 Convergent structure and function: BIPEDALISM (adaptation for [1] maximum oxygen uptake during sustained running or hopping, [2] predator detection at a distance, [3] freeing the forelimbs for functions other than locomotion)

Convergent lineages:

2.1 Bolosaurs (Amniota: Sauropsida: Anapsida: Bolosauridae: *Eudibamus cursoris* †Permian)

2.2 Ornithosuchids (Diapsida: Archosauromorpha: Crurotarsi: Ornithosuchidae; *Ornithosuchus longidens* †Triassic)

2.3 Dinosauromorphs (Archosauromorpha: Archosauria: Ornithodira: Dinosauromorpha: Lagosuchidae; *Marasuchus illoensis* †Triassic)

2.4 Dinosaurs (Archosauria: Ornithodira: Dinosauria: Herrerasauridae; *Herrerasaurus ischigualastensis* †Triassic)

2.5 Red kangaroo (Amniota: Synapsida: Therapsida: Mammalia: Marsupialia: Diprotodontia: Macropodidae; *Macropus rufus*)

2.6 Fawn hopping mouse (Mammalia: Eutheria: Euarchontoglires: Rodentia: Muridae; *Notomys cervinus*)

2.7 California kangaroo rat (Eutheria: Euarchontoglires: Rodentia: Heteromyidae; *Dipodomys californicus*)

2.8 Springhaas (Eutheria: Euarchontoglires: Rodentia: Pedetidae; *Pedetes capensis*)

2.9 Humans (Eutheria: Euarchontoglires: Primates: Catarrhini: Hominidae: Hominini; *Ardipithecus ramidus* †Pliocene)

Note: The geological age of extinct species is marked with a †. For data sources, see text.

are much more diverse than nontetrapod. However, one particular form of tetrapod walking locomotion is curiously quite restricted: that of bipedalism. The first tetrapods evolved in the Late Devonian, and were quadrupedal amphibians (nonamniote tetrapods). The plesiomorphic quadrupedal condition is the norm for almost all living tetrapods; very few have made the transition to standing up and walking solely on their hind limbs.

The adaptive significance of bipedalism is still open to debate. Standing erect on the hind limbs clearly gives an animal the ability to see much farther in the distance, thus giving it an early-warning capability in predator detection. Walking only on the hind limbs also gives an animal the capability to use the forelimbs for functions other than locomotion: both theropod maniraptoran dinosaurs and primate mammals convergently evolved prehensile hands in their forelimbs. Another possible bipedal adaptation is allowing greater expansion of the lungs and rib cage during fast running or hopping, for greater oxygen uptake. Many species of quadrupedal lizard will stand up on their hind limbs in order to rapidly run away from danger. This behavior is most notorious in the brown basilisk lizard, *Basiliscus basiliscus*, of Central and South America, where it is known as the *Lagartija Jesucristo* (Jesus Christ lizard) because of its ability to run so fast on its hind limbs that it can actually run across the surface of water without sinking.

Bipedal locomotion has convergently arisen multiple times in both reptiles and mammals (table 2.3). The oldest currently known bipedal reptile is the Late Paleozoic anapsid species *Eudibamus cursoris* (Berman et al. 2000). Although it was a very ancient type of reptile, it was not only bipedal but digitigrade, standing on its toes. Following the end-Permian mass extinction, bipedal locomotion reevolved independently in two major groups of the archosaurs, the advanced reptiles, in the Triassic. The ornithosuchids, or "birdlike crocodiles," were bipedal predators much more closely related to living crocodiles than other archosaurians (Benton 2005). Much more significantly, in terms of subsequent evolution, is the convergent appearance of bipedalism in the advanced ornithodires and the earliest dinosaurs in the Triassic. *Marasuchus* was a bipedal dinosauromorph predator—not quite a dinosaur—and the earliest dinosaurs *Herrerasaurus* and *Eoraptor* were bipedal predators (Benton 2005). In further dinosaurian evolution, the earliest of the other major branch of the dinosaurs, the ornithischians, were bipedal fabrosaurids, such as *Lesothosaurus* (Benton 2005). Thus, bipedalism may well be a synapomorphy for the entire Dinosauria, with subsequent quadrupedalism in dinosaurs evolving as a secondary trait. All of the living dinosaurs, the birds, are bipedal.

Bipedalism has convergently arisen in two major groups of mammals: the marsupials and the placentals (table 2.3). The fascinating bipedal marsupials of Australia are well known, particularly the largest, the red kangaroo. Within the placentals, bipedalism has evolved independently several times in the rodents, and notably in the primates. The fawn hopping mouse is typical of Australian hopping-mice placental species that coexist with marsupial bipedal species. Independently, in North America, the Californian kangaroo rat (which is not a rat at all) and related species have evolved bipedal locomotion. And in South Africa, yet another independent group of rodents has evolved bipedalism, the Springhaas *Pedetes capensis*, which has been described as the "kangaroo rabbit" although it is neither a rabbit nor a kangaroo. Last, bipedalism arose independently in a group of primates in Africa, the Hominini, and is a synapomorphy for all of the various human species.

Why is bipedalism so rare in the mammals in contrast to the dinosaurs? The dinosaurian ecosystem was enormously successful; it persisted for some 150 million years, and it contained numerous bipedal animal groups. All of the theropods, the predators of the dinosaurian ecosystem, were bipedal. Numerous herbivore groups, both saurischian prosauropods and ornithischian ornithopods, were bipedal, and many of these were quite large animals, such as the hadrosaurs. Why are predatory bipedal lions or tigers nonexistent in mammalian ecosystems? Why are there no large, herbivorous bipedal horses or buffalos?

The rarity of bipedal mammalian animals is probably a function of developmental constraint, as opposed to functional constraint, for the earliest mammals and their ancestors were quadrupedal. The persistence of this symplesiomorphic trait in mammalian evolution may be due more to "developmental inertia," in that the quadrupedal condition is the inherited norm, than to any possible nonfunctionality of bipedal mammalian forms. In contrast, bipedalism is a synapomorphy for the Dinosauria—their earliest forms, both carnivores and herbivores, were bipedal. Only when they evolved such large and heavy animals as the gigantic sauropods, the massively armored ankylosaurs, and ceratopsids with enormously large skulls did dinosaurs depart from this developmental norm because bipedalism was no longer functional; at that point, they secondarily reevolved quadrupedal forms.

To conclude this section on walking-locomotion morphologies, I would like to take another thought-experiment excursion into theoretical morphology. The mythical centaur is described as a vertebrate animal with four walking legs present on a horselike posterior, and two arms present on a humanlike anterior. In considering this nonexistent form, Maclaurin and Sterelny (2008, 104) ask: "No six-limbed vertebrates have ever evolved from four-limbed ancestors. Is this evidence of the developmental impossibility of centaurs? How can we tell from the fact that the elements in a trait cluster *did not* diverge independently of one another, that they *could not* evolve independently of one another?" This is an interesting question, and it illustrates the difficulty of proving developmental constraint. But, as in our previous thought experiment on the nonexistence of dragons, we can argue for developmental constraint on the grounds of improbability. In order to evolve a centaur-form, a group of tetrapods would have to first develop an additional pair of legs (not impossible, but highly improbable for vertebrates) and, second, develop arms from the new pair of legs. It is interesting to note here that the insects have in fact developed a centaur-form: the praying mantis, *Mantis religiosa*. Evolved from ancestors for whom the six-legged condition is the developmental norm, the mantid has four walking appendages on its posterior and two arms on the upright thorax below the head, like a centaur. And not only have the insects evolved centaurs, but they did it twice! Centaur morphologies very similar to true mantids (Hexapoda: Mantodea: Mantidae) have been independently evolved by the "false mantid" mantidflies (Hexapoda: Neuroptera: Mantispidae), a group of insects more related to ant lions than mantids. The green mantidfly of southeastern North American, *Zeugomantispa minuta*, looks virtually identical in form to *Mantis religiosa* but is smaller, being only about one-fifth the length of the mantid. Thus, even centaurs have been convergently evolved, but only by insects.

So far we have considered animal locomotion under water (swimming), aloft in the air (flying), and on the surface of the Earth (walking). Now let us consider animal locomotion under the surface of the Earth: burrowing. The number of independent convergences on burrowing morphologies is amazing (table 2.4), particularly when it would seem that these morphologies are related to a rather restrictive and highly specialized way of life.

The evolution of the four walking limbs of the earliest amphibians from their fish ancestors' lobed fins is a complicated process spanning

Table 2.4
Convergent evolution of animal burrowing morphologies

1 Convergent structure and function: SNAKE-SHAPED BODIES (penetrator adaptations: streamlined, pointed heads for soil penetration; elongated, thin cylindrical bodies to minimize drag resistance in burrows; loss of fore and hind limbs no longer used in locomotion)

Convergent lineages:

1.1 Aistopods (Tetrapoda: Lepospondyli: Aistopoda: Phlegethontiidae; *Phlegethonia linearis* †Carboniferous)

1.2 Greater yellow-banded caecilian (Tetrapoda: Batrachomorpha: Lissamphibia: Gymnophiona: Cecelidae; *Ichthyophis glutinosus*)

1.3 Florida worm lizard (Tetrapoda: Reptiliomorpha: Amniota: Sauropsida: Diapsida: Lepidosauromorpha: Lepidosauria: Squamata: Scleroglossa: Amphisbaenea: Amphisbaenidae; *Rhineura floridana*)

1.4 Burton's legless lizard (Lepidosauria: Squamata: Scleroglossa: Gekkonta: Pygopodidiae; *Lialis burtonis*)

1.5 Brazilian bachian lizard (Squamata: Scleroglossa: Autarchoglossa: Scincomorpha: Gymnophthalmidae; *Bachia oxyrhinus*)

Table 2.4
(continued)

1.6 California legless lizard (Squamata: Scleroglossa: Autarchoglossa: Anguimorpha: Anniellidae; *Anniella pulchra*)
1.7 Eastern glass lizard (Squamata: Scleroglossa: Autarchoglossa: Anguimorpha: Anguidae; *Ophisaurus ventralis*)
1.8 Scarlet snake (Squamata: Scleroglossa: Autarchoglossa: Anguimorpha: Serpentes: Colubridae; *Cemophora coccinea*)

2 Convergent structure and function: MOLE-SHAPED BODIES (excavator adaptations: compact cylindrical bodies to fit within burrows, powerful forelimbs or hind limbs with enlarged claws or shovel structures for digging, rudimentary eyes)

Convergent lineages:
2.1 Mole cricket (Bilateria: Protostomia: Ecdysozoa: Arthropoda: Mandibulata: Hexapoda: Orthoptera: Gryllotalpidae; *Gryllotalpa gryllotalpa*)
2.2 Mexican burrowing toad (Bilateria: Deuterostomia: Chordata: Osteichthyes: Sarcopterygii: Tetrapoda: Batrachomorpha: Lissamphibia: Batrachia: Anura: Rhinophrynidae; *Rhinophrynus dorsalis*)
2.3 Couch's spadefoot toad (Batrachomorpha: Lissamphibia: Batrachia: Anura: Pelobatidae; *Scaphiopus couchii*)
2.4 Guinea shovel-snout frog (Lissamphibia: Batrachia: Anura: Hemisotidae; *Hemisus guineensis*)
2.5 Mole salamander (Lissamphibia: Batrachia: Urodela: Ambystomidae; *Ambystoma talpoideum*)
2.6 Southern marsupial mole (Tetrapoda: Reptiliomorpha: Amniota: Synapsida: Therapsida: Mammalia: Marsupialia: Notoryctemorpha: Notoryctidae; *Notoryctes typhlops*)
2.7 Pink fairy armadillo (Mammalia: Eutheria: Xenarthra: Dasypodidae; *Chlamydophorus truncatus*)
2.8 Cape golden mole (Mammalia: Eutheria: Afrotheria: Afrosoricida: Chrysochloridae; *Chrysochloris asiatica*)
2.9 European mole (Mammalia: Eutheria: Laurasiatheria: Eulipotyphles: Talpidae; *Talpa europaea*)
2.10 Ancient palaeanodonts (Eutheria: Laurasiatheria: Pholidota: Palaeanodonta: Epoicotheridae; *Epoicotherium unicum*, *Xenocranium pileorivale* †Oligocene)
2.11 Common vole (Eutheria: Euarchontoglires: Rodentia: Myomorpha: Arvicolidae; *Microtus arvalis*)
2.12 Lesser mole rat (Eutheria: Euarchontoglires: Rodentia: Myomorpha: Spalacidae; *Spalax leucodon*)
2.13 Siberian zokor (Eutheria: Euarchontoglires: Rodentia: Myomorpha: Cricertidae: Mylospalacinae; *Myospalax myospalax*)
2.14 Asian bamboo rat (Eutheria: Euarchontoglires: Rodentia: Myomorpha: Rhizomyidae; *Rhizomys sumatraensis*)
2.15 Chilean coruro (Eutheria: Euarchontoglires: Rodentia: Hystricomorpha: Octodontidae; *Spalacopus cyanus*)
2.16 Rio Negro tuco-tuco (Eutheria: Euarchontoglires: Rodentia: Hystricomorpha: Ctenomyidae; *Ctenomys rionegrensis*)
2.17 African mole rat (Eutheria: Euarchontoglires: Rodentia: Hystricomorpha: Bathyergidae; *Cryptomys hottentotus*)
2.18 Plains pocket gopher (Eutheria: Euarchontoglires: Rodentia: Sciuromorpha: Geomyidae; *Geomys bursarius*)

Note: The geological age of extinct species is marked with a †. For data sources, see text.

millions of years in the Late Devonian and Early Carboniferous. So it is astonishing to note that some ancient amphibians promptly devolved these same hard-earned appendages and became limbless in the later Carboniferous. These are the aistopods, like *Phlegethonia linearis*, which has a morphology virtually identical to a modern snake (Benton 2005) but is not a snake. Amphibian convergence on penetrator-form burrowing morphologies did not end in the Carboniferous, for an entire group of legless amphibians exist today, the Gymnophiona, whose caecilian members strongly resemble earthworms. In addition, they have secondarily evolved a hydrostatic system for burrowing and locomotion that is convergent on the ancient plesiomorphic hydrostatic system seen in true earthworms (O'Reilly et al. 1997).

These amphibians are not alone, as the amphisbaenid Florida worm lizard is also identical in overall appearance to an earthworm, even to its pinkish-purple color, yet it is a lepidosaurian reptile. Lee and Shine (1998) have demonstrated that the amphisbaenid worm lizards are more closely related to gekko lizards than to the snakes. But other groups of lizards have evolved snakelike penetrator morphologies, like the ancient amphibian aistopods. These modern reptiles include the California legless lizard of western North America, the legless glass lizards of eastern North America, the pygopodid legless lizards of Australia, and the recently discovered gymnophthalmid legless lizard *Bachia oxyrhinus* of Brazil, South America. The true penetrator-form burrowing snakes, like the very pretty scarlet snake of southeastern North America, are more closely related to modern varanid lizards, such as the very impressive predator *Varanus komodensis*, the Komodo dragon (Lee and Shine 1998), and more distantly related to the ancient marine snakelike mosasaurs (see eel-form swimmers in table 2.1).

When burrowing animals are mentioned, most people think of the mole, with its short, compacted, cylindrical body and powerful digging forelimbs—typical excavator-form burrowers. However, this same form can be found in the insects, as seen in the mole crickets. And, just as the mammalian mole is considered by many to be a pest, the mole cricket is injurious to garden plants and crops. Excavator-form burrowers are also found in the amphibians. For example, ambystomid mole salamanders have short, compacted, cylindrical bodies and blunt heads for burrowing. The amphibian toads have convergently evolved excavator-forms twice, but in a backward fashion: both the Mexican burrowing toads and the spadefoot toads burrow with their powerful hind limbs, not their forelimbs, and in both groups the hind feet of the animals have been enlarged and stiffened into spadelike digging structures. Last, excavator forms are also found in the frogs, as in the Guinea shovel-snout frog.

In modern placental mammals, the mole-form has independently evolved in three very distantly related groups: the laurasiatherian true talpid moles, so common in Europe and North America; the afrotherian golden chrysochlorid moles; and the xenarthran armadillos (table 2.4). In the Oligocene, some 30 million years ago, another group of placental mammals independently evolved the mole-form, the ancient epoicotherids (Rose and Emry 2005). An even more distantly related mole-form has been convergently evolved by the marsupial mammals, as is seen in the modern southern marsupial mole.

Last, multiple convergent evolutions of excavator-form morphologies have taken place in another group of placental mammals, the rodents, as documented by Nevo (1999). Mole-forms have been convergently evolved in four groups of myomorph rodents around the world: the voles, the lesser mole rats, the Siberian zokors, and the Asian bamboo rats. In South America, two groups of hystricomorph rodents have evolved mole-forms: the coruros, endemic to Chile, and the tuco-tucos, found far to the southeast in Uruguay. In Africa, another group of hystricomorph rodents have evolved mole-forms: the African mole rats. And last, in North America, the sciuromorph rodents have produced the pocket gophers.

From the perspective of theoretical morphology, it is interesting to note that burrowing mammals have never produced penetrator-forms—there exist no mammalian snake-forms. Yet it is easy to envision an elongated, furry, snakelike mammal, one that has secondarily lost its legs. Of all the mammals, the weasels and ferrets—with their elongated bodies and small legs—are perhaps the best candidates for producing a hypothetical snake-form mammal. In contrast, the reptiles have never produced excavator mole-forms; many reptiles excavate (turtles come to mind) but usually only sporadically and shallowly, as when they are preparing to hibernate. The burrowing reptiles' preference for penetrator-forms and the burrowing mammals' preference for excavator-forms again raise the question of the role of functional versus developmental constraint in explaining the absence of theoretically possible morphologies in evolution.

In the animals, mode of locomotion is usually independent of how the animal feeds; that is, whether it is a carnivore or an herbivore. Some burrowing mammals are carnivores, like talpid moles, while others are herbivores, like tuco-tucos (Nevo 1999). Some flying birds are carnivores, like hawks; others are herbivores, like parrots. Similary, some flying bats are carnivores, like the insectivorous microbats, while others are herbivores, like the frugivorous megabats. In addition to locomotion, the particular mode of feeding of an animal imposes additional functional constraints, which again are reflected in subsequent convergent evolution, as we shall see in the next section of the chapter.

Carnivores: Prey Detection

As heterotrophic organisms, animals are incapable of synthesizing their own food, unlike plants, which we shall consider in the next chapter. Because all animals need food, it is no surprise that many different animal groups have repeatedly, convergently, evolved the same successful forms and structures used in food acquisition.

In order to survive, carnivores must be able to detect, capture, and kill prey animals. One very obvious way to detect prey is to be able to see the prey animal, that is, to have eyes. The convergent evolution of eyes themselves will be considered later in this chapter; of interest here are the modifications of eyes found in the carnivores. An oft-repeated rule for animals is that carnivorous animals have binocular vision—their eyes are located forward enough on their skulls to give the animal overlapping fields of vision, which gives the animal three-dimensional depth perception and thus enables it to precisely locate a prey animal in space. In contrast, herbivorous animals usually have eyes located on either side of their heads, giving them an almost 360-degree field of vision for the detection of a predator in front of them, to either side, or even behind them. The disadvantage of herbivorous-animal vision is that the animal has either no binocular vision at all or only very restricted binocular vision directly in front of it. In some carnivores, highly developed binocular vision has the disadvantage that their field of vision is restricted to the front of the animal, so that they must turn their heads in order to see to their sides or to look behind themselves.

Exceptions to this rule are numerous. For example, the primates have highly developed binocular vision that is, in general, unrelated to a carnivorous mode of life. Instead, many primates need binocular vision because they are arboreal—they need precise depth perception in order to jump from tree branch to tree branch without missing the branch, and perhaps falling to their deaths to the forest floor below.

Binocular vision is a very ancient condition for chordate animals. Even though their eyes are placed on either side of their laterally flattened bodies, many fish still possess a limited degree of binocular vision directly in front of their bodies, where their visual fields briefly overlap. For this reason, the rainbow trout, *Oncorhynchus mykiss*, can accurately stalk and precisely bite a floating insect resting on the water surface above it.

Thus, in many cases the possession of binocular vision in a carnivore is simply a plesiomorphic trait that the animal inherited from a distant ancestor, and not a newly derived or convergent trait. What is often convergent, however, is the *degree of binocular vision*—the degree to which the eyes have been moved forward on the skull. Consider the owl and the cat. These are very different animals—one is an avian dinosaur and the other is a placental mammal, and they represent two branches of the evolutionary tree that diverged over 340 million years ago (the sauropsids and the synapsids). Yet both animals have eyes rotated so far anterior on their skulls that they face directly forward, a convergent condition also found in humans. These animals have very highly developed three-dimensional depth perception, as we do.

However, if we closely examine the eyes of the owl, cat, and human, we immediately spot a major difference—the cat has eyes with vertical-slit pupils. The vertical-slit pupil has been repeatedly and independently evolved many times throughout the evolutionary history of animals (table 2.5). In the seas, three separate groups of modern cartilaginous fishes have convergently evolved eyes with vertical-slit pupils. The angel sharks (species of *Squatina*) are bottom dwellers with flattened bodies, and lie in wait for passing prey that they then ambush. They often operate in dim light or at night, where it is difficult for prey animals to detect them lying on the sea bottom. The carcharhinid sharks, such as the whitetip shark, have eyes with vertical-slit pupils and gold irises that are astonishingly similar to those often found in black cats. Last, the myliobatid rays, such as the beautiful spotted eagle ray, have independently evolved the same type of pupil.

On land, the vertical-slit pupil has repeatedly evolved independently in the eyes of amphibians, reptiles, birds, and mammals (table 2.5). In the amphibians, the great majority of frogs and toads have eyes with horizontal-slit pupils (a trait we shall consider in more detail later). However, some tree frogs have evolved eyes with vertical-slit pupils, a pupil type that has also been convergently evolved in the peculiar tailed frogs, which are aquatic carnivores that inhabit streams and creeks. Vertical-slit pupils have evolved independently in two families of fossorial toads: the rhinophrynid burrowing toads and the pelobatid spadefoot

Table 2.5
Convergent evolution of predator eye structures and vision systems

1 Convergent structure and function: VERTICAL-SLIT PUPILS IN EYES (allows [1] full usage of the diameter of the lens of the eye in bright light as well as in very low light intensities, with well-focused images in all light intensities; [2] particular detection of motion in the horizontal plane)

Convergent lineages:

1.1 Pacific angel shark (Vertebrata: Chondrichthyes: Elasmobranchii: Squatinidae; *Squatina californica*)

1.2 Whitetip shark (Chondrichthyes: Elasmobranchii: Carcharhinidae; *Charcharhinus longimanus*)

1.3 Spotted eagle ray (Chondrichthyes: Elasmobranchii: Myliobatidae; *Aetobatus narinari*)

1.4 Brownbelly leaf frog (Vertebrata: Osteichthyes: Sarcopterygii: Tetrapoda: Batrachomorpha: Lissamphibia: Batrachia: Anura: Hylidae; *Phyllomedusa tarsius*)

1.5 Tailed frog (Batrachomorpha: Lissamphibia: Batrachia: Anura: Ascaphidae; *Ascaphus truei*)

1.6 Mexican burrowing toad (Lissamphibia: Batrachia: Anura: Rhinophrynidae; *Rhinophrynus dorsalis*)

1.7 Couch's spadefoot toad (Lissamphibia: Batrachia: Anura: Pelobatidae; *Scaphiopus couchii*)

1.8 Helmeted gecko (Tetrapoda: Reptiliomorpha: Amniota: Sauropsida: Diapsida: Lepidosauromorpha: Lepidosauria: Squamata: Scleroglossa: Gekkonta: Gekkonidae; *Tarentola chazaliae*)

1.9 Burton's legless lizard (Lepidosauria: Squamata: Gekkonta: Pygopodidae; *Lialis burtonis*)

1.10 Granite night lizard (Lepidosauria: Squamata: Scleroglossa: Autarchoglossa: Scincomorpha: Xantusidae; *Xantusia henshawi*)

1.11 Indian python (Squamata: Scleroglossa: Autarchoglossa: Anguimorpha: Serpentes: Boidae; *Python molurus*)

1.12 Timber rattlesnake (Squamata: Scleroglossa: Autarchoglossa: Anguimorpha: Serpentes: Viperidae: Crotalinae; *Crotalus horridus*)

1.13 Nile crocodile (Diapsida: Archosauromorpha: Archosauria: Crurotarsi: Crocodylia: Crocodylidae; *Crocodylus niloticus*)

1.14 Black skimmer (Archosauria: Ornithodira: Dinosauria: Saurischia: Theropoda: Maniraptora: Aves: Charadriiformes: Rhynchopidae; *Rhynchops niger*)

1.15 Small cats (Amniota: Synapsida: Therapsida: Mammalia: Eutheria: Laurasiatheria: Carnivora: Feliformia: Felidae; *Felis sylvestris catus*)

1.16 Red fox (Mammalia: Eutheria: Laurasiatheria: Carnivora: Caniformia: Canidae; *Vulpes vulpes*)

1.17 Harp seal (Eutheria: Laurasiatheria: Carnivora: Caniformia: Phocidae; *Pagophilus groenlandicus*)

1.18 Slow loris (Eutheria: Euarchontoglires: Primates: Lorisiformes; *Nycticebus coucang*)

2 Convergent structure and function: HORIZONTAL-SLIT PUPILS IN EYES (allows [1] full usage of the diameter of the lens of the eye in bright light as well as in very low light intensities, with well-focused images in all light intensities; [2] particular detection of motion in the vertical plane)

Convergent lineages:

2.1 Common octopus (Bilateria: Protostomia: Lophotrochozoa: Mollusca: Cephalopoda: Coleoidea: Octopodidae; *Octopus vulgaris*)

Table 2.5
(continued)

2.2 Common toad (Bilateria: Deuterostomia: Chordata: Osteichthyes: Sarcopterygii: Tetrapoda: Batrachomorpha: Lissamphibia: Batrachia: Anura: Bufonidae; *Bufo bufo*)
2.3 Bullfrog (Batrachomorpha: Lissamphibia: Batrachia: Anura: Ranidae; *Rana catesbeiana*)
2.4 Green tree frog (Lissamphibia: Batrachia: Anura: Hylidae; *Hyla cinerea*)
2.5 Asian palm civet (Tetrapoda: Reptiliomorpha: Amniota: Synapsida: Therapsida: Mammalia: Eutheria: Laurasiatheria: Carnivora: Viveridae; *Paradoxurus hermaphroditus*)
2.6 Bottlenose dolphin (Mammalia: Eutheria: Laurasiatheria: Cetartiodactyla: Cetacea: Odontoceti: Delphinidae; *Tursiops truncatus*)

3 Convergent structure and function: ENLARGED EYES (increases retinal surface area for low-light-intensity vision in nocturnal predators)
Convergent lineages:
3.1 Common octopus (Bilateria: Protostomia: Lophotrochozoa: Mollusca: Cephalopoda: Coleoidea: Octopodidae; *Octopus vulgaris*)
3.2 Squirrelfish (Bilateria: Deuterostomia: Chordata: Osteichthyes: Actinopterygii: Beryciformes: Holocentridae; *Holocentrus adscensionis*)
3.3 Great horned owl (Osteichthyes: Sarcopterygii: Reptiliomorpha: Amniota: Sauropsida: Diapsida: Archosauromorpha: Ornithodira: Dinosauria: Saurischia: Theropoda: Maniraptora: Aves: Strigiformes: Strigidae; *Bubo virginianus*)
3.4 Tasmanian devil (Amniota: Synapsida: Therapsida: Mammalia: Marsupialia: Dasyuromorpha: Dasyuridae; *Sarcophilus harrisii*)
3.5 Small cats (Mammalia: Eutheria: Laurasiatheria: Carnivora: Feliformia: Felidae; *Felis sylvestris catus*)
3.6 Lesser bushbaby (Mammalia: Eutheria: Euarchontoglires: Primates: Lorisiformes: Galagidae; *Galago senegalensis*)
3.7 Philippine tarsier (Eutheria: Euarchontoglires: Primates: Tarsiiformes: Tarsiidae; *Tarsius syrichta*)

4 Convergent structure and function: INFRARED "VISION" ([1] pit organs in snakes that detect infrared light, used for detecting heat from warm-blooded, endothermic prey animals even in total darkness; [2] infrared receptors in insects, used to detect forest fires)
Convergent lineages:
4.1 Pyrophyllic beetle (Bilateria: Protostomia: Ecdysozoa: Arthropoda: Mandibulata: Hexapoda: Coleptera: Buprestidae: Buprestinae; *Melanophila acuminata*)
4.2 Australian fire beetle (Arthropoda: Mandibulata: Hexapoda: Coleptera: Buprestidae: Chrysobothrinae; *Merimna atrata*)
4.3 Australian flat bug (Arthropoda: Mandibulata: Hexapoda: Heteroptera: Aradidae; *Aradus albicornis*)
4.4 Indian python (Bilateria: Deuterostomia: Chordata: Osteichthyes: Sarcopterygii: Reptiliomorpha: Amniota: Sauropsida: Diapsida: Lepidosauromorpha: Squamata: Scleroglossa: Autarchoglossa: Anguimorpha: Serpentes: Boidae; *Python molurus*)
4.5 Timber rattlesnake (Lepidosauromorpha: Squamata: Scleroglossa: Autarchoglossa: Anguimorpha: Serpentes: Viperidae: Crotalinae; *Crotalus horridus*)

5 Convergent structure and function: ELECTRIC-FIELD "VISION" (electroreceptive organs allowing the detection of moving prey by the electrical-field activity associated with muscle contractions)
Convergent lineages:
5.1 Cephalaspid osteostracans (Vertebrata: Osteostraci: Cephalaspidae; *Hemicyclaspis murchisoni* †Devonian)
5.2 Marbled electric ray (Vertebrata: Chondrichthyes: Elasmobranchii: Batoidea: Torpediniformes: Torpedinidae; *Torpedo marmorata*)

Table 2.5
(continued)

5.3 Winter skate (Chondrichthyes: Elasmobranchii: Batoidea: Rajiformes: Rajidae; *Raja ocellata*)
5.4 Pacific stargazer (Vertebrata: Osteichthyes: Actinopterygii: Teleostei: Perciformes: Uranoscopidae; *Astroscopus zephyreus*)
5.5 Banded knifefish (Actinopterygii: Teleostei: Gymnotiformes: Gymnotidae; *Gymnotus carapo*)
5.6 Peters' elephant-nose fish (Actinopterygii: Teleostei: Osteoglossiformes: Mormyridae; *Gnathonemus petersii*)
5.7 African electric catfish (Actinopterygii: Teleostei: Siluriformes: Malapteruridae; *Malapterurus electricus*)
5.8 Duckbill platypus (Osteichtyes: Sarcopterygii: Reptiliomorpha: Amniota: Synapsida: Therapsida: Mammalia: Monotremata: Ornithorhynchidae; *Ornithorhynchus anatinus*)
5.9 Australian echidna (Mammalia: Monotremata: Tachyglossidae; *Tachyglossus aculeatus*)

Note: The geological age of extinct species is marked with a †. For data sources, see text.

toads. These toads burrow into the ground during the day, and emerge to hunt at night.

In the lepidosaurian reptiles, the great majority of both lizards and snakes have eyes with round pupils. But among the lizards, we find vertical-slit pupils in the helmeted geckos (whose bizarre eyes are unique in many ways) and in all the nocturnal species of the gecko family. On opposite sides of the planet, the xantusid night lizards of North America and many species of the pygopodid legless lizards of Australia, like Burton's legless lizard, have eyes with vertical-slit pupils. Among the snakes, the two deadliest snake groups have convergently evolved eyes with vertical-slit pupils: the boid constrictors and the venomous pit vipers. Both of these predator groups have also evolved infrared-sensitive pit organs, and are able to hunt warm-blooded animals in total darkness (a trait that we shall consider in more detail in a moment).

The advanced archosaurian reptiles, such as the crocodiles and alligators, are aquatic carnivores that have independently evolved eyes with vertical-slit pupils. Since the great majority of the archosaurian dinosaurs are extinct, we may never know what type of eyes they possessed. This fact does not prevent the makers of science fiction movies from usually portraying extinct theropod predators, such as the velociraptors, with eyes with vertical-slit pupils. In this context, it is interesting to note that only one species of the modern avian dinosaurs has eyes with vertical-slit pupils: the black skimmer *Rhynchops niger* (Zusi and Bridge 1981). The many species of modern owls, which are avian nocturnal predators, all have eyes with round pupils. Zusi and Bridge (1981) suggest that the wide range of light intensities encountered by the black skimmer, which seeks prey during both the dark of night and brilliant daylight, led it to evolve vertical-slit pupils to protect its eyes during daylight fishing.

In the placental mammals, eyes with vertical-slit pupils are most characteristic of the small cats, such as the European wildcat *Felis sylvestris* or the North American bay lynx *Lynx rufus*. All the large cats that hunt during daylight, such as the African lion *Panthera leo*, have eyes with round pupils. However, eyes with vertical-slit pupils have been convergently evolved within the canid predators as well, even though these animals are so unlike the cats in many ways. One example is the modern red fox, *Vulpes vulpes*. The small foxes are in essence honorary cats, converging on a felid way of life. Many of the marine carnivores, such as the harp seal, have also evolved eyes with pupils that are round when the animal is hunting in the water, but that contract to vertical slits when in the air. Last, eyes with vertical-slit pupils have independently evolved within the primates themselves, and are present in the lorisiform nocturnal carnivore the slow loris, *Nycticebus coucang* (Malmström and Kröger 2006).

In summary, the vertical-slit pupil has repeatedly evolved in carnivores that are nocturnal, hunting at night, or crepuscular, hunting in the dim light of twilight or just before sunrise, as well as in those that hunt in environments that normally have low light intensities, as in many aquatic habitats. Malmström and Kröger (2006) have argued that in animals with multifocal lenses, the slit-pupil system allows the full usage of the diameter of the lens of the eye in bright light as well as in very low light intensities, with well-focused images at all light intensities. In contrast, the round-pupil system shades the peripheral zones of the lens of the eye as it is constricted, leading to the loss of well-focused images at wavelengths of light that are normally focused in those regions of the lens. Zusi and Bridge (1981) report that, in the eyes of small cats, the slit-pupil system allows a greater reduction in pupil size than is possible with a round-pupil sphincter system; this is particularly important to animals that have very large pupils when the pupillary opening is at its maximum dilation. Thus, for animals with large pupils, the slit-pupil system offers much better protection to the retina of the eye in bright light conditions than a round-pupil system.

The catch in the reasoning of these two studies is this: it does not matter how the slit in the slit-pupil system is oriented in order to obtain both of these superior-focus and retinal-protection advantages. That is, a horizontal-slit pupil would work just as well as a vertical-slit pupil. And indeed we shall see that many animals—both carnivores and herbivores—have convergently evolved eyes with horizontal-slit pupils. Why, then, do we find the vertical-slit pupil only in carnivore eyes, and never in herbivores? The answer appears to involve the type of motion detection that is important to a predator. The vertical-slit-pupil system in the eye is particularly good at detecting motion in a horizontal plane. And since most prey animals are moving in a near-horizontal plane, parallel to the surface of the ground or to the sea floor, predators with vertical-slit pupils in their eyes have an advantage in spotting prey motion, even in very low light intensities.

Although the frogs and toads are carnivores, the great majority of frogs and toads have eyes with horizontal-slit pupils, not vertical. These include the true toads (bufonids), most of the true frogs (ranids), and most of the true tree frogs (hylids); see table 2.5. In these animals, the horizontal-slit-pupil system is particularly good at detecting motion in a vertical plane. Because many of their prey species are flying insects, which move in three dimensions off the surface of the Earth, the horizontal-slit pupils of toads and frogs give these predators an advantage in spotting flying prey. Many also are diurnal predators, hunting during the daylight hours, as opposed to nocturnal or crepuscular predators.

Otherwise, the horizontal-slit pupil is rare in predators. The catlike Asian palm civet has convergently evolved horizontal-slit-pupil eyes, but its adaptive significance is open to question, as these animals hunt in a fashion similar to true cats, which have vertical-slit-pupil eyes. Curiously, eyes with horizontal-slit pupils have also convergently arisen in a few marine predators like the bottlenose dolphin and the octopus, animals with radically different evolutionary backgrounds but with similar eyes. As with the palm civets, it is unclear what selective advantage these eye systems may possess, if any, in the habitats frequented by the bottlenose dolphin and the octopus.

In addition to pupil type, the size of the eye itself is subject to convergent evolution depending on the hunting behavior of the animal. Nocturnal predators have much larger eyes, relative to the size of their skulls, than diurnal predators. Within the avian clade, the diurnal-hunting hawk has small eyes, while the nocturnal-hunting owl has enormous eyes. Within the felid clade, the diurnal-hunting lion has small eyes, while the nocturnal-hunting small cats have enormous eyes. I have listed only a few notable examples of this convergent trait in table 2.5, as it is rife throughout the animal kingdom. What is not common among animals is an extreme, but highly advantageous, form of nocturnal "vision": the ability to "see" infrared light.

The evolution of infrared vision would obviously give a predator an extremely deadly advantage over warm-blooded, endothermic prey animals in that the predator would be able to detect the prey animal even in total darkness simply by detecting its body heat. Yet no predator has evolved eyes that are capable of seeing the infrared spectrum of light. It may well be that there are limits on the maximum wavelength of light that can be detected by the vertebrate eye (the same appears to be true for the compound eyes of the insects, as we shall see in a moment). In the classic work *On Growth and Form*, Thompson (1942) suggested that there is a minimum size to the vertebrate eye, below which the eye simply cannot function given the wavelengths of visible light (around 400 nm to 700 nm for humans), and Purnell (1995) has marshaled evidence from divergent vertebrates that indicates this limit is in the range of 1.0 to 1.5 millimeters in diameter. Thus, at the maximum end of the visible light spectrum, it may well be impossible to evolve an eye that can detect electromagnetic radiation with wavelengths greater than 800 nanometers, the beginning of the infrared spectrum.

However, two convergent groups of snakes have evolved pit organs capable of detecting infrared radiation (table 2.5). In the crotaline pit vipers, a heat-sensitive facial pit is located between the eye and the nostril on each side of the head. In boid constrictors, heat-sensitive pits are located in the lip scales. The sensory information from the pit organs is processed in the same region of the snake's brain as the sensory information from the eyes; thus, this may be the only way in which vertebrates may "see" infrared light. This type of convergent "vision" may also be attainable only by ectothermic predators, as an endothemic predator's own body heat may overwhelm, and hence render useless, the heat-at-a-distance detection ability of such an infrared-perception organ. Although at first glance the convergent evolution of infrared "vision" via heat-sensitive facial pits in two separate groups of snakes is remarkable, the selective advantage of being able to detect endothermic prey in total darkness leads one to wonder why more ectothermic predators have not convergently evolved this capability.

There exists another group of animals that have evolved the capability of "seeing" infrared light—the bizarre pyrophilous beetles and flat bugs—but they also cannot detect infrared light with their compound eyes. Twice within the buprestid beetle family (Evans 1966; Schmitz et al. 2008) and once within the aradid flat bug family (Schmitz et al.

2008), three groups of insects have convergently evolved specialized infrared-receptor organs (table 2.5). These insects actually hunt for forest fires, seeking out burnt wood in order to feed on fast-growing postfire fungi. As such, they are not strictly predators because their intended prey are fungi, not other animals. But, as the fungi are much more closely related to metazoans than either of these two groups are to the plants (see appendix of this book), I include them here.

The fact that both the camera-eyed vertebrates and the compound-eyed arthropods independently evolved the capability of detecting infrared radiation—but not with their eyes—argues for functional constraint rather than developmental constraint in the absence of infrared vision in animals. Surely, if it were functionally possible, some group of animals with these very different types of eyes would have evolved a modified eye structure that could detect electromagnetic radiation with wavelengths greater than 800 nm in the past 600 million years of evolution. But could the absence of infrared vision also be considered a developmental constraint peculiar to Earth-type life? In an interesting case of deep homology, it is now known that all animal eyes—both camera and compound—contain highly conserved transcription factor *Pax-6* gene homologs (a fact that we shall consider in more detail later in the chapter). Might some alien forms of life elsewhere in the universe, life forms that evolved eyes with a different genetic coding and totally different eye structure for detecting electromagnetic radiation, easily see long-wavelength light as well as short-wavelength light? We shall return to this question in chapter 7, when we consider the possibility that there might exist organic forms that function perfectly well, but that nevertheless cannot be developed by Earth life.

An even more exotic prey-detection system is the electric fishes' ability to "see" an electric field itself, and not just electromagnetic radiation, in that the same regions of the brain that process visual information also process electrosensory information in these fish (Bastian 1982; Moller 1995). Seven different lineages of fish have convergently evolved the capability of detecting an electric field (table 2.5). These include two different groups of rays and skates in the cartilaginous fishes, four different groups of modern bony fishes, and one group of ancient agnathan fishes. This latter group, the cephalaspid osteostracans, is particularly interesting. These primitive jawless fishes possessed three sensory fields on their bony head shields, two lateral and one on the top of the head. Thick bundles of nerves led from these sensory fields back to the brain, but appear to have led to the auditory region of the brain (Stensiö 1963) rather than to the visual region. These ancient fishes therefore may have evolved the capability to "hear" the static of an electric field, rather than "seeing" the electric field.

Modern electric fishes are predators, and use their electroreceptive organs to locate moving prey by detecting the electric-field activity associated with muscle contractions in the prey organisms. Many are able to "see" the electric field of their prey even in total darkness, similar to the ability of the pit viper to "see" the infrared radiation emitted by endothermic prey species in total darkness. For example, the African electric catfish is a voracious nocturnal predator, but in contrast to most nocturnal predators, it has very small eyes (even smaller than those of diurnal fishes). Instead of using visual detection to hunt its prey, it uses electrosensory detection.

Water is an excellent conductor of electricity, and seven groups of fishes have independently evolved the capability to detect an electric field in water. But these fishes are not alone: two separate groups of primitive monotreme mammals—the duckbill platypus and the Australian echidna—have also evolved the capability of detecting electric fields (table 2.5). The duckbill platypus is a semiaquatic predator capable of hunting in total darkness under water (Gregory et al. 1987; Proske and Gregory 2003); it has thus converged on the same hunting strategy used by the electric fishes. The Australian echidna is a land-dwelling animal, however, and electric field intensities in air are very weak compared to those that can be developed in water. The echidna uses the electroreceptors on its beak to detect an electric field in the moist soil of its rainforest habitat, produced by moving earthworms (Manger et al. 1997).

Charles Darwin found the convergent evolution of electroreception in the electric fishes to be so unusual and so improbable that he included it in *On the Origin of Species* in his list of difficulties for the theory of natural selection to explain: "The electric organs offer another and more serious difficulty; for they occur in only about a dozen fishes, of which several are widely remote in their affinities. Generally when the same organ appears in several members of the same class, especially if in members having very different habits of life, we may attribute its presence to inheritance from a common ancestor; and its absence in some of the members to its loss through disuse or natural selection. But if the electric organs had been inherited from one ancient progenitor thus provided, we might have expected that all electric fishes would have been specially related to each other" (Darwin 1859, 193). Today we know that the convergent distribution of electrosensory organs in fish is much more widespread than Darwin realized, and that they have convergently appeared even in the mammals. Rather than posing a difficulty for the theory of natural selection, the convergent evolution of electrosensory organs in distantly related animals is seen today as a prime example of functional constraint and the process of natural selection in action (Zakon and Unguez 1999, Hopkins 2008).

Many animals on Earth have developed the capability of detecting a magnetic field as well, but I know of no animal that uses this capability to locate prey (presumably because prey animals do not produce detectable magnetic fields). Rather than hunting, animals use magnetosensory capabilities to orient themselves relative to the magnetic field of the Earth, particularly in migratory species.

In addition to electromagnetic waves, many carnivores can detect pressure waves in water or air when locating prey. The possession of the tympanic-membrane system of hearing, or sound detection, in land animals is a trait that can be traced back to the early tetrapods in the Late Devonian and Carboniferous (more on the convergent evolution of this trait later in the chapter). However, tympanal hearing systems have been convergently modified in surprising ways by more derived tetrapod predators, particularly nocturnal predators. For example, as flying nocturnal predators, owls need to be able to locate prey animals in all three spatial dimensions; otherwise, they might overshoot or undershoot the prey animal when they swoop down out of the sky. Norberg (1977) has argued that asymmetric ear systems for three-dimensional stereophonic hearing have been convergently evolved by at least five separate groups of owls, independently of one another (table 2.6). In the tytonid owls, asymmetric ears have been independently evolved by species of the genus *Tyto* and the genus *Pholidus*. In the strigid owls, asymmetric ears of similar structure are found in owl species in the genera *Bubo*, *Strix* (such as the Eurasian tawny owl, *S. aluco*), and *Ciccaba* (such as the mottled owl, *C. virgata*). Asymmetric ears of different structure are found in strigid owl species of the genera *Asio*, *Rhinoptynx* (such as the striped owl, *R. clamator*), and *Pseudoscops* (such as the Jamaican owl, *P. grammicus*). Last, yet another group of strigid owls, consisting of species of the genus *Aegolius*, have asymmetric ears that are different from the two other strigid owl groups.

An even more radical alteration of the hearing system occurs in animals that have essentially evolved an organic form of sonar (table 2.6). These animals can produce very high-pitched ultrasonic sound waves that are bounced off the surrounding environment—including

Table 2.6
Convergent evolution of predator ear structures and auditory systems

1 Convergent structure and function: ASYMMETRIC EARS (allows three-dimensional stereophonic hearing—distance perception in both the vertical and horizontal plane—for precise prey location)

Convergent lineages:

1.1 Barn owl (Archosauromorpha: Archosauria: Ornithodira: Dinosauria: Saurischia: Theropoda: Maniraptora: Aves: Strigiformes: Tytonidae; *Tyto alba*)

1.2 Asian bay owl (Dinosauria: Saurischia: Theropoda: Maniraptora: Aves: Strigiformes: Tytonidae; *Phodilus badius*)

1.3 Great horned owl (Dinosauria: Saurischia: Theropoda: Maniraptora: Aves: Strigiformes: Strigidae; *Bubo virginianus*)

1.4 Long-eared owl (Dinosauria: Saurischia: Theropoda: Maniraptora: Aves: Strigiformes: Strigidae; *Asio otus*)

1.5 Boreal owl (Dinosauria: Saurischia: Theropoda: Maniraptora: Aves: Strigiformes: Strigidae; *Aegolius funereus*)

2 Convergent structure and function: ULTRASONIC HEARING ("sonar" adaptation: three-dimensional depth perception by echolocation for precise prey location)

Convergent lineages:

2.1 Edible-nest swiftlet (Amniota: Sauropsida: Diapsida: Archosauromorpha: Archosauria: Ornithodira: Dinosauria: Saurischia: Theropoda: Maniraptora: Aves: Apodiformes: Apodidae; *Aerodramus fuciphagus*)

2.2 Oilbird (Dinosauria: Saurischia: Theropoda: Maniraptora: Aves: Caprimulgiformes: Steatornithidae; *Steatornis caripensis*)

2.3 Lesser hedgehog tenrec (Amniota: Synapsida: Therapsida: Mammalia: Eutheria: Afrotheria: Afrosoricida: Tenrecidae; *Echinops telfairi*)

2.4 Eurasian shrew (Mammalia: Eutheria: Laurasiatheria: Eulipotyphles: Soricidae; *Sorex araneus*)

2.5 Mouse-eared bat (Eutheria: Laurasiatheria: Chiroptera: Microchiroptera: Vespertilionidae; *Myotis myotis*)

2.6 Bottlenose dolphin (Eutheria: Laurasiatheria: Cetartiodactyla: Cetacea: Odontoceti: Delphinidae; *Tursiops truncatus*)

Note: For data sources see text.

potential prey animals—and have ears that can hear the ultrasonic echoes, or return waves. Echolocation thus allows these animals not only to hunt prey animals by ultrasound but to safely locomote without sight. The most spectacular example of echolocation is that seen in bats flying after insects in the dim light of an early summer evening (while we can still see them). Two separate groups of birds have also convergently evolved ultrasonic hearing and echolocation capabilities: the apodid swiftlets of the tropical Indo-Pacific and the steatornithid oilbirds of South America. The edible-nest swiftlets are crepuscular-to-nocturnal flying insectivores that inhabit dark caves, and thus are ecological equivalents to bats. Yet the swiftlet is an avian dinosaur, whereas the bat is a placental mammal. The closest common ancestor for the two is found back in the Carboniferous, over 340 million years ago, yet they have convergently evolved the ability to hear ultrasound. The South American oilbirds are included here as an anomaly—they have independently evolved echolocation, as they are birds that are nocturnal and inhabit caves, but they are frugivores, not carnivores. Thus, they do not use echolocation to hunt.

Ground-dwelling organisms have also convergently evolved echolocation. Many of the fascinating little tenrecs of Madigascar, such as the lesser hedgehog tenrec, have evolved the capability to locate their insect prey using echolocation (Gould 1965). They produce very high-pitched squeaks and then analyze the pattern of the ultrasound echoes, a capability that has been convergently evolved by species of two genera of insectivorous shrews: *Sorex* and *Blarina*. Nevertheless, these two groups of mammals are very distantly related: the shrews are laurasiatherians, while the tenrecs are afrotherians. Last, ocean-dwelling animals have convergently evolved echolocation. The predaceous toothed cetaceans, such as the bottlenose dolphin, produce high-frequency clicks and can use the subsequent submarine echoes to echolocate as well as to detect prey.

Carnivores: Prey Capture

The possession of teeth in animals is a symplesiomorphic trait that can be traced back to the early gnathostomes in the Devonian, and the possession of claws is a symplesiomorphic trait that can be traced back to the early tetrapods in the Carboniferous. However, both teeth and claws have been repeatedly, convergently, modified to produce highly efficient killing structures in animal predators. The most widely cited example of remarkable convergence in predator dentary evolution is the convergent modification of normally conical fang teeth into saber teeth in three separate groups of mammals in the Cenozoic (table 2.7). Elongation of the fang teeth in creodont oxyaenid carnivores, such as *Machaeroides simpsoni*, began during the Paleocene and is seen as a precursor to the evolution of saber-tooth predators in the placental mammals (Turner 1997). The Creodonta are a sister lineage to the true Carnivora, and by the late Eocene the first saber-tooth predators evolve in the nimravids, or false cats, such as *Hoplophoneus mentalis*. Saber-tooth nimravid predators persisted throughout the Oligocene, producing the lion-sized saber-tooth *Barbourofelis fricki* in the Miocene (Turner 1997). The first true cats evolve in the Oligocene, and by the late Miocene are convergently producing saber-tooth species, culminating in the Pleistocene saber-tooth great cat, *Smilodon fatalis*. As such, it would appear that the evolution of saber-tooth predators in the nimravids and felids is an interesting, but hardly remarkable, case of parallel evolution in the Carnivora. Not so. During the same interval of time in which felid saber-tooth cats roamed North America, Europe, and Asia, the marsupial mammals of South America evolved a saber-tooth predator, *Thylacosmilus atrox*, that is almost identical in morphology to *Smilodon fatalis* (Turner 1997, 136). Although we have to go back in time 100 million years or so, to the mid-Cretaceous, to find a common ancestor between the placental and marsupial mammals, the two groups converged on the same predator morphology in the Pliocene and Pleistocene.

Elongated, saber-shaped teeth are an adaptation for producing shearing bites, bites that can tear off whole chunks of prey flesh and that produce gaping wounds, leading to massive blood loss and shock in the prey animal. Shearing bites to the neck can rip out the entire neck of the prey animal below the cervical vertebrae, or a saber-tooth bite can crush and collapse the windpipe of large prey animals, quickly causing unconsciousness due to oxygen deprivation (Turner 1997, 125). The extinction of the saber-tooth predators in the Pleistocene appears to be ecologically linked to the extinction of the large-prey species that formed their chief source of food.

The modern cats are well known for another highly derived trait: retractable claws. Most predators, like bears and wolves, have claws that can cause serious damage to a prey animal, but that are very dull compared to the razor-sharp claws of the cat. The cat's claws are usually retracted, and thus do not come into contact with the ground when they are walking; hence they are not dulled by constant wear and abrasion, like the claws of the dog. When the cat deploys its claws, its entire paw expands as the toe digits are extended, and the sickle-shaped claws may be used for either slashing the prey animal or clinging to it while the cat's fang teeth are in use.

Interestingly, retractable claws were first evolved by the dinosaurs, not by the cats (table 2.7). Both the dromaeosaurids and the troodontids evolved a retractable claw on the second digits of their hind feet. Like the cat's claws, these claws were sickle-shaped and were retracted when not in use. The dromaeosaurids and troodontids were related maniraptoran theropods, but their retractable claws were independently evolved in parallel and differ in several aspects of their anatomy (Varricchio 1997). A third example of convergent evolution of this trait is seen in the very plesiomorphic Late Cretaceous bird *Rahonavis ostromi*,

Table 2.7
Convergent evolution of predator killing structures

1 Convergent structure and function: SABER TEETH (fang teeth elongated and laterally flattened for shearing bites to produce gaping wounds)

Convergent lineages:

1.1 Marsupial saber-tooth "cat" (Mammalia: Marsupialia: Sparassodonta: Thylacosmilidae; *Thylacosmilus atrox* †Pliocene)

1.2 False saber-tooth "cat" (Mammalia: Eutheria: Laurasiatheria: Carnivora: Feliformia: Nimravidae; *Barbourofelis fricki* †Miocene)

1.3 True saber-tooth cat (Eutheria: Laurasiatheria: Carnivora: Feliformia: Felidae; *Smilodon fatalis* †Pleistocene)

2 Convergent structure and function: RETRACTABLE CLAWS (protection of claws when not in use, enhances sharpness of sickle claws for ripping prey)

Convergent lineages:

2.1 Dromaeosaurs (Amniota: Sauropsida: Diapsida: Archosauromorpha: Archosauria: Ornithodira: Dinosauria: Saurischia: Theropoda: Maniraptora: Dromaeosauridae; *Velociraptor mongoliensis* †Cretaceous)

2.2 Troodontids (Dinosauria: Saurischia: Theropoda: Maniraptora: Troodontidae; *Saurornithoides mongoliensis* †Cretaceous)

2.3 Rahonaves (Dinosauria: Saurischia: Theropoda: Maniraptora: Aves: incertae sedis; *Rahonavis ostromi* †Cretaceous)

2.4 Red-legged seriema (Dinosauria: Saurischia: Theropoda: Maniraptora: Aves: Gruiformes: Cariamae: Cariamidae; *Cariama cristata*)

2.5 Nimravid "cats" (Amniota: Synapsida: Therapsida: Mammalia: Eutheria: Laurasiatheria: Carnivora: Feliformia: Nimravidae; *Hoplophoneus mentalis* †Eocene)

2.6 True cats (Mammalia: Eutheria: Laurasiatheria: Carnivora: Feliformia: Felidae; *Proailurus lemanensis* †Oligocene)

2.7 Banded palm civet (Eutheria: Laurasiatheria: Carnivora: Feliformia: Viverridae; *Hemigalus derbyanus*)

3 Convergent structure and function: RAPTORIAL BEAK (laterally compressed, hooked beak for piercing and tearing flesh)

Convergent lineages:

3.1 Common octopus (Bilateria: Protostomia: Lophotrochozoa: Mollusca: Cephalopoda: Coleoidea: Octopodidae; *Octopus vulgaris*)

3.2 Snapping turtle (Bilateria: Deuterostomia: Chordata: Osteichthyes: Sarcopterygii: Reptiliomorpha: Amniota: Sauropsida: Anapsida: Testudines: Chelydridae; *Macroclemys temmincki*)

3.3 Peregrine falcon (Sauropsida: Diapsida: Archosauromorpha: Archosauria: Ornithodira: Dinosauria: Saurischia: Theropoda: Maniraptora: Aves: Falconiformes: Falconidae; *Falco peregrinus*)

3.4 Red-tailed hawk (Dinosauria: Saurischia: Theropoda: Maniraptora: Aves: Accipitriformes: Accipitridae; *Buteo jamaicensis*)

3.5 Secretary bird (Dinosauria: Saurischia: Theropoda: Maniraptora: Aves: Accipitriformes: Sagittaridae; *Sagittarius serpentarius*)

3.6 Osprey (Dinosauria: Saurischia: Theropoda: Maniraptora: Aves: Accipitriformes: Pandionidae; *Pandion haliaetus*)

3.7 Turkey vulture (Dinosauria: Saurischia: Theropoda: Maniraptora: Aves: Ciconiiformes: Cathartidae; *Cathartes aura*)

3.8 Great horned owl (Dinosauria: Saurischia: Theropoda: Maniraptora: Aves: Strigiformes: Strigidae; *Bubo virginianus*)

3.9 Red-legged seriema (Dinosauria: Saurischia: Theropoda: Maniraptora: Aves: Ralliformes: Cariamae: Cariamidae; *Cariama cristata*)

Table 2.7
(continued)

4 Convergent structure and function: POISON-INJECTING FANGS (hollow fangs connected with poison glands for injection of poison into prey)

Convergent lineages:

4.1 Black widow spider (Bilateria: Protostomia: Ecdysozoa: Arthropoda: Cheliceriformes: Arachnida: Araneae: Theridiidae; *Latrodectus mactans*)

4.2 Amazonian giant centipede (Arthropoda: Mandibulata: Myriapoda: Chilopoda: Scolopendridae; *Scolopendra gigantea*)

4.3 European ant lion larvae (Arthropoda: Mandibulata: Hexapoda: Neuroptera: Myrmeleontidae; *Euroleon nostras*)

4.4 Four-spotted owlfly larvae (Arthropoda: Mandibulata: Hexapoda: Neuroptera: Ascalaphidae; *Ululodes quadripunctatus*)

4.5 Timber rattlesnake (Bilateria: Deuterostomia: Chordata: Osteichthyes: Sarcopterygii: Reptiliomorpha: Amniota: Sauropsida: Diapsida: Lepidosauromorpha: Squamata: Scleroglossa: Autarchoglossa: Anguimorpha: Serpentes: Viperidae: Crotalinae; *Crotalus horridus*)

5 Convergent structure and function: POISONOUS STINGERS (stingers coated with poison or connected with poison glands for injection of poison into prey)

Convergent lineages:

5.1 Portuguese man-of-war (Metazoa: Cnidaria: Siphonophora: Physaliidae; *Physalia physalis*)

5.2 Striated cone-shell snails (Metazoa: Bilateria: Protostomia: Lophotrochozoa: Mollusca: Gastropoda: Prosobranchiata: Neogastropoda: Conidae; *Conus striatus*)

5.3 Babylonian auger-shell snails (Mollusca: Gastropoda: Prosobranchiata: Neogastropoda: Terebridae; *Terebra babylonia*)

5.4 Giant desert hairy scorpion (Protostomia: Ecdysozoa: Arthropoda: Cheliceriformes: Arachnida: Scorpiones: Iuridae; *Hadrurus arizonensis*)

5.5 Giant cicada-killer wasp (Arthropoda: Mandibulata: Hexapoda: Hymenoptera: Sphecidae; *Sphecius speciosus*)

5.6 Giant short-tailed stingray (Bilateria: Deuterostomia: Chordata: Chondrichthyes: Elasmobranchii: Batoidea: Dasyatidae; *Dasyatis brevicaudata*)

6 Convergent structure and function: ELECTRIC-FIELD GENERATION (generation of an electric field to stun or kill prey)

Convergent lineages:

6.1 Marbled electric ray (Vertebrata: Chondrichthyes: Elasmobranchii: Batoidea: Torpediniformes: Torpedinidae; *Torpedo marmorata*)

6.2 Pacific stargazer (Vertebrata: Osteichthyes: Actinopterygii: Teleostei: Perciformes: Uranoscopidae; *Astroscopus zephyreus*)

6.3 South American electric eel (Actinopterygii: Teleostei: Gymnotiformes: Electrophoridae; *Electrophorus electricus*)

6.4 African electric catfish (Actinopterygii: Teleostei: Siluriformes: Malapteruridae; *Malapterurus electricus*)

Note: The geological age of extinct species is marked with a †. For data sources, see text.

which looked very similar to the Late Jurassic bird *Archaeopteryx lithographica* (it still possessed three fingers with claws on its wing, and had caudal vertebrae along with tail feathers), but differed markedly in possessing retracted, enlarged sickle-shaped claws on the second digits of its feet. This last convergence has been questioned by some, who argue that *Rahonavis* was not a bird but a dromaeosaurid with wings, and thus conclude that its sickle claw was not a convergent character. That proposal, however, would require the convergent evolution of the wing within the maniraptors. And indeed, Mayr et al. (2005) have proposed that birds—winged maniraptors—independently evolved twice, although that conclusion remains controversial (Mayr and Peters 2006).

The dromaeosaurids, troodontids, and rahonaves did not survive the mass extinction at the end of the Cretaceous, but some of their living maniraptoran relatives—the seriema birds—have convergently evolved a retracted, sickle-shaped claw in the second digits of their feet that is eerily similar to that seen in the second hind digits of the troodontids. This convergent retracted sickle claw is well developed in the feet of the modern red-legged seriema, *Cariama cristata*, of South America. The seriemas are long-legged, mostly terrestrial predators that are ecologically convergent with the secretary birds of Africa. Only two species of seriemas survive today, but these cariamaen birds are part of a long lineage of terrestrial avian predators that includes the Eocene "terror bird" *Phorusrhacos longissimus*, a flightless raptor that stood almost 3 meters tall.

Is the reappearance of the retracted sickle claw in the Cenozoic cariamaen birds an example of convergence due to functional or to developmental constraint? The number of ways a claw might be modified is limited, and a sickle claw serves a clear function. Yet why not develop the retracted sickle claw on the third digit, or the fourth digit, of the foot? Is there a developmental predisposition within the cariamaen birds to modify the second digit of the foot, like their ancient troodontid relatives, and not the third or fourth? Or does the sickle claw simply function best if it is placed in the second digit of the foot?

Within the placental mammals, fully retractable claws evolved in parallel in three separate lineages during the Cenozoic: the nimravid false cats, the true cats, and the civets (table 2.7). Not all of the civets have retractable claws and, of those that do, the number of digits on the foot that have retractable claws is variable from species to species. Thus, the pattern of convergent evolution of retractable claws in the civets is strikingly different from the pattern seen in the troodontids and the cariamaen birds. On the other hand, the pattern of convergent evolution in the true cats and the nimravid false cats is identical—both groups developed fully retractable claws on all of their digits.

Not all predators have teeth like the sharks in the sea and the cats on the land. The cephalopods, such as the modern octopus, are ancient molluscan predators that grasp their prey with multiple tentacles, and bite with a beak that is surprisingly similar in form to the raptorial beak seen in a hawk (table 2.7). But because the cephalopods evolved in the Late Cambrian, some 340 million years before the evolution of the first avian dinosaurs in the Late Jurassic, it would be more correct to state that the hawks have beaks like the octopus rather than vice versa. Raptorial beaks appear to be a synapomorphy for the entire Cephalopoda, where the beaks seen in modern coleoid cephalopods are more similar to the extinct ammonites than to the modern living ectocochleate cephalopod *Nautilus pompilius*.

Raptorial beaks have been convergently evolved by gnathostome vertebrates that have secondarily lost an ancient characteristic trait of most jawed animals—their teeth. The raptorial beak is well developed in many types of turtles, themselves very ancient sauropsids, as seen in the modern snapping turtle. But the raptorial beak is most characteristically, and convergently, developed in a more derived group of sauropsids, the theropod dinosaurs, as seen in the modern hawks (table 2.7).

The accipitriform raptors include a very diverse group of bird species: hawks, eagles, kites, and Old World vultures (Accipitridae), the secretary birds (Sagittaridae), and the ospreys (Pandionidae). The hawks, eagles, and kites hunt on the wing, whereas the secretary birds are largely terrestrial hunters. The ospreys are specialized piscivores that catch fish with the talons of their feet (the single species *Pandion haliaetus* is considered to be so different from the other accipitriform birds that it constitutes the entire family Pandionidae). And the Old World vultures are mostly scavengers instead of hunters. Although these birds differ greatly in their morphologies and ecologies, the raptorial beak may be a synapomorphy for the entire Accipitriformes, rather than three separate convergences in the three separate families of accipitriforms (table 2.7).

The falconiform raptors, the falcons and caracaras, are more distantly related to the accipitriforms, and a more substantial argument may be made for the convergent development of the raptorial beak in this group. The New World vultures, such as the familiar turkey vulture in the skies of North America, are ciconiiform birds even more distantly related to the accipitriforms (and the ecologically convergent Old World vultures), and have convergently evolved the raptorial beak. The strigiform owls, specialized nocturnal predators, are more closely related to hummingbirds than they are to hawks (Lecointre and Le Guyader 2006), yet they too have convergently developed the raptorial beak. And last, the modern seriemas and their ancient "terror bird" cousins, *Phorusrhacos longissimus*, are basal neoavians very distantly related to all the other modern bird groups (Benton 2005), yet they too have convergently evolved the raptorial beak. In all, some five to seven different groups of raptorial birds have developed beaks convergent on those seen in the ancient turtles and cephalopods.

Some carnivores have evolved even more specialized structures for killing prey animals than the convergent teeth, claws, and beaks that we have considered so far. These animals have evolved fangs like hypodermic needles: fangs that are capable of injecting poison directly into the body of the prey animal. Poison-injecting fangs have been independently evolved in widely separated groups of arthropods (table 2.7). The most familiar to us are the spiders and the centipedes. These two predator groups are very distantly related; the spiders are cheliceriforms and the centipedes are mandibulates. Yet they have evolved very similar poison-injecting fangs, as have the larvae of two additional groups of neuropteran hexapods: the ant lions and the owlflies. Although one might suppose that the development of such a specialized structure as a hypodermic fang must surely be confined to the arthropods and their exoskeletal mode of growth, this is not the case. The vertebrate snakes, such as the timber rattlesnakes and their pit viper kin, have convergently evolved hollow, poison-injecting fangs.

Other predators have taken a backward approach to poisoning their prey: rather than attacking with poisonous fangs on the front of their bodies, they use poisonous stingers located on the back of their bodies (table 2.7). Of these, the most familiar to us are the scorpions and the wasps. Once again, although these two groups are both arthropods, they are very distantly related: the scorpions are cheliceriforms and the wasps are mandibulates. The wasps are a very highly diverse group of flying predators, yet some have even devolved their wings and become ground dwellers. One particularly notable example is the velvet ant, also known as the cow killer ant, which is not an ant at all (as anyone who tries to pick one up will painfully discover). Just as with poison-injecting fangs, poisonous stingers are not confined to the arthropods. The vertebrate stingrays, a group of cartilaginous fishes, have also convergently evolved a poisonous stinger located on the rear of the animals.

Stingers are not confined to the rear end of animals (table 2.7). The cone-shell snails, voracious predators of other marine invertebrates, have a long tonguelike proboscis with tiny radulae, or teeth. Most snails use these barbed tongues to drill holes into the bodies of their prey, but the cone-shell snails have evolved radulae that can inject an extremely deadly neurotoxin. This poison is so deadly it can kill animals as large as humans, animals much larger than the snail can actually use as a food resource. A related group of snails, the auger-shell snails, have also evolved stinger tongues, but the venom they inject is not nearly as poisonous as that of the cone-shell snails.

Stingers are also not confined to highly derived, protostomous and deuterostomous bilaterian animals. Near the base of the metazoan phylogenetic tree, the cnidaria have evolved barbed, poisonous stingers that are contained in specialized explosive cells, the cnidocytes. The tentacles of the cnidarians are lined with cnidocytes, which project the stingers into the body of the prey animal, injecting a paralyzing poison. The cnidarian then contracts its tentacles to pull the prey to its mouth. Some cnidarians, like the Portuguese man-of-war jellyfish, are mobile predators of fish and zooplankton, while others, such as the corals, are sessile predators that simply wait for prey animals to come within striking distance.

Perhaps the most exotic prey-capture capability has been independently evolved by four of the groups of electric fishes: the ability to generate an electric field strong enough to stun, and some cases even kill, the prey animal (table 2.7). These fishes have gone beyond the already unusual ability to detect an electric field in prey hunting that we considered previously (table 2.5). For example, the freshwater electric eels and electric catfish can use their electric organs to discharge amplitudes between 300 to 600 volts, and the marine electric rays and stargazers discharge amplitudes between 15 to 50 volts (Moller 1995). The electric eels and electric catfish are thus in many cases able to directly kill their target frogs and fish with an electric shock, while the electric rays and stargazers stun their prey into immobility or disorientation, and then proceed to attack. Interestingly, the electric catfish no longer possess the poisonous stingers on their fins (a trait we shall consider in more detail later in the chapter) that are found in so many other groups of catfish—since they can directly use their electric-field generative ability to defend themselves, the stingers have been lost.

Herbivores: Plant Processing

Land plants cover extensive areas of the surface of the Earth; thus, life would at first glance seem to be much easier for a plant-eating herbivore than for a carnivore. That is, plant food is generally abundant, and plants cannot run away, as hunted prey can. The major problem for herbivores is that plant food is difficult to digest, and essential nutrients are difficult to obtain in sufficient quantities from plants. For example, consider many plant seeds and grains. They are rich in nutrients, but they are also very hard, particularly when dried, and can crack your teeth if you try to chew them. Yet think of the common domestic chicken, *Gallus gallus*, an animal that survives on a diet of seeds and grains but has no teeth. The chicken has evolved an alternative method of grinding up hard plant material—it has a gastric mill, or gizzard. The chicken actually deliberately swallows sharp stone fragments, which are then held in the muscular walls of its gizzard and are used to grind up seeds and grains.

Gastric mills have been convergently evolved by an astonishing variety of animals, from earthworms to mammals (table 2.8). The gizzard of the domestic chicken is most familiar to us, and some consider it to be a delicacy. But gizzards are not just found in the galliform birds or just in herbivorous or granivorous birds. The gaviiform red-throated loon, *Gavia stellata*, is more closely related to a penguin than a chicken, yet it also possesses a gastric mill that it uses to grind up the bones of frogs and the exoskeletons of crustaceans that form part of its diet. The plesiomorphic paleognathaean ostrich, *Struthio camelus*, a modern large flightless bird, also possesses a gastric mill that it uses to grind the grasses, roots, and leaves that it normally eats, plus the occasional insect, lizard, or small mammal.

Curiously, we even have a fossil record for gastric mills. The sharp stones located within the gizzard are termed "gastroliths," or stomach stones, and they become worn and rounded in a characteristic fashion with usage. Gastroliths are occasionally found within the skeletal remains of fossil birds and other animals, indicating that these animals possessed gastric mills in life. Thus, we know that the Cretaceous maniraptoran dinosaur *Caudipteryx zoui* had a gizzard—and it was not a bird, although closely related (note, however, that Maryańska et al. [2002] consider the oviraptorosaurians to be secondarily flightless birds, and so the gastric mill of *Caudipteryx zoui* may be an avian synapomorphy rather than convergent within the maniraptorans).

Table 2.8
Convergent evolution of gastric mills in herbivores and other animals that require mechanical assistance to digestion

Convergent structure and function: GASTRIC MILLS (muscular gizzards with embedded stones or teeth for grinding/processing plant material)

Convergent lineages:

1 Common European earthworm (Bilateria: Protostomia: Lophotrochozoa: Annelida: Oligochaeta: Lumbricidae; *Lumbricus terrestris*)

2 Great pond snail (Lophotrochozoa: Mollusca: Gastropoda: Pulmonata: Lymnaeidae; *Lymnaea stagnalis*)

3 Priapulid worm (Protostomia: Ecdysozoa: Introverta: Priapulida: Priapulidae; *Priapulus caudatus*)

4 Hermit crab (Arthropoda: Mandibulata: Malacostraca: Decapoda: Paguridae; *Eupagurus bernhardus*)

5 Madagascar hissing cockroach (Arthropoda: Mandibulata: Hexapoda: Blattodea: Blattellidae; *Gromphadorhina portentosa*)

6 Striped mullet (Bilateria: Deuterostomia: Chordata: Osteichthyes: Actinopterygii: Perciformes: Mugilidae; *Mugil cephalus*)

7 American gizzard shad (Osteichthyes: Actinopterygii: Clupeiformes: Clupeidae; *Dorosoma cepedianum*)

8 Nile crocodile (Osteichtyes: Sarcopterygii: Reptiliomorpha: Amniota: Sauropsida: Diapsida: Archosauromorpha: Crurotarsi: Crocodylia: Crocodylidae; *Crocodylus niloticus*)

9 Sauropodomorph dinosaur (Archosauromorpha: Ornithodira: Dinosauria: Saurischia: Sauropodomorpha: Plateosauridae; *Plateosaurus engelhardi* †Triassic)

10 Caudipteryx oviraptor (Dinosauria: Saurischia: Theropoda: Maniraptora: Oviraptorosauria: Caudipterygidae; *Caudipteryx zoui* †Cretaceous)

11 Ostrich (Dinosauria: Saurischia: Theropoda: Maniraptora: Aves: Paleognathae: Struthionidae; *Struthio camelus*)

12 Psittacosaurid dinosaur (Dinosauria: Ornithischia: Cerapoda: Marginocephalia: Psittacosauridae; *Psittacosaurus mongoliensis* †Cretaceous)

13 Giant scaly anteater (Amniota: Synapsida: Therapsida: Mammalia: Eutheria: Laurasiatheria: Pholidota: Manidae; *Manis gigantea*)

Note: The geological age of extinct species is marked with a †. For data sources, see text.

From the fossil record, we know that gastric mills were convergently evolved by the second major branch of the saurischian dinosaurs, the herbivorous sauropodomorphs. Characteristic gastroliths have been found with the skeletal remains of plesiomorphic prosauropods, like *Plateosaurus engelhardti*, of the Triassic and at the very base of the sauropodomorph clade. Gastroliths are later found in the rib cages of giant herbivorous sauropod dinosaurs, like *Seismosaurus*, that apparently swallowed all of their plant food whole and ground it up internally in their gastric mill (McIntosh 1997). As many as 50 polished gastroliths have been found in several psittacosaur skeletons, indicating that gastric mills were convergently evolved in the herbivorous ornithischian dinosaurs as well (Sereno 1997).

Distant cousins to the dinosaurs, the modern crocodiles and alligators, have also evolved gastric mills, but not for grinding tough plant material or hard seeds. These animals are carnivores, but they swallow their food whole, including the prey's very hard bones. The crocodiles thus have convergently evolved a gastric mill for the same reason, but not for the same food type, as the extinct sauropodomorphs. The bones of their prey are cracked and ground up by their gastric mills, and then sent onward to their very high-acid-content stomachs for further processing.

Gastric mills are not a convergent phenomenon to be found only in the archosaurian clade of tetrapods. Among the mammals, the giant scaly anteater has evolved a gastric mill to grind the carapaces of the ants that form the bulk of its diet. It actually has no teeth in its mouth and swallows stones just as granivorous birds do, depending upon its gizzard to process the hard exoskeletons of its food. The gizzard of the giant scaly anteater contains an additional element to aid in its grinding function—stomach teeth. This gizzard innovation, too, has been convergently evolved at least twice in the arthropods. The stomatogastric systems of the decapod malacostracans contain gizzards with three hard, chitinous stomach teeth that assist in the grinding function of the gastric mill. Diverse decapods such as the red swamp crayfish, *Procambarus clarkii*, a common freshwater dweller in streams of southeastern North America, and the familiar hermit crab, *Eupagurus bernhardus*, all possess gizzards with teeth. Many of the hexapod insects, such as the spectacular Madagascar hissing cockroach, have convergently evolved gizzards with chitinous teeth similar to those found in the decapods.

Gastric mills have been convergently evolved in molluscs as well (table 2.8). Many snails, such as the great pond snail, deliberately ingest sand for their gizzards, and use these gizzards to process tough plant material. Even the earthworms have evolved gizzards to grind the detritus and plant fragments that form their diet. The peculiar priapulid worms in the oceans, more carnivores than detritivores, have also convergently evolved gizzards to grind up the carapaces of their prey, as well as ingested detritus from sediment. The priapulid worms are ecdysozoans, more closely related to arthropods than to the lophotrochozoan annelid worms, yet the two distantly related groups of worms independently evolved similar gastric mills.

Last, gastric mills have been convergently evolved at least twice in the fishes. The striped mullet and the American gizzard shad both ingest sand for their gizzards, and use these gizzards to grind the small invertebrates and detritus that form their diets. The striped mullet is a perciform fish and the gizzard shad is a clupeiform fish, and their gizzards are independently evolved.

The alternative to grinding up plant material in an internal gizzard is to grind it up in the mouth itself. In the modern world, mammals are the undisputed masters of processing food in the mouth. The characteristic heterodont dentition of the mammals is a veritable toolbox, in that mammals have specialized teeth for slicing, piercing, shearing, crushing, and grinding. Moreover, these specialized teeth meet in a precise arrangement when the jaws of the mammal are closed, an arrangement known as dental occlusion. Mammals represent an extreme in the development of dental occlusion, where the complex crowns of the molar teeth in the upper and lower jaw fit together in a precise mortar-and-pestle fashion, even to the extent of facilitating tooth wear on the crowns to produce wear facets resulting from tooth-to-tooth contact (DeMar and Bolt 1981; Benton 2005, 292). In particular, the mammals possess tribosphenic molars in which a cusp in the upper-jaw molar, the protocone, fits into an opposing basin in the lower-jaw molar, the talonid (Benton 2005, 307). These complex molars are capable of both shearing and grinding occlusal functions, and it was long thought that they were a unique autapomorphy of the therian clade of extinct and extant marsupial and placental mammals. However, we now know that an extinct group of australosphenid mammals, the Ausktribosphenida, convergently evolved the tribosphenic molar (table 2.9). Luo et al. (2001) argue that the tribosphenic molar evolved vicariantly in the southern continent of Gondwana (as seen in *Ausktribosphenos nyktos*), and in the northern continent of Laurasia (as seen in marsupials and placentals) during the Cretaceous. In addition, yet another group of even more distantly related Jurassic mammals, the Shuotheriidae, evolved a reversed, or "pseudo-tribosphenic," molar in which the position of a pseudo-talonid basin is shifted from the posterior part of the molar to the anterior in order to receive the pseudo-protocone cusp (Luo et al. 2007). The functional end result is the same as that for the tribosphenic molar, and Luo et al. (2007) argue that early mammalian dental evolution may be much more iterative and parallel than currently recognized.

Hunter and Jernvall (1995) point out that many of the more derived tribosphenic mammals have added an additional cusp, a hypocone, to the original three cusps found in the upper molar. They argue that the acquisition of the hypocone is a key evolutionary adaptation for herbivory, and that it has been convergently evolved over twenty separate times within the tribosphenic mammals (see table 1 in Hunter and Jernvall

Table 2.9
Convergent evolution of teeth and dental systems in herbivores and other animals that require mechanical assistance to digestion

1 Convergent structure and function: TRIBOSPHENIC MOLARS (complex molars capable of both shearing and grinding occlusal functions)

Convergent lineages:

1.1 Shuotheriid mammal (Mammalia: Yinotheria: Shuotheriidae; *Pseudotribos robustus* †Jurassic)

1.2 Ausktribosphenid mammal (Mammalia: Australosphenida: Ausktribosphenida: Ausktribosphenidae; *Ausktribosphenos nyktos* †Cretaceous)

1.3 Virginia opossum (Mammalia: Theria: Marsupialia: Didelphimorphia: Didelphidae; *Didelphis virginiana*)

2 Convergent structure and function: OCCLUSAL DENTITION (dentition capable of interlocking occlusion for crushing and chewing food)

Convergent lineages:

2.1 Idiognathodontid conodont (Chordata: Craniata: Conodonta: Ozarkodinida: Idiognathodontidae; *Idiognathodus claviformis* †Carboniferous)

2.2 Queensland lungfish (Chordata: Vertebrata: Osteichthyes: Sarcopterygii: Dipnoi: Ceratodontidae; *Neoceratodus forsteri*)

2.3 Procolophonoid reptile (Sarcopterygii: Tetrapoda: Reptiliomorpha: Amniota: Sauropsida: Anapsida: Procolophonoidea; *Procolophon trigoniceps* †Triassic)

2.4 Trilophosaurid reptile (Sauropsida: Diapsida: Archosauromorpha: Trilophosauridae; *Trilophosaurus buettneri* †Triassic)

2.5 Chimaerasuchid notosuchian (Archosauromorpha: Crurotarsi: Crocodylomorpha: Notosuchia: Chimaerasuchidae; *Chimaerasuchus paradoxus* †Cretaceous)

2.6 Pakasuchid notosuchian (Archosauromorpha: Crurotarsi: Corcodylomorpha: Notosuchia: Pakasuchidae; *Pakasuchus kapilimai* †Cretaceous)

2.7 Diademondontid cynodont (Amniota: Synapsida: Therapsida: Cynodontia: Diademodontidae; *Scalenodon angustifrons* †Triassic)

2.8 Traversodontid cynodont (Therapsida: Cynodontia: Traversodontidae; *Massetognathus pascuali* †Triassic)

2.9 Tritylodontid cynodont (Therapsida: Cynodontia: Tritylodontidae; *Kayentatherium wellesi* †Triassic)

2.10 Triassic mammal (Therapsida: Cynodontia: Mammalia: *Adelobasileus cromptoni* †Late Triassic)

3 Convergent structure and function: DENTAL BATTERIES (multiple rows of teeth in jaw to form a grinding/shearing surface for processing plant material)

Convergent lineages:

3.1 Hadrosaurid dinosaur (Archosauria: Ornithodira: Dinosauria: Ornithischia: Cerapoda: Ornithopoda: Hadrosauridae; *Parasaurolophus walkeri* †Cretaceous)

3.2 Ceratopsid dinosaur (Dinosauria: Ornithischia: Cerapoda: Marginocephalia: Ceratopsia: Ceratopsidae; *Chasmosaurus belli* †Cretaceous)

Note: The geological age of extinct species is marked with a †. For data sources, see text.

1995). Jernvall (2000) argues that only small developmental changes are needed to produce major changes in cusp numbers and sizes in mammalian teeth, and that this explains the frequent convergent evolution of new cusps.

Just as the tribosphenic molar in mammals is convergent, dental occlusion itself is not a unique trait of the mammals. In fact, mammals are only one of four separate groups of cynodont therapsids that convergently evolved complex dental occlusion in the Triassic (table 2.9). The tritylodontids, such as *Kayentatherium wellesi*, are the closest relatives to the most ancient mammal known to us, *Adelobasileus cromptoni* (Benton 2005, 290), but they independently evolved dental occlusion. The traversodontids are more distantly related, and the diademondontids even more so, yet they too independently evolved dental occlusion (Rowe 1993; Martinez et al. 1996). Interestingly, all three of these cynodont groups were secondarily herbivorous, in that most of the cynodonts were carnivores or omnivores. Thus, in the Triassic, evolution produced four separate and different cynodont experiments in complex dental occlusion and extensive food processing in the mouth, only one of which survives to the present day (the mammals).

Dental occlusion is also not a unique trait of the synapsid amniotes. The other branch of the amniote animals, the reptilian sauropsids, also evolved independent lineages of animals with at least partial dental occlusion. This is particularly interesting in that the sauropsids usually replace their teeth continually throughout life, whereas the mammals have only two sets of teeth—the early deciduous teeth, and the later set of permanent teeth. Occlusion is particularly difficult to maintain in a jaw that contains teeth of differing ages, and hence different shapes and positions, as is commonly the case in sauropsids.

Even so, the convergent evolution of dental occlusion has been reported in some extinct archosaurs and primitive sauropsids. Wu et al. (1995, 678) describe a Cretaceous notosuchian crocodylomorph from China, *Chimaerasuchus paradoxus*, that has teeth "very similar to that of the postcanine teeth of tritylodontid synapsids and represents a particularly striking example of convergent evolution." This animal was secondarily herbivorous, as most of the crocodile-like animals were carnivores, and thus also represents an ecological convergence to the tritylodontid cynodonts, which were also secondary herbivores. A second, independent, evolution of molariform teeth and occlusal precision in Cretaceous notosuchians is reported from Tanzania for *Pakasuchus kapilimai* (table 2.9). This species belongs to a separate notosuchian clade than the

Chinese *Chimaerasuchus paradoxus*, yet it also evolved molariform teeth "paralleling the level of occlusal complexity seen in mammals" (O'Connor et al. 2010, 748). Noting that ancient mammals were rare on the southern supercontinent Gondwana in the Cretaceous, in contrast to the northern supercontinent Laurasia, O'Connor et al. (2010, 748) suggest that these "notosuchians probably filled niches and inhabited ecomorphospaces that were otherwise occupied by mammals on northern continents." (The phenomenon of ecological-niche convergence by other disparate phylogenetic lineages will be considered in detail in chapter 4.) Finally, the convergent evolution of "complex, multi-cusped, mammal-like teeth" has also been reported in another Cretaceous crocodylomorph from Africa (Clark et al. 1989, 1064), and the evolution of dental occlusion has been reported in even more distantly related archosauromorph diapsids, *Trilophosaurus buettneri* and *Tricuspisaurus thomasi* (Robinson 1956; DeMar and Bolt 1981), and in a procolophonoid anapsid (Carroll and Lindsay 1985), all from the Triassic (table 2.9).

Leaving terrestrial ecosystems, we find that occlusal dentition has only rarely been reported for aquatic animals. The Queensland lungfish develops a series of tooth plates within its mouth, which are cusped and which occlude to form a crushing surface that it uses to eat shelled invertebrates such as snails and prawns (Kemp 1977). Far back in time, the Carboniferous conodont *Idiognathodus claviformis* apparently crushed food with the complex, interpenetrative oral surfaces of its molariform Pa elements. These were toothlike structures reminiscent of gastric mill teeth in that, because the conodonts had no jaws, these element structures were located in the throat posterior to other teethlike elements near the opening of the mouth. Still, microwear patterns on their oral surfaces indicate that these conodonts "developed dental occlusion of mammal-like complexity" (Donoghue and Purnell 1999, 58).

In the end, mammals still remain the evolutionary masters of chewing efficiency. Or perhaps they are not? There exist two groups of herbivorous dinosaurs that clearly rivaled—if not surpassed—the mammals in specialized masticatory adaptation. These were the hadrosaurids and the ceratopsids. Of all the sauropsids, only the ornithopod ornithischians were able to chew their food with the efficiency of a mammal, although they did so in a different fashion (Benton 2005, 207). The highly derived hadrosaurid ornithopods further increased their chewing efficiency by evolving multiple rows of teeth in each jaw, not just the normal one row. These multiple rows of teeth, known as dental batteries, contained 40 to 60 teeth per battery, with the result that the typical hadrosaur contained hundreds of teeth in its mouth. Moreover, the hadrosaurs continued to grow and replace teeth in the dental batteries throughout their lifespan (a plesiomorphic sauropsid feature), so a typical animal produced thousands of teeth, replacing them constantly as the older teeth were worn down. The hadrosaurs have been argued to have been "Mesozoic ungulates" (Carrano et al. 1999), ecologically and morphologically convergent on modern ungulate mammals (we shall consider ecological convergence in detail in chapter 4). Unlike the hadrosaurs, modern grazing ungulates that eat tough grass, like the horse or cow, have only one set of permanent teeth. Because these teeth have very high crowns, they can last the lifetime of the animal, as the tooth is worn away progressively from the top down (Janis 1990).

Dental batteries are not unique to the hadrosaurs. The ceratopsid ornithischians also evolved multiple rows of teeth, but their dental batteries are different in structure from those of the hadrosaurs. The surfaces of the ceratopsid dental batteries are more bladelike, rather than the rasplike batteries of the hadrosaurs, and were used more for shearing than for grinding. The ceratopsids are descendants of the psittacosaurs, animals that simply swallowed their food whole and processed it internally in a gastric mill (Sereno 1997). Thus, the peculiar evolution of multiple rows of teeth in the jaw occurred twice, independently, in the ornithischian dinosaurs (table 2.9).

A major problem for herbivores is the digestion of cellulose. Sometimes mechanical crushing and shredding of cellulose is simply not sufficient, and so chemical fermentation is used as an alternative digestive route. Multiple lineages of mammals have convergently evolved specialized digestive structures to house and protect anaerobic bacteria, and to allow these bacteria to ferment cellulose within the structures. Specifically, two types of cellulose fermentation systems have evolved: stomachal-fermentation systems, in which specialized stomach compartments house the anaerobic bacteria, and cecal-fermentation systems, in which an intestinal cecum houses the anaerobic bacteria (Lecointre and Le Guyader 2006). Stomachal-fermentation systems have convergently arisen in three groups of cetartiodactyls—the ruminants, the camels, and the hippopotamuses—one group of primates—the colobus monkeys—and even in one type of herbivorous bird (table 2.10). Lecointre and Le Guyader (2006) argue that the stomachal-fermentation system evolved independently and in parallel in the three cetartiodactyl groups, and that it is not a synapomorphy for these groups. The four-chambered stomachs of the many ruminant groups are quite different from those of the

Table 2.10
Convergent evolution of chemical digestive structures and systems in herbivores

1 Convergent structure and function: STOMACHAL-FERMENTATION SYSTEM (specialized stomach compartments house anaerobic bacteria for fermentation of cellulose in digestion)

Convergent lineages:

1.1 Hoatzin (Amniota: Sauropsida: Diapsida: Archosauromorpha: Ornithodira: Dinosauria: Saurischia: Theropoda: Aves: Opisthocomiformes: Opisthocomidae; *Opisthocomus hoatzin*)

1.2 Dromedary camel (Amniota: Synapsida: Therapsida: Mammalia: Eutheria: Laurasiatheria: Cetartiodactyla: Tylopoda: Camelidae; *Camelus dromedarius*)

1.3 Red deer (Mammalia: Eutheria: Laurasiatheria: Cetartiodactyla: Ruminantia: Cervidae; *Cervus elaphus*)

1.4 Hippopotamus (Eutheria: Laurasiatheria: Cetartiodactyla: Hippopotamidae; *Hippopotamus amphibius*)

1.5 Western red colobus monkey (Eutheria: Euarchontoglires: Primates: Cercopithecidae; *Colobus badius*)

2 Convergent structure and function: CECAL-FERMENTATION SYSTEM (intestinal cecum houses anaerobic bacteria for fermentation of cellulose in digestion)

Convergent lineages:

2.1 Yellow-spotted hyrax (Mammalia: Eutheria: Afrotheria: Hyracoidea: Procaviidae; *Heterohyrax brucei*)

2.2 Mountain zebra (Eutheria: Laurasiatheria: Perissodactyla: Equidae; *Equus zebra*)

2.3 European rabbit (Eutheria: Euarchontoglires: Lagomorpha: Leporidae; *Oryctolagus cuniculus*)

3 Convergent structure and function: RUMINANT-STOMACH SYSTEM (multiple-stomach food processing to allow delayed mastication, thus decreasing actual feeding time and predator exposure)

Convergent lineages:

3.1 Red kangaroo (Mammalia: Marsupialia: Diprotodontia: Macropodidae; *Macropus rufus*)

3.2 Red deer (Mammalia: Eutheria: Laurasiatheria: Cetartiodactyla: Ruminantia: Cervidae; *Cervus elaphus*)

3.3 Dromedary camel (Eutheria: Laurasiatheria: Cetartiodactyla: Tylopoda: Camelidae; *Camelus dromedarius*)

Note: For data sources, see text.

tylopod camels, guanocos, and their kin, and yet more different from the stomachs of the hippopotamuses. The colobus monkeys are primates, very different animals from the cetartiodactyls, yet they too have convergently evolved a stomachal-fermentation system to deal with their exclusively leaf-eating herbivorous diet. And last, the peculiar hoatzin or "stink bird" is an avian dinosaur, very far removed from the clade of the mammals, yet it also has independently evolved a stomachal-fermenting system (table 2.10).

The cecal-fermentation system also has been convergently evolved by a phylogenetically diverse group of mammals (table 2.10): euarchontoglirean lagomorphs (rabbits, hares, and pikas), afrotherian hyracoids (hyraxes), and laurasiatherian perissodactyls (horses, rhinos, and tapirs). The lagomorphs are particularly interesting in that they recycle the cellulose-rich part of their diet: the food passes first through the cecum, where the bacteria degrade the cellulose and the cecum absorbs the released nutrients, leaving behind a soft fecal pellet enriched in vitamins and bacteria. These soft pellets are defecated, reswallowed, and redigested by the rabbit, so that the pellet's vitamins are absorbed and the symbiotic bacteria are digested as food. This peculiar digestive process, termed "caecotrophy" (Lecointre and Le Guyader 2006, 438), is convergent—at least in its digestive effects—on the regurgitation and redigestion process used by the ruminant cetartiodactyls.

Rumination has been convergently evolved by three separate groups of mammals (table 2.10): twice independently in two groups of placental cetartiodactyls (ruminants and camels), and a third time in the very distantly related marsupial mammals, the kangaroos (Lecointre and Le Guyader 2006, 469). Ruminants minimize feeding time and predator exposure by eating their food rapidly, and thus usually chew it poorly. Later, in more protected settings, the partially digested food is regurgitated, remasticated, and reswallowed for further digestive processing.

Before moving on to consider convergent antipredator adaptations, let us consider the pattern of evolution of one last feeding structure: animal beaks. We are all familiar with the distinctive beaks of modern ducks and parrots. Parrots use their pincerlike, hooked beaks to manipulate, crack open, and extract food from seeds and nuts. Ducks use their flattened beaks to filter food material from water. Both of these modern animals are highly derived theropod dinosaurs. In the Cretaceous another group of dinosaurs, the psittacosaurs and their ceratopsian descendants, independently evolved beaks that are astonishingly similar to parrot beaks, yet these dinosaurs were highly derived marginocephalian ornithischians, not theropods (table 2.11). Analyses of tooth wear in a recently discovered, exceptionally well-preserved psittacosaur from Mongolia has led Sereno et al. (2009) to the conclusion that the diet of this psittacosaur was identical to that of a parrot, that is, it was a "nucivore" (nut eater).

The closest common ancestor for these two groups of animals (psittacosaurs and parrots) lived over 240 million years ago, in the Middle Triassic. In the Triassic there existed another group of animals that convergently evolved parrotlike beaks, the rhynchosaurid reptiles. Parrotlike beaks are not confined to the sauropsid amniotes, however, as in the Permian a group of dicynodontid therapsids in the synapsid amniote

Table 2.11
Convergent evolution of parrot and duck beaks

1 Convergent structure and function: PARROT BEAK (laterally flattened, dorsoventrally flared, pincerlike hooked beak for holding, cracking open, and extracting food from seeds and nuts)

Convergent lineages:

1.1 Rhynchosaurid reptile (Amniota: Sauropsida: Diapsida: Archosauromorpha: Rhynchosauria: Rhynchosauridae; *Hyperodapedon gordoni* †Triassic)

1.2 Scarlet macaw (Archosauromorpha: Archosauria: Ornithodira: Dinosauria: Saurischia: Therapoda: Maniraptora: Aves: Psittaciformes: Psittacidae; *Ara macao*)

1.3 Psittacosaurid dinosaur (Dinosauria: Ornithischia: Cerapoda: Marginocephalia: Psittacosauridae; *Psittacosaurus mongoliensis* †Cretaceous)

1.4 Dicynodontid therapsid (Amniota: Synapsida: Therapsida: Dicynodontia: Dicynodontidae; *Pristerodon mackayi* †Permian)

2 Convergent structure and function: DUCK BEAK (dorsoventrally flattened, laterally flared, horn beak for filtering food from water)

Convergent lineages:

2.1 Wood duck (Amniota: Sauropsida: Diapsida: Archosauromorpha: Ornithodira: Dinosauria: Saurischia: Theropoda: Maniraptora: Aves: Anseriformes: Anatidae; *Aix sponsa*)

2.2 Hadrosaurid dinosaur (Dinosauria: Ornithischia: Cerapoda: Ornithopoda: Hadrosauridae; *Anatotitan copei* †Cretaceous)

2.3 Duckbill platypus (Amniota: Synapsida: Therapsida: Mammalia: Monotremata: Ornithorhynchidae; *Ornithorhynchus anatinus*)

Note: The geological age of extinct species is marked with a †. For data sources, see text.

lineage also independently evolved parrotlike beaks. All of these animals were herbivorous.

There are no living synapsids that still possess parrotlike beaks, but there are some that possess beaks that are identical to those of ducks—the duckbill platypuses (table 2.11). As occurred with parrot beaks, a group of distantly related ornithischian dinosaurs, the hadrosaurs, independently evolved beaks like those of theropod ducks. Mechanically, the mammalian platypus uses its beak exactly as a duck does, filtering small aquatic animals out of the water with the transverse ridges of its beak, ridges that are similar to the teethlike serrations of the duck's beak. The hadrosaurs, however, were not aquatic animals. Their beak usage was probably more like that of modern geese, which use their ducklike beaks to crop grass and other plant material.

Defense: Antipredator Adaptations

Many of the same structures that predators use to kill prey can also be used to defend themselves from other predators. In this section of the chapter, I will concentrate on the convergent evolution of structures and systems that are primarily for defense. Although a few predators have evolved eyes with horizontal-slit pupils (as we have previously discussed), eyes of this type are very widespread in herbivores. Most familiar, perhaps, are the odd-looking eyes of domestic goats, although the same type of eye is present in sheep, cows, deer, elks, hippopotamuses, and so on. It is a near universal trait of the cetartiodactyl mammals. However, horses and their kin also have eyes with horizontal-slit pupils; thus, this characteristic has clearly evolved independently in the perissodactyl mammals. A third group of mammals, the afrotherian manatees, have convergently evolved the horizontal-slit pupil in the eye as well (table 2.12). In all cases, this type of eye is used by herbivores for predator detection at a distance, in all directions, and thus for predator avoidance.

A typical herbivore strategy of predator avoidance is simply to run away—as fast as possible. The evolution of fast running has anatomical consequences, a functional constraint that leads directly to convergent evolution. Bakker (1983) has demonstrated that six morphological

Table 2.12
Convergent evolution of herbivore predator-avoidance systems

1 Convergent structure and function: HORIZONTAL-SLIT PUPILS IN EYES (allows [1] full usage of the diameter of the lens of the eye in bright light as well as in very low light intensities, with well-focused images in all light intensities, and [2] in combination with near 360-degree vision, continuous viewing of almost total horizon for predator detection in all light intensities)

Convergent lineages:

1.1 African manatee (Mammalia: Eutheria: Afrotheria: Sirenia: Trichechidae; *Trichechus senegalensis*)

1.2 Mountain zebra (Mammalia: Eutheria: Laurasiatheria: Perissodactyla: Equidae; *Equus zebra*)

1.3 Domestic goat (Eutheria: Laurasiatheria: Cetartiodactyla: Ruminantia: Bovidae; *Capra aegagrus*)

2 Convergent structure and function: DIGIT-NUMBER REDUCTION (reduction of muscle mass in the distal parts of limbs, elongation of the bones in the feet, loss of digits in the foot, and the aquisition of hooves to protect the digits from shock in running, to produce fast-running spindly-limbed animals)

Convergent lineages:

2.1 Modern horse (Mammalia: Eutheria: Laurasiatheria: Perissodactyla: Equidae; *Equus caballus*)

2.2 Litoptern "horse" (Eutheria: Laurasiatheria: Meridiungulata: Litopterna: Proterotheriidae; *Thoatherium minusculum* †Miocene)

2.3 Red deer (Eutheria: Laurasiatheria: Cetartiodactyla: Ruminantia: Cervidae; *Cervus elaphus*)

2.4 Guanaco (Eutheria: Laurasiatheria: Cetartiodactyla: Tylopoda: Camelidae; *Lama guanicoe*)

Note: The geological age of extinct species is marked with a †. For data sources, see text.

transitions occur in the evolution of fast runners from slow, and that these morphological changes have repeatedly occurred in distantly related animal groups that convergently evolved fast runners. These evolutionary changes involve lengthening and attenuation of the distal bone shafts in the limbs while the proximal bones of the limbs become shorter, lengthening and attenuation of the metapodial bones in the feet as the animals locomote first on their toes and then on the tips of their toes, shortening of the phalangeal bones in the foot, and transformation of claws into hooves to protect the ends of the toes from shock and damage while running. Last, the number of hoofed toes in the foot are reduced, leading in some cases to the animal standing upon a single hoofed toe. The spindly legs of red deer are a familiar example of this morphological transition: the upper bone of the hind limb (femur) is short and surrounded by a compact mass of muscle (which controls the limb) held next to the body of the deer, the lower limb bone shafts (tibia and fibula) are elongated away from the deer's body and contain much less muscle mass, the foot is enormously elongated into a thin shaft of very low mass, and the animal stands on the tips of only two hoofed toes. This form of limb, so common in many ruminant cetartiodactyls, has been independently evolved in the tylopod cetartiodactyls as well, even to the extent of reduction of the number of digits in the foot to two (table 2.12). The more distantly related perissodactyls have not only independently evolved the same type of running limb but have further reduced the number of hoofed digits in the foot to one, as is seen in the modern horse. In the Miocene, a group of distantly related South American meridiungulates convergently evolved limbs like those seen in modern horses. The ancient one-toed litopterns that ran across the plains of South America looked like horses, but were not (table 2.12).

Another antipredator adaptation is to acquire armor. An armored body can be of the passive, tanklike form or the actively offensive, spine armor form. Very distantly related animals have repeatedly evolved tanklike body armor: the saurposid turtles, the dinosaurian ankylosaurs, and the mammalian glyptodonts, armadillos, and pangolins (table 2.13). When confronted by a predator, these animals retreat within their armor and passively wait until the predator gives up trying to attack them and moves on to more vulnerable prey. The morphological convergence between the Cretaceous ankylosaurid dinosaurs and the Pleistocene glyptodontid mammals is astonishing—not just in the similarity of the body armor, but even down to the offensive club located at the ends of their tails, which these animals used to strike out at a predator while

Table 2.13
Convergent evolution of body armor in animals

1 Convergent structure and function: TANKLIKE BODY ARMOR (shell of rigid to semirigid plates that enclose soft body tissues for defense against predators)

Convergent lineages:

1.1 Eastern box turtle (Amniota: Sauropsida: Anapsida: Testudines: Emydidae; *Terrapene carolina*)

1.2 Ankylosaur (Diapsida: Archosauromorpha: Ornithodira: Dinosauria: Ornithischia: Thyreophora: Ankylosauridae; *Ankylosaurus magniventris* †Cretaceous)

1.3 Glyptodont (Amniota: Synapsida: Therapsida: Mammalia: Eutheria: Xenarthra: Cingulata: Glyptodontidae; *Doedicurus clavicaudatus* †Pleistocene)

1.4 Nine-banded armadillo (Mammalia: Eutheria: Xenarthra: Dasypodidae; *Dasypus novemcinctus*)

1.5 Giant scaly anteater (Mammalia: Eutheria: Laurasiatheria: Pholidota: Manidae; *Manis gigantea*)

2 Convergent structure and function: SPINOSE BODY ARMOR (external covering of sharp spines that enclose soft body tissues for defense against predators)

Convergent lineages:

2.1 Australian echidna (Mammalia: Monotremata: Tachyglossidae; *Tachyglossus aculeatus*)

2.2 Greater hedgehog tenrec (Mammalia: Eutheria: Afrotheria: Afrosoricida: Tenrecidae; *Setifer setosus*)

2.3 European hedgehog (Eutheria: Laurasiatheria: Eulipotyphles: Erinaceidae; *Erinaceus europaeus*)

2.4 North American porcupine (Eutheria: Euarchontoglires: Rodentia: Caviomorpha: Erethizontidae; *Erethizon dorsatum*)

2.5 Crested porcupine (Eutheria: Euarchontoglires: Rodentia: Hystricomorpha: Hystricidae; *Hystrix cristata*)

3 Convergent structure and function: CHEMICAL BODY ARMOR (poison concentrated in skin or body tissues for defense against predators)

Convergent lineages:

3.1 Mamor nudibranch snail (Bilateria: Protostomia: Lophotrochozoa: Mollusca: Gastropoda: Opisthobranchia: Nudibranchia: Dendrodorididae; *Dendrodoris grandiflora*)

3.2 Monarch butterfly (Protostomia: Ecdysozoa: Arthropoda: Mandibulata: Hexapoda: Lepidoptera: Danaidae; *Dananus plexippus*)

3.3 Margined blister beetle (Arthropoda: Mandibulata: Hexapoda: Coleoptera: Meloidae; *Epicauta pestifera*)

3.4 Pyralis firefly (Arthropoda: Mandibulata: Hexapoda: Coleoptera: Lampyridae; *Photinus pyralis*)

3.5 Puffer fish (Bilateria: Deuterostomia: Chordata: Osteichthyes: Actinoptergyii: Tetraodontiformes: Tetraodontidae; *Takifugu vermicularis*)

3.6 Bumblebee poison-arrow frog (Osteichtyes: Sarcopterygii: Tetrapoda: Batrachomorpha: Lissamphibia: Batrachia: Anura: Dendrobatidae; *Dendrobates leucomelas*)

3.7 Fire salamander (Lissamphibia: Batrachia: Urodela: Salamandridae; *Salamandra salamandra*)

3.8 Crested auklet (Tetrapoda: Reptiliomorpha: Amniota: Sauropsida: Diapsida: Archosauromorpha: Ornithodira: Dinosauria: Saurischia: Theropoda: Maniraptora: Aves: Charadriiformes: Alcidae; *Aethia cristatella*)

3.9 Hooded pitohui (Dinosauria: Saurischia: Theropoda: Maniraptora: Aves: Passeriformes: Pachycephalidae; *Pitohui dichrous*)

Note: The geological age of extinct species is marked with a †. For data sources, see text.

crouched down under their armor. Both creatures had to live with deadly predators—the ankylosaurids were preyed upon by the tyrannosaurs, and the glyptodontids had to deal with saber-tooth cats—and they evolved the same morphological solution, even though they are vastly different animals.

Besides passively protecting the bearer, spinose body armor also presents the danger of injury to the predator. Animals bristling with sharp spines have repeated evolved in widely separated lineages—monotreme echidnas, laurasiatherian hedgehogs, and afrotherian tenrec "hedgehogs"—and twice independently have evolved among the rodents—the North American porcupines and the African porcupines (table 2.13).

A third type of body armor is not mechanical but chemical. Animals with chemical defenses are also some of the prettiest, most colorful, and most deadly in nature (table 2.13). Their brightly colored bodies are a visible warning to predators that their body tissues are loaded with poison. In the oceans, the wildly colorful nudibranch snails not only manufacture their own poison, such as sesquiterpene toxin (Cimino et al. 1985), but can store and concentrate poisons found in their environment. The bizarre puffer fish have deadly tetrodotoxin in their skin and organs; if eaten, the predator swiftly dies.

On land, diverse arthropods such as pretty orange-and-black monarch butterflies store and accumulate cardenolide poison from the milkweeds on which they feed (we shall consider poisonous plants in detail in the next chapter); blister beetles can burn with cantharidin toxin if eaten; and brilliant flashing fireflies contain lucibufagin poison—and insectivores avoid them all (Agosta 1996). In the amphibians, the beautifully colored poison-arrow frogs are deadly, and toxin from their skin has been used by humans to poison the tips of arrows (hence their name). Fire salamanders have independently evolved poison glands along the tops of their backs. Even a few birds have convergently evolved chemical body armor; the skin and feathers of the hooded pitohui are poisonous (Agosta 1996), and the crested auklet secretes antiectoparasite toxins (Douglas et al. 2005), creating its own mosquito repellant.

A last line of antipredator defense is to fight back. In the seas, five different groups of bony fishes have evolved poisonous stinger spines for defense (table 2.14). The scorpion fishes are diverse and deadly: some are very pretty, like the red lionfish that is popular in aquariums even though it is dangerous; others are nondescript and cryptic, like the stonefish *Synanceia horrida*, which can kill a human if inadvertently stepped

Table 2.14
Convergent evolution of specialized defensive structures in animals

1 Convergent structure and function: STINGER SPINES (poison-bearing spines for defensive use against predators)

Convergent lineages:

1.1 Red lionfish (Osteichthyes: Actinopterygii: Teleostei: Scorpaeniformes: Scorpaenidae; *Pterois volitans*)

1.2 Weever fish (Actinopterygii: Teleostei: Perciformes: Trachinidae; *Trachinus vipera*)

1.3 Coral rabbitfish (Actinopterygii: Teleostei: Perciformes: Siganidae; *Siganus corallinus*)

1.4 Poison toadfish (Actinopterygii: Teleostei: Batrachoidiformes: Batrachoididae; *Thalassophryne megalops*)

1.5 Striped eel catfish (Actinopterygii: Teleostei: Siluriformes: Plotosidae; *Plotosus lineatus*)

1.6 Duckbill platypus (Osteichthyes: Sarcopterygii: Tetrapoda: Reptiliomorpha: Amniota: Synapsida: Therapsida: Mammalia: Monotremata: Ornithorhynchidae; *Ornithorhynchus anatinus*)

2 Convergent structure and function: STINGERS (poison stingers for defensive use against predators)

Convergent lineages:

2.1 Honeybee (Arthropoda: Mandibulata: Hexapoda: Hymenoptera: Apidae; *Apis mellifera*)

2.2 Fire ant (Arthropoda: Mandibulata: Hexapoda: Hymenoptera: Formicidae; *Solenopsis geminata*)

2.3 Long-horned beetle (Arthropoda: Mandibulata: Hexapoda: Coleoptera: Cerambycidae; *Onychocerus albitarsis*)

3 Convergent structure and function: CHEMICAL SPRAY (poisonous chemical sprays for defensive use against predators)

Convergent lineages:

3.1 Common millipede (Bilateria: Protostomia: Ecdysozoa: Arthropoda: Mandibulata: Myriapoda: Diplopoda: Spirobolidae; *Narceus annularis*)

3.2 Flat-backed millipede (Arthropoda: Mandibulata: Myriapoda: Diplopoda: Polydesmidae; *Polydesmus angustus*)

3.3 Neotropical arboreal termite (Arthropoda: Mandibulata: Hexapoda: Isoptera: Termitidae; *Nasutitermes corniger*)

3.4 Bombardier beetle (Arthropoda: Mandibulata: Hexapoda: Coleoptera: Carabidae; *Brachinus explodens*)

3.5 Green stinkbug (Arthropoda: Mandibulata: Hexapoda: Hemiptera: Pentatomidae; *Acrosternum hilare*)

3.6 Striped skunk (Bilateria: Deuterostomia: Chordata: Osteichthyes: Sarcopterygii: Reptiliomorpha: Amniota: Synapsida: Therapsida: Mammalia: Eutheria: Laurasiatheria: Carnivora: Mephitidae; *Mephitis mephitis*)

3.7 European polecat (Eutheria: Laurasiatheria: Carnivora: Mustelidae; *Mustela putorius*)

Note: For data sources, see text.

on. The weever fishes, rabbitfishes, toadfishes, and catfishes have all convergently evolved this line of defense—as has the duckbill platypus on land, with stinger spurs on its hind feet.

We have previously considered predators that sting; three groups of arthropods have convergently evolved stingers for defense. These are the bees, some ants, and some long-horned beetles (table 2.14). The social bees in particular will defend their nests by stinging, which is an act of self-sacrifice as the bee dies afterward. Most ants bite to defend their nests, but a few have independently modified their ovipositors to produce stingers, in parallel evolution with their relatives the bees. The long-horned beetle *Onychocerus albitarsis* has evolved hypodermic antennae; that is, it uses its antennae as stingers with "a delivery system almost identical to that found in the stinger of a deadly buthid scorpion" (Berkov et al. 2008, 257).

Rather than use a stinger, which requires close contact with the predator, other animals have convergently evolved systems to spray poisonous chemicals at the predator from a distance (table 2.14). In the mammals, the New World skunks and Old World polecats are legendary as well as interesting, in that most mammals do not use chemical defenses. In the arthropods, the bombardier beetle is most famous because it not only sprays noxious liquids with quinone (Agosta 1996) but heats the liquid to boiling before spraying. The common millipede also will spray quinone if disturbed, whereas the polydesmid millipedes spray deadly hydrogen cyanide.

Why is chemical defense so rare in terrestrial mammals and birds, but so common in terrestrial panarthropods and amphibians? From the perspective of theoretical morphology, one can easily visualize a nonexistent poison-skinned mouse or sparrow, convergent on a poison-arrow frog, that has evolved a chemical defense to deter a cat or hawk predator. Yet the poison-arrow frog exists, whereas the poison-skinned mouse or sparrow does not. Does a developmental constraint exist for the generation and maintenance of skin poisons in endothermic animals? Or is the constraint functional, in that other forms of defense, perhaps less metabolically costly and complicated, are adequate in endothermic animals?

Organ Systems

We have previously considered the convergent evolution of pupil shapes in the eyes of carnivores (table 2.5) and herbivores (table 2.12). But what about the eye itself? In the classic study of the convergent evolution of the eye, Salvini-Plawen and Mayr (1977) presented evidence for the definite independent evolution of light-detection organs in 40 separate animal lineages, and the probable independent origination of light-detection organs in 20 additional lineages, bringing the total to 60 separate convergences. They also document the convergent evolution of the process of eye evolution itself: from single photoreceptor cells scattered across the body of an animal, to distinct aggregations of photoreceptor cells into eyespots, to ocellar pits, to ocellar cups, to simple eyes, and to complex compound and camera eyes. This same evolutionary sequence can be observed in multiple independent lineages of animals.

In table 2.15 I have conservatively taken the 40 definite cases of convergent eye evolution given by Salvini-Plawen and Mayr (1977), reanalyzed each case in terms of a phylogenetic classification of life (see appendix), and added additional convergent cases that modern phylogenetic analyses have revealed. These additional factors have altered the reasoning of Salvini-Plawen and Mayr (1977) somewhat; for example, because Salvini-Plawen and Mayr considered the arthropods to be descendants of the annelids, they considered the possible homology of light-detection structures found in both groups of animals. We now know that the arthropods are ecdysozoan protostomes more closely related to introvert worms like the priapulids and nematodes, whereas the annelids are lophotrochozoan protostomes more closely related to molluscs (Lecointre and Le Guyader 2006), and thus the two are not closely related. These considerations have resulted in bringing the number of definite convergences to 49 (table 2.15).

It is clear that eyes and other light-detection organs are ancient inventions of the animals. Simple eyespots, more complex ocellar cups, and even complex camera eyes can be found in some of the most plesiomorphic animals still alive, the most simple of the eumetazoa, the cnidarians. Complex camera eyes in our own lineage, the Chordata, were developed over 500 million years ago in the craniate conodonts of the Cambrian (Purnell 1995).

Light-detection organs have been repeatedly evolved, lost, and reevolved to varying degrees of complexity throughout the past 600 million years. Thus, we find simple, isolated photoreceptor cells scattered across the tissues of bilaterian animals as widely divergent as lophotrochozoan protostome rock chitons, ecdysozoan protostome earthworms, and even in the tails of the ammocoete larvae of deuterostome sea lampreys (table 2.15). These are all highly derived animals compared

Table 2.15
Convergent evolution of eyes and light-detection systems in animals

1 Convergent structure and function: PHOTORECEPTOR CELLS (cells with enlarged surface areas containing photopigment for light detection)

Convergent lineages:

1.1 Common European earthworm (Bilateria: Protostomia: Lophotrochozoa: Annelida: Oligochaeta: Lumbricidae; *Lumbricus terrestris*)

1.2 Common rock chiton (Lophotrochozoa: Mollusca: Polyplacophora: Chitonidae; *Acanthopleura spiniger*)

1.3 Soft-shelled clam (Mollusca: Bivalvia: Myoida: Myidae; *Mya arenaria*)

1.4 Marine nematode (Protostomia: Ecdysozoa: Introverta: Nematozoa: Nematoda: Oncholaimidae; *Oncholaimus vesicarius*)

1.5 Sea lamprey (Bilateria: Deuterostomia: Chordata: Vertebrata: Petromyzontiformes: Petromyzontidae; *Petromyzon marinus*)

2 Convergent structure and function: EYESPOTS (photoreceptor cells aggregated into distinct spot regions for increased light detection abilities)

Convergent lineages:

2.1 Pandeid jellyfish (Metazoa: Cnidaria: Hydrozoa: Anthomedusae: Pandeidae; *Leuckartiara octona*)

2.2 Asplanchnid rotifer (Metazoa: Bilateria: Protostomia: Lophotrochozoa: Syndermata: Rotifera: Asplanchnidae; *Asplanchna brightwelli*)

2.3 Green spoon worm larvae (Lophotrochozoa: Annelida: Echiuria: Bonelliidae; *Bonellia viridis*)

2.4 Palola worm (Annelida: Polychaeta: Eunicemorpha: Eunicidae; *Palola viridis*)

2.5 Sabellid bristle worm (Annelida: Polychaeta: Sabellidae; *Notaulax rectangulata*)

2.6 Serpulid bristle worm (Lophotrochozoa: Sipuncula: Sipunculidea: Sipunculidae; *Sipunculus nudus*)

2.7 Planarian flatworm (Lophotrochozoa: Parenchymia: Platyhelminthes: Turbellaria: Microstomidae; *Microstomum lineare*)

2.8 Lampert's sea cucumber (Bilateria: Deuterostomia: Echinodermata: Holothuroidea: Apoda: Synaptidae; *Synaptula lamperti*)

3 Convergent structure and function: OCELLAR PIT (photoreceptor cells located in an open pit, connected to an optic nerve, for directional light detection)

Convergent lineages:

3.1 Spiral tufted bryozoa larvae (Bilateria: Protostomia: Lophotrochozoa: Bryozoa: Cheilostomata: Bugulidae; *Bugula neritina*)

3.2 Megathyrid lamp shell larvae (Lophotrochozoa: Brachiopoda: Terebratulida: Megathyrididae; *Argyrotheca cistellula*)

3.3 Loxosomatid goblet worm larvae (Lophotrochozoa: Spiralia: Entoprocta: Loxosomatidae; *Loxosomella cochlear*)

3.4 Common starfish (Bilateria: Deuterostomia: Echinodermata: Asteroidea: Forcipulatida: Asteriidae; *Asterias rubens*)

3.5 Tornarian larvae of *Balanoglossus proterogonius* (Deuterostomia: Phryngotremata: Hemichordata: Enteropneusta: Tornaria: *Tornaria ancoratae*)

4 Convergent structure and function: OCELLAR CUP (photoreceptor cells located in partially enclosed cup-shaped structure, connected to an optic nerve, for increased precision in directional light detection)

Convergent lineages:

4.1 Bougainvillid jellyfish (Metazoa: Cnidaria: Hydrozoa: Anthomedusae: Bougainvillidae; *Bougainvillea principis*)

Table 2.15
(continued)

4.2 Tiaropsid jellyfish (Cnidaria: Hydrozoa: Leptomedusae: Tiaropsidae; *Tiaropsis multicirrata*)
4.3 Moon jellyfish (Cnidaria: Scyphozoa: Semaeostomae: Ulmaridae; *Aurelia aurita*)
4.4 Pectinariid trumpet worm (Metazoa: Bilateria: Protostomia: Lophotrochozoa: Annelida: Polychaeta: Terebellomorpha: Pectinariidae; *Pectinaria gouldii*)
4.5 Opheliid bristle worm (Annelida: Polychaeta: Opheliidae; *Armandia brevis*)
4.6 Nephthyid bristle worm (Annelida: Polychaeta: Phyllodocemorpha: Nephthyidae; *Nephthys ciliata*)
4.7 Lineid ribbon worm (Lophotrochozoa: Parenchymia: Nemertea: Lineidae; *Lineus ruber*)
4.8 Echinoderid kinorhynch (Protostomia: Ecdysozoa: Introverta: Cephalorhyncha: Kinorhyncha: Echinoderidae; *Echinoderes aquilonius*)
4.9 Salpid sea squirt (Bilateria: Deuterostomia: Chordata: Urochordata: Thaliacea: Salpidae; *Salpa cylindrica*)
4.10 Florida lancet (Chordata: Myomerozoa: Cephalochordata: Branchiostomidae; *Branchiostoma floridae*)

5 Convergent structure and function: SIMPLE EYE (globular eyes with pinhole openings, closed openings, or closed openings with a simple lens for increased precision in directional light detection)
Convergent lineages:
5.1 Onitochiton (Protostomia: Lophotrochozoa: Mollusca: Polyplacophora: Chitonidae; *Onitochiton neglectus*)
5.2 Queen scallop (Mollusca: Bivalvia: Pectenacea: Pectenidae; *Pecten maximus*)
5.3 Pterotracheid snail (Mollusca: Gastropoda: Caenogastropoda: Littorinimorpha: Pterotracheidae; *Pterotrachea mutica*)
5.4 Common garden snail (Mollusca: Gastropoda: Pulmonata: Helicidae; *Helix aspersa*)
5.5 Paragordian hair worm (Protostomia: Ecdysozoa: Introverta: Nematozoa: Nematomorpha: Gordiidae; *Paragordius varius*)
5.6 Peripatid velvet worm (Panarthropoda: Onychophora: Peripatidae; *Peripatus antiguensis*)
5.7 Sagittid arrow worm (Protostomia: Chaetognatha: Aphragmophora: Sagittidae; *Sagitta enflata*)

6 Convergent structure and function: COMPOUND EYE (complex eye with multiple lenses for image resolving)
Convergent lineages:
6.1 Sabellid bristle worm (Protostomia: Lophotrochozoa: Annelida: Polychaeta: Sabellidae; *Bispira volutacornis*)
6.2 Sabellid bristle worm (Annelida: Polychaeta: Sabellidae; *Branchiomma bombyx*)
6.3 Sabellid bristle worm (Annelida: Polychaeta: Sabellidae; *Megalomma vesiculosum*)
6.4 Sabellid bristle worm (Annelida: Polychaeta: Sabellidae; *Pseudopotamilla occelata*)
6.5 Noah's ark shell (Mollusca: Bivalvia: Arcoida: Arcidae; *Arca noae*)
6.6 Mydocopid ostracode (Protostomia: Ecdysozoa: Arthropoda: Mandibulata: Crustacea: Ostracoda: Mydocopida: *Azygocypridina lowryi*)
6.7 Praying mantis (Arthropoda: Mandibulata: Hexapoda: Mantodea: Mantidae; *Mantis religiosa*)
6.8 Eukrohnid arrow worm (Protostomia: Chaetognatha: Phragmophora: Eukrohniidae; *Eukrohnia hamata*)

Table 2.15
(continued)

7 Convergent structure and function: CAMERA EYE (complex eye with a single lens for image resolving)

Convergent lineages:

7.1 Carybdeid box jellyfish (Metazoa: Cnidaria: Cubozoa: Carybdeidae; *Carybdea marsupialis*)

7.2 Alciopod annelid worm (Metazoa: Bilateria: Protostomia: Lophotrochozoa: Annelida: Polychaeta: Alciopidae; *Torrea candida*)

7.3 Common octopus (Lophotrochozoa: Mollusca: Cephalopoda: Coleoidea: Octopodidae: *Octopus vulgaris*)

7.4 Jumping spider (Protostomia: Ecdysozoa: Arthropoda: Cheliceriformes: Arachnida: Salticidae; *Portia fimbriata*)

7.5 Ogre-faced spider (Arthropoda: Cheliceriformes: Arachnida: Deinopidae; *Dinopsis subrufus*)

7.6 Dawn conodont (Bilateria: Deuterostomia: Chordata: Myomerozoa: Craniata: Conodonta: Cordylodontidae; *Eoconodontus notchpeakensis* †Cambrian)

Note: For data sources, see text.

to the nonbilaterian cnidarians, where one might expect to find only simple photoreceptor cells and eyespots—but that is not so: the simple carybdeid box jellyfish have convergently evolved camera eyes (Martin 2002).

The independent evolution of complex camera eyes in the chordates (humans and our kin) and in the molluscs (the octopus and their kin) is a classic case of amazing convergent evolution, oft cited along with the convergent evolution of wings in tetrapods. Yet we now know that camera eyes have also been evolved by alciopid annelid worms (Wald and Raypart 1977) and by two separate groups of spiders (Williams and McIntyre 1980; Laughlin 1980), in addition to some box jellyfish. The ogre-faced spiders are nocturnal predators, and have the largest single-lens eyes described among all the arthropods (Laughlin 1980), a large-eyed predatory convergence seen repeatedly in vertebrate nocturnal carnivores (table 2.5). The telescopic eyes of the jumping spider *Portia fimbriata* are convergent on those of raptorial birds, who use a similar morphological trick to increase the focal length of their eyes: "a parallel to the pit [of the eye] of *P. fimbriata* is the deep convexiclivate fovea found in some vertebrates, notably falconiform birds. . . . [B]oth a group of vertebrates and an invertebrate have therefore adopted the same strategy to improve visual acuity despite a restricted cephalic space" (Williams and McIntyre 1980, 580).

Phylogenetic analyses by Fitzhugh (1989) have revealed that compound eyes have independently evolved four separate times in sabellid bristle worms, not just once. Molecular analyses by Oakley and Cunningham (2002) have revealed that the compound eyes of mydocopid ostracodes have an origin independent from those in other arthropods. Thus, Salvini-Plawen and Mayr's (1977) upper estimate of 60 independent origins of light-detection organs may ultimately prove to be closer to the actual mark. Modern analyses do not always increase the number of convergences discovered, however. Fitzhugh's (1991) phylogenetic analysis revealed that the pygidial eyes of the sabellid bristle worms *Fabriciola* and *Pseudofabriciola* were a synapomophy, and not independently derived.

It is interesting to note that the complex camera eye has been independently evolved in deuterostome chordates, protostome molluscs and arthropods, and nonbilaterian cnidarians (table 2.15), but that complex compound eyes have appeared only in the protostomes. Compound eyes have independently evolved eight different times within the three major types of protostomes (lophotrochozoans, ecdysozoans, and chaetognaths), but not outside of the protostome clade, to my knowledge. Is there a developmental constraint limiting compound eyes to the protostome lineage? This question leads to the consideration of the development of the eye itself.

Modern developmental studies have revealed that all animal eyes contain highly conserved transcription factor *Pax-6* gene homologs, and the ubiquitous presence of the photopigment opsin proteins (Salvini-Plawen 2008). This discovery has led to the proposal of two different models for eye evolution: (1) prototype photoreceptor structures and their developmental genes evolved only once, and further evolution has elaborated them along independent lines, or (2) photoreceptor structures evolved independently multiple times, each time co-opting homologous genes for use in developing those structures (Oakley and Cunningham 2002, 1429). The first model is, in essence, a model of parallel evolution of the eye, and the second model is a model of convergent evolution of the eye. Both models thus can be used to argue for functional and developmental constraints in the evolution of the eye—eyes function only if they are constructed in a limited number of ways, and there are a limited number of genes available that can be used for their development.

Both Oakley and Cunningham (2002) and Salvini-Plawen (2008) argue for the second model. Oakley and Cunningham (2002) point out that the first model requires a photoreceptor structure to be present in all common ancestors, whereas their analyses indicate that compound eyes were not present in ostracode ancestors. Conway Morris (2003) has also argued for the second model, and points out that in mammals alone the *Pax-6* gene is redeployed multiple times for other functions, as it is associated with the development of the brain, nose, pituitary gland, pancreas, and the gut. The *Pax-6* gene is, in essence, equivalent to the homologous tetrapod limb that has convergently evolved into wings for flying in three separate groups of vertebrates (table 2.2), and into fins and paddle-form swimming appendages in seven other groups of vertebrates (table 2.1).

We have previously considered the convergent evolution of asymmetric positioning and ultrasonic hearing in the ears of carnivores (table 2.6). But what about the ear itself? As an organ for detecting pressure waves in air, or for hearing, the ear is an adaptation of land animals. The ancestors of land animals, the fish, possess lateral line systems for detecting pressure waves in water, which is much denser than air, and these lateral line systems are still present in many aquatic amphibians, the tetrapod descendants of fish. Experimentation has shown that lateral line systems are extremely sensitive to minute differences in water pressure (Dijkgraaf 1967), and that fish and aquatic amphibians have well-developed directional "hearing" under water (Kuroki 1967).

Lateral line systems are not a unique synapomorphy of the vertebrates, however. Both molluscs and arthropods have evolved species that have convergently evolved lateral line systems (table 2.16). In the molluscs, many cephalopods have evolved lines of ciliated tissue on their heads and arms, and these epidermal lines have been shown to function like the lateral line systems of fish, allowing the cephalopods to have directional hearing underwater. In addition to the convergent evolution of the complex camera eye (table 2.15), these epidermal lateral line systems are "another example of convergent evolution between a sophisticated cephalopod and vertebrate sensory system" (Budelmann and Bleckmann 1988, 1). In the marine arthropods, some pelagic decapod crustaceans have evolved very long antennae that are held parallel to the long axis of the body of the animal on either side, and these structures also function as lateral line systems for pressure-wave detection (Denton and Gray 1985).

The thin gaseous mixture of the Earth's atmosphere has very low pressure compared to underwater habitats; consequently, land tetrapods have evolved tympanal ear systems in order to hear in air, and have lost the lateral line system. However, some tetrapods that have secondarily returned to aquatic habitats have convergently reevolved a lateral line

Table 2.16
Convergent evolution of ears and sound-detection systems in animals

1 Convergent structure and function: LATERAL LINE SYSTEMS (lateral line row of pores with nerve endings for detecting minute changes in water pressure, used for directional hearing in aqueous habitats)

Convergent lineages:

1.1 Shallow-water brief squid (Bilateria: Protostomia: Lophotrochozoa: Mollusca: Cephalopoda: Coleoidea: Loliginidae; *Lolliguncula brevis*)

1.2 Penaeid shrimp (Protostomia: Ecdysozoa: Arthropoda: Mandibulata: Malacostraca: Decapoda: Penaeidae; *Acetes sibogae*)

1.3 Astrapid jawless fish (Bilateria: Deuterostomia: Chordata: Vertebrata: Pteraspidomorphi: Astrapida: Astrapidae; *Astrapis desiderata* †Ordovician)

1.4 Florida manatee (Vertebrata: Osteichthyes: Sarcopterygii: Reptiliomorpha: Amniota: Synapsida: Therapsida: Mammalia: Eutheria: Afrotheria: Sirenia: Trichechidae; *Trichechus manatus*)

2 Convergent structure and function: TYMPANAL EAR SYSTEMS (tympanum-based hearing system for detecting minute changes in air pressure, used for directional hearing in terrestrial habitats)

Convergent lineages:

2.1 Field cricket (Bilateria: Protostomia: Ecdysozoa: Arthropoda: Mandibulata: Hexapoda: Orthoptera: Gryllidae; *Gryllus pennsylvanicus*)

2.2 Cricket-parasite tachinid fly (Arthropoda: Mandibulata: Hexapoda: Diptera: Tachinidae; *Ormia ochracea*)

2.3 Green lacewing (Arthropoda: Mandibulata: Hexapoda: Neuroptera: Chrysopidae; *Chrysopa carnea*)

2.4 Grand western cicada (Arthropoda: Mandibulata: Hexapoda: Hemiptera: Cicadidae; *Tibicen dorsata*)

2.5 Praying mantis (Arthropoda: Mandibulata: Hexapoda: Mantodea: Mantidae; *Mantis religiosa*)

2.6 Tiger beetle (Arthropoda: Mandibulata: Hexapoda: Coleptera: *Cincindela marutha*)

2.7 Ornate tiger moth (Arthropoda: Mandibulata: Hexapoda: Lepidoptera: Noctuoidea: Arctiidae; *Apantesis ornata*)

2.8 Mallow moth (Arthropoda: Mandibulata: Hexapoda: Lepidoptera: Geometroidea: Geometridae; *Larentia clavaria*)

2.9 Red postman butterfly (Arthropoda: Mandibulata: Hexapoda: Lepidoptera: Papilionoidea: Nymphalidae; *Heliconius erato*)

2.10 Butterfly moth (Arthropoda: Mandibulata: Hexapoda: Lepidoptera: Hedyloidea; Hedylidae; *Macrosoma nigrimacula*)

2.11 Balanerpeton amphibian (Bilateria: Deuterostomia: Chordata: Osteichthyes: Sarcopterygii: Tetrapoda: Batrachomorpha: Dendrerpetonidae; *Balanerpeton woodi* †Carboniferous)

2.12 Eastern box turtle (Tetrapoda: Reptiliomorpha: Amniota: Sauropsida: Anapsida: Testudines: Emydidae; *Terrapene carolina*)

2.13 Granite night lizard (Diapsida: Lepidosauromorpha: Squamata: Scleroglossa: Autarchoglossa: Scincomorpha: Xantusidae; *Xantusia henshawi*)

2.14 Nile crocodile (Diapsida: Archosauromorpha: Crurotarsi: Crocodylia: Crocodylidae; *Crocodylus niloticus*)

2.15 Duckbill platypus (Amniota: Synapsida: Therapsida: Mammalia: Monotremata: Ornithorhynchidae; *Ornithorhynchus anatinus*)

2.16 Mouse-eared bat (Mammalia: Eutheria: Laurasiatheria: Chiroptera: Microchiroptera: Vespertilionidae; *Myotis myotis*)

Note: The geological age of extinct species is marked with a †. For data sources, see text.

system. In addition to losing their tetrapods and reevolving fins, the manatees have acquired a body covering of sinus tactile hairs (which are usually confined to the head region) that function as a lateral line system underwater (Reep 2002).

Two major groups of animals have invaded the land from the water: the panarthropods and the tetrapods. Among the tetrapods, many of the early amphibians still possessed lateral line systems—evidence that they continued to spend a great deal of time in the water. With the evolution of the amniotes, tetrapod invasion of the land began in earnest. It was long thought that tetrapod tympanal ear systems were inherited from the first fully terrestrial amphibian ancestors, and thus that ears evolved very early and only once in the tetrapod lineage. But this is not so. The phylogenetic analyses of Clack (2002) have revealed that tympanal ear systems evolved independently in at least five different tetrapod lineages: once in the ancient batrachomorph amphibians and still present in their lissamphibian descendants (frogs, salamanders, and kin), three additional separate times in the sauropsid amniotes (ancestors of living turtles, lizards, and archosaurs), and independently in the synapsid amniotes (ancestors of mammals). It is now known that the mammalian tympanal ear, with its characteristic inner-ear stapes, incus, and malleus bones, evolved twice independently in the synapsid lineage: once in the monotremes, and independently in the therians (table 2.16). This fact was revealed by the discovery of the basal monotreme species *Teinolophus trusleri* from the Early Cretaceous of Australia, in which the three inner-ear bones had not yet separated from the mandible of the animal (Rich et al. 2005). Therefore, the divergence of the australosphenids and monotremes from the therian lineage occurred before the evolution of the characteristic mammalian typanal ear, and the separation of the three inner-ear bones from the mandible and the evolution the mammalian inner ear occurred independently in the monotreme and therian lineages.

We can hear birds sing, and birds can hear humans sing, but our ears do not have a common origin. As summarized by Clack (2002, 300): "The evolution of hearing in air was a slow and complicated process; far from being achieved and perfected in the earliest tetrapods, it was separately invented many times by different groups—and a long time after tetrapods first gained the land." Clack (2002, 300) surmises that a natural-selective impetus for the evolution of tympanal ears may have been the evolution of noisy, buzzing insects in the Carboniferous: "the timing of the evolution of the characteristics of an ear capable of receiving airborne sound well certainly corresponds to the huge radiation of insects that can be seen in the fossil record of the Late Permian." Tetrapods that could hear insects, as well as see them, would certainly be more successful in hunting insects for food than a tetrapod that could not hear.

Insectivory-triggered ear evolution is not unique to the tetrapods. Within the hexapods themselves, today there exist the cricket-parasite tachinid flies, which have evolved tympanal ears that allow them specifically to hear the crickets that they hunt for food (Robert et al. 1992; Robert and Willi 2000). Tympanal ears have independently evolved in seven different orders of hexapods, and four times independently within the lepidopteran order itself (Hoy and Robert 1996), for a total of at least ten convergent originations (table 2.16). The hexapods show much greater inventiveness in the evolution of their ears than tetrapods, as they variously have placed the tympanum on their abdomens, their thoraxes, their wing bases, and so on. The early evolution of ears in hexapods may have had more to do with their social interactions—calling, courtship, and territorial signals—than with predation, with subsequent ear evolution as antipredator adaptations, such as ultrasonic hearing in moths as a means of evading bats (Hoy and Robert 1996).

In addition to sight and hearing, another major animal sensory capability is that of smell. To my knowledge, very little attention has been given to the phylogenetic distribution of different olfactory organs as a means to determining how often these organs have been independently evolved. Most mentions of convergence in olfactory organs are of the notably unusual convergences, as in star-nosed moles (Conway Morris 2003). Although possible convergences in the olfactory organs of hexapods and vertebrates have been discussed, at present "a definitive answer is not yet possible, but given the differences in the molecular genetics of olfactory receptor coding, the evidence points strongly to convergence" (Conway Morris 2003, 180).

Related to the chemosensory process of detecting molecules in air, or smelling, is the process of breathing itself in the thin atmosphere of the terrestrial realm. The two major groups of animals that have invaded the land from the water have evolved two different organ systems to deal with gas exchange. The panarthropods use a tracheal system of tubes to carry oxygen from the outside air deep into their tissues, and to remove carbon dioxide waste from those tissues and expel it to the outside environment. The tetrapods use lungs instead, with oxygen absorbed across the surface area of the lung tissue and diffused into the blood of the circulatory system, where it is then transported to interior tissues. Carbon dioxide waste is transported in the reverse direction; it is expelled across the surface area of the lung tissue, and exhaled to the outside environment. Thus, the circulatory system of tetrapods serves the dual function of transporting both oxygen and nutrients to interior tissues and removing both carbon dioxide and chemical metabolic wastes from those tissues. The tracheal system of the panarthropods is used only for gas exchange; chemical metabolic wastes are removed by other excretory systems.

As such, these two breathing systems would be an example of divergent evolution in two separate groups of animals, and of no particular interest in the consideration of convergent evolution. However, it appears that the panarthropods have evolved the tracheal system of breathing not just once but five separate times: in the Onychophora (land-dwelling velvet worms), in Cheliceriformes (land-dwelling scorpions and arachnids), and three times in the Mandibulata—the land-dwelling myriapods, malacostracan isopods, and the hexapods (Klok et al. 2002). Klok et al. (2002) give further evidence that the ability to open and close the spiracles of the trachael system (to produce discontinuous gas exchange) has independently evolved four separate times in separated species of tracheated panarthropods. In all cases, the evolution of the tracheal system of breathing is linked to the developmental invagination of the respiratory surfaces of the aquatic panarthropod forms (Pritchard et al. 1993).

Convergent evolution has also occurred in the lung-breathing tetrapods. The plesiomorphic condition is to have lungs like bellows and to breathe in a biflow system, where oxygen is inhaled and carbon dioxide is exhaled. In a biflow breathing system, the same pathway for air transport is used both in inhaling and in exhaling. The birds, however, have a much more efficient uniflow system of breathing, where air is first inhaled into a system of interior air sacs, then moved from the air sacs to the lungs, then exhaled from the lungs to the outside environment. The pathway of air through the bird's breathing system thus moves only in one direction, in a looped pathway rather than in and out along the same pathway, as with other tetrapods. This efficient uniflow system of breathing and oxygen extraction allows birds to easily breathe in regions of low-oxygen partial pressure, as is encountered high in the sky, whereas humans at the same altitude may need the assistance of breathing masks and oxygen cylinders.

It was long thought that the uniflow system of breathing was an autapomorphy of the birds, and related to the evolution of flight in dinosaurs. However, exceptionally well-preserved fossils of the plesiomorphic theropod *Majungatholus atopus* reveals that the animal had the avian pulmonary air-sac system, and that the avian style of breathing evolved very early in the evolution of the theropod clade of dinosaurs (O'Connor and Claessens 2005). Now, a more careful examination of other skeletons of the ornithodires in general reveals that the pterosaurs also had a pulmonary air-sac system and uniflow respiration (Claessens et al. 2009). In essence, the pterosaurs evolved the uniflow system of breathing 70 million years before the birds did. As the pulmonary air-sac system is demonstrably absent in dinosauromorphs and in the herrerasaurids (the basal dinosaurs), the breathing system clearly evolved convergently twice within the archosaurs—once in the pterosaurs, and once in the theropod dinosaurs.

It is interesting that the panarthropods have never evolved lungs. This may well be another example of developmental constraint on the number of evolutionary possibilities open to a group of animals, and a limit to convergent evolution. The tracheal system of breathing limits the size of an animal, and only in the Carboniferous—when oxygen partial pressure in the atmosphere was higher than at present—did very large insects evolve, such as meganeurid dragonflies with wingspans of 75 centimeters.

The small size of insects has allowed these animals to converge on a thermoregulatory system that is usually thought to be possessed only by the birds and mammals—endothermy. Like all ectotherms, insects use a variety of behavioral mechanisms to warm up, such as basking in the sun or shivering. Due to the insects' small size, the heat produced by their muscular actions alone is large relative to the volume of their bodies, and the insects have evolved a series of organ systems to retain their heat once it is acquired: insulating air sacs around their thoracic muscles, an insulating air sac between their thorax and their abdomen, a circulatory system with countercurrent heat exchange, and external body insulation, like the scales of moths and the setae fuzz of bees (Heinrich 1993, 1996).

Large ectotherms, like amphibians and reptiles, have a much larger volume of body tissue to warm up, and a much smaller body surface to absorb heat by basking. The larger the animal, the more serious the problem, because volume increases as a cubic function of dimension, whereas surface area increases only as a square function. One way to solve this problem would be to somehow increase the surface area of the body, in order to absorb more heat while basking. At least six different groups of tetrapods have increased their body surface areas by vastly elongating the neural spines of their vertebrae, thus creating a large "sail fin" structure on their backs (table 2.17). Interestingly, both the derived synapsid pelycosaurs and the ancient batrachomorph amphibians convergently evolved this surface-area-increase structure in the Permian. Sail fins were convergently evolved twice, at the same time, by the pelycosaurs, as the sail fins of the carnivorous sphenacodontids are structurally very different from those of the herbivorous edaphosaurids (Bennett 1996). The fact that the large pelycosaurs evolved sail fins is taken as evidence that these synapsids were still ectotherms, and that endothermy had not yet evolved in the synapsid lineage.

Even more interesting, three separate groups of dinosaurs independently evolved sail fins (table 2.17). As I shall discuss in a moment, the dinosaurs have been argued to have been endotherms, so why would they evolve sail fins like an ectotherm? If the dinosaurs were endotherms, it may well be that the sail fin structure was primarily used to shed excess heat and to cool the body down, rather than to absorb heat.

Two modern groups of highly active, energetic animals are known for their endothermic metabolisms—the birds and the mammals. But when did endothermy evolve in these two separate lineages? The first bird, *Archaeopteryx lithographica*, dates from the Late Jurassic. However, the discovery of fossil non-avian coelurosaurian theropods with feathers but no wings indicates that feathers evolved first as body insulation to retain heat, and that the coelurosaurs were endothermic. Is endothermy thus a theropod trait? The fact that many small ornithopod ornithischians, such as the hypsilophodonts, have fibrolamellar bone (an endothermic characteristic) has been used to argue that they also were endothermic. Did endothermy convergently arise twice in the dinosaurs, once in the theropod saurischians and once in the ornithopod ornithischians, or is endothermy a synapomorphy of the entire clade of dinosaurs? Bakker (1986) has been the most vocal advocate of the latter position, though others still dispute the uniform distribution of endothermy within all of the dinosaurs (see the discussion in Benton 2005, 219–223).

The ornithodiran pterosaurs have also been argued to have been endothermic (see Benton 2005, 226–228). As discussed above, the uniflow system of breathing using lungs and pulmonary air sacs was independently evolved by the pterosaurs and the birds (Claessens et al. 2009), and the evidence is strong that endothermy also independently evolved in the pterosaurs and the dinosaurs (table 2.17). Endothermy clearly arose independently in the synapsid amniote lineage, far removed from the sauropsid dinosaur and pterosaur lineages. Most date the evolution

Table 2.17
Convergent evolution of animal thermoregulation systems

1 Convergent structure and function: HEAT TRANSFER SURFACE (organic surface-area structures for absorbing heat from the external environment or shedding excess heat to the external environment)

Convergent lineages:

1.1 Dissorophid amphibians (Tetrapoda: Batrachomorpha: Dissorophidae; *Platyhystrix rugosus* †Permian)

1.2 Spinosaur (Tetrapoda: Reptiliomorpha: Amniota: Sauropsida: Diapsida: Archosauromorpha: Ornithodira: Dinosauria: Saurischia: Theropoda: Carnosauria: Spinosauridae; *Spinosaurus maroccanus* †Cretaceous)

1.3 Amargasaur (Dinosauria: Saurischia: Sauropodomorpha: Dicraeosauridae; *Amargasaurus cazaui* †Cretaceous)

1.4 Ouranosaur (Dinosauria: Ornithischia: Cerapoda: Ornithischia: Iguanodontidae; *Ouranosaurus nigeriensis* †Cretaceous)

1.5 Edaphosaurid synapsid (Amniota: Synapsida: Edaphosauridae; *Edaphosaurus cruciger* †Permian)

1.6 Sphenocodontid synapsid (Synapsida: Sphenacodontidae; *Dimetrodon grandis* †Permian)

2 Convergent system and function: ENDOTHERMY (active internal generation and maintenance of body heat via metabolism)

Convergent lineages:

2.1 Salmon shark (Vertebrata: Chondrichthyes: Elasmobranchii: Lamnidae; *Lamna ditropis*)

2.2 Common thresher shark (Chondrichthyes: Elasmobranchii: Alopidae; *Alopias vulpinus*)

2.3 Tuna (Vertebrata: Osteichthyes: Actinopterygii: Teleostei: Scombridae; *Thunnus alalunga*)

2.4 Butterfly mackerel (Actinopterygii: Teleostei: Scombridae; *Gasterochisma melampus*)

2.5 Swordfish (Actinopterygii: Teleostei: Xiphiidae; *Xiphias gladius*)

2.6 Pterosaur "hairy devil" (Osteichthyes: Sarcopterygii: Reptiliomorpha: Amniota: Sauropsida: Diapsida: Archosauromorpha: Ornithodira: Pterosauria: Rhamphorhynchidae; *Sordes pilosus* †Jurassic)

2.7 King penguin (Archosauromorpha: Ornithodira: Dinosauria: Saurischia: Theropoda: Aves: Sphenisciformes: Spheniscidae; *Aptenodytes patagonica*)

2.8 Polar bear (Amniota: Synapsida: Therapsida: Mammalia: Eutheria: Laurasiatheria: Carnivora: Ursidae; *Thalarctos maritimus*)

Note: The geological age of extinct species is marked with a †. For data sources, see text.

of endothermic metabolisms in the synapsids to the evolution of the mammalian clade itself; thus it is intriguing that endothermic metabolisms arose three times, independently, in amniote tetrapods (pterosaurs, dinosaurs, and mammals) at the same time—in the Late Triassic.

Yet endothermy is not, as we might conclude, an advanced, derived condition of active terrestrial animals, because it has also convergently arisen in the fishes. Block et al. (1993) have demonstrated that endothermic metabolisms have convergently evolved five separate times in the fishes, twice in the cartilaginous fishes and three times in the bony fishes (table 2.17). In the fishes, it has been argued that the evolution of endothermy is not primarily a result of the adaptive benefits of energetic lifestyles with sustained high activity levels, but rather the niche expansion of these fish groups into cold-water regions of the oceans (Watson 1993; Block et al. 1993).

Last in our consideration of organ systems, it has been recently proposed (Schierwater et al. 2009) that the nervous system itself is convergent, and has independently evolved in two separate metazoan lineages (table 2.18). In the current classification used in this book (see appendix), the nervous system evolved in the cnidarians, and is thus a synapomorphy for subsequent metazoan evolution (table 2.18). In the cnidarians, the nerve cells form a network without a central nervous system. In the

Table 2.18
The alternative classification of metazoans proposed by Schierwater et al. (2009)

Current classification	Alternative classification
Metazoa	Metazoa
– Placozoa	– Diploblasta
– Demospongiae	– – Placozoa
– Hexactinellida	– – – Porifera
– Calcarea	– – – – Calcarea
– Eumetazoa	– – – – Demospongiae
– – Ctenophora	– – – – Hexactinellida
– – Cnidaria*	– – – Ctenophora
– – Bilateria	– – – Cnidaria*
– – – Protostomia	– Bilateria*
– – – Deuterostomia	– – Protostomia
	– – Deuterostomia

Note: The alternative classification of Schierwater et al. (2009) would require the independent evolution of nervous systems in the cnidarians and the bilaterians (marked by asterisks in the table).

bilaterians, a central nerve system is present, organized around a cephalic ganglion from which the central nerve cord emerges; it is considered to have evolved from the simpler cnidarian nervous system (Lecointre and Le Guyader 2006).

Schierwater et al. (2009) have argued that two basal monophyletic groups of metazoans exist: the Diploblasta and the Bilateria (table 2.18). In the current classification, an evolutionary grade exists at the base of the metazoan clade, running from the simple placozoans through to the more complex cnidarians (table 2.18). All of these animals have a roughly radial symmetry, and are sometimes referred to as the "Radiata," a taxon that has no formal status, as it is paraphyletic. In the alternative classification, the placozoans are proposed to be the basal group of a monophyletic clade consisting of three phyla: the Porifera (which are paraphyletic in the current classification, table 2.18), the Ctenophora, and the Cnidaria. In this alternative classification, the cnidarians are not ancestral to the bilaterians; therefore, the evolution of the network nervous system of the cnidarians would be independent of the evolution of the central nervous system of the bilaterians, as these two groups of animals occur in two separate metazoan lineages (table 2.18).

Needless to the say, the argument of Schierwater et al. (2009) remains controversial at present. The peculiar placozoans play a pivotal part in the alternative classification (table 2.18). These organisms are known from a single species, *Trichoplax adhaerens*, which was first discovered in a saltwater aquarium in Austria and subsequently has been found in marine waters worldwide. Lecointre and Le Guyader (2006) suggest that placozoan morphological simplicity may be secondary, and that they are derived from more complex eumetazoans, based upon preliminary molecular studies. Schierwater et al. (2009, 37) argue that placozoan morphological simplicity is primary, and that these organisms possess "a living fossil genome" that shows that "complex genetic tool kits arise before morphological complexity" (in that they were present prior to the Diploblasta-Bilateria bifurcation in evolution) and that "these kits may form similar morphological structures in parallel," such as nervous systems.

Reproduction

"Classic textbook examples of convergence, such as the evolution of flight, and fusiform body shapes in aquatic vertebrates, pale beside those relating to viviparity and matrotrophy" (Blackburn 1992, 320). This observation, surprising as it might appear at first glance, is true, as we shall see in this final section of the chapter. Viviparous animals give live birth, and matrotrophic females provide nutrients for the internally developing embryo. Both of these conditions, complex though they may be, have repeatedly and independently evolved in an astonishing number of animals.

In the plesiomorphic condition for animals, the female produces the egg, the male produces sperm, and the fertilization of the egg by sperm takes place external to the female's body. In marine invertebrates, the egg may contain only enough nutrient for development to the larval stage, after which the larva is on its own and must find its own food to survive, a condition known as planktotrophy. Alternatively, the female can develop a larger egg, one that contains enough yolk nutrient for substantial larval development, a condition known as lecithotrophy. In terms of bioeconomics, it costs the female more energy to produce lecithotrophic larvae than planktotrophic, but in terms of reproductive success lecithotrophic larvae have a better chance of survival than planktotrophic.

In external fertilization, the female fish lay a gelatinous mass of eggs over which the male fish sprays a milky cloud of sperm. This is the plesiomorphic reproductive condition for many oviparous, egg-laying fish and some aquatic amphibians. A more certain form of fertilization, and one that is less costly in terms of sperm numbers, is internal, in which the sperm is introduced into the body of the female, fertilization of the eggs occurs internally, and the female then lays the fertilized eggs. Simple internal fertilization requires the alignment of the female and male cloacal vents so that sperm may be transferred between them. Many male fish, such as sharks, have evolved clasping appendages from the pelvic fins which can be used to hold onto the female during mating, maintaining contact between the male and female cloacae. Terrestrial tetrapods are able to hold one another by their appendages, and males have evolved the ability to somewhat protrude their cloaca so that it fits a slight distance within the cloaca of the female. Some primitive sauropsids have added barbs to their cloaca, which can be used to hold the male's everted cloaca within the female's during fertilization.

In the most efficient mode of internal fertilization, the male evolves an inflatable copulatory organ that can deposit the sperm as close to the eggs within the female's body as possible, and that can be deflated when it is not in use so that it is not a liability in locomotion: in other words, a penis (Kelly 2002). The hydrostatic penis has been convergently evolved

Table 2.19
Convergent evolution of male internal fertilization organs

Convergent structure and function: HYDROSTATIC PENIS (inflatable copulatory organ for passing sperm directly into the body of the female)

Convergent lineages:

1 Golden apple snail (Bilateria: Protostomia: Lophotrochozoa: Mollusca: Gastropoda: Caenogastropoda: Ampullariidae; *Pomacea bridgesii*)

2 Cayenne caecilian (Bilateria: Deuterostomia: Chordata: Osteichthyes: Sarcopterygii: Tetrapoda: Batrachomorpha: Lissamphibia: Gymnophiona: Typhlonectidae; *Typhlonectes compressicauda*)

3 North American tailed frog (Lissamphibia: Batrachia: Anura: Ascaphidae; *Ascaphus truei*)

4 Snapping turtle (Tetrapoda: Reptiliomorpha: Amniota: Sauropsida: Anapsida: Testudines: Chelydridae; *Macroclemys temmincki*)

5 Green anole (Sauropsida: Diapsida: Lepidosauromorpha: Squamata: Iguania: Polychrotidae; *Anolis carolinensis*)

6 Nile crocodile (Archosauromorpha: Crurotarsi: Crocodylia: Crocodylidae; *Crocodylus niloticus*)

7 Ostrich (Archosauromorpha: Ornithodira: Dinosauria: Saurischia: Theropoda: Maniraptora: Aves: Paleognathae: Struthionidae; *Struthio camelus*)

8 Argentine lake duck (Dinosauria: Saurischia: Theropoda: Maniraptora: Aves: Neognathae: Anseriformes: Anatidae; *Oxyura vittata*)

9 Nine-banded armadillo (Amniota: Synapsida: Therapsida: Mammalia: Eutheria: Xenarthra: Dasypodidae; *Dasypus novemcinctus*)

Note: For data sources, see text.

at least six separate times by amniote animals: four separate times within the sauropsids, and once within the synapsids (table 2.19). A single penis is present in anapsid turtles and archosaurian crocodiles but, curiously, two penises, or "hemipenes," were evolved by the lepidosaurian lizards and passed down to their snake descendants. Although the lizards and snakes have two penises, they only use one at a time in intercourse. Equally curious, the penis was lost in highly derived archosaurs like the avian dinosaurs, which reverted back to the use of the cloaca-versus-cloaca method of fertilization. Only a very few birds, such as the ostrich and the duck, have secondarily reevolved a penis organ (King 1981; Kelly 2002).

In the nonamniote tetrapods, males of the caecilian amphibians have an evertable rear part of the cloaca, or phallodeum, that they use to transfer sperm into the cloaca of the female (Wake 1993). Among the frogs, species members of only one genus, the tailed frog *Ascaphus*, have evolved a similar penislike extension of the cloaca. Of the invertebrate animals, the snails are unique in that both dioecious and hermaphroditic species of gastropods have evolved a penis. Some other marine invertebrates, such as the male octopus, secondarily use one of their tentacles as a pseudopenis, or gonopod, to transfer sperm packets to the female. Most curious, the female dwarf seahorse *Hippocampus zosterae*, a peculiar bony fish, uses a gonopod-like organ to transfer eggs from her body into the body of the male seahorse, where they are fertilized (Masonjones and Lewis 1996), in what could be argued to be the convergent evolution of a pseudopenis by a female animal.

Internal fertilization is a precursor to the evolution of viviparity, in which the female does not lay the eggs but rather gives live birth to the offspring. As in fertilization modes, there is a spectrum of modes or degrees of viviparity. The simplest is lecithotrophic viviparity, in which the female produces eggs with large yolks but does not lay them; instead, the embryo develops within the body of the female (this reproductive mode is also sometimes called ovoviviparity). A very simple placenta is present for gas exchange between the female and the eggs, as the developing embryos must respire. The opposite extreme is obligate placentotrophy, in which the female produces small eggs with little or no yolk and the necessary nutrient is directly provided by the female across the placenta to the developing embryos (Thompson and Speake 2006). The spectrum between these two extrema is usually divided into four modes or types of matrotrophy (Blackburn 1993): Type I (lecithotrophic viviparity), Type II, Type III, and Type IV (obligate placentotrophy).

As mammals, many of us have the impression that giving live birth is a unique mammalian trait. This is not true—the convergent evolution of viviparity is rampant in all kinds of animals, from very simple to very complex (tables 2.20, 2.21, 2.22). It is often also thought that viviparity is associated only with highly derived land animals, the amniotes who have adapted to life in harsh terrestrial habitats, but this also is not true, as many marine animals have evolved viviparity (table 2.20). Viviparity is not even a vertebrate trait, for it has independently evolved in both main branches of the invertebrate protostomes, from lophotrochozoan marine bristle worms (Baskin and Golding 1970) to ecdysozoan velvet worms, and in invertebrate deuterstomes as well, such as marine starfish and sea cucumbers, where it has independently evolved at least four separate times (Byrne 2005).

The evolution of viviparity is also ancient, as fossils of the Devonian placoderm fish *Materpiscis* (mother fish) have revealed (Long et al. 2008). In the living sharks, descendants of the ancient placoderms, phylogenetic analyses have revealed that viviparity has independently evolved in nine separate lineages (Dulvy and Reynolds 1997). In table 2.20, I have listed examples of basal or near-basal species for each independent evolution of viviparity in sharks from the cladogram of Dulvy and Reynolds (1997); subsequent evolution within some of these clades has resulted in the presence of viviparity in 40 percent of all living shark species. Fossils of extinct Carboniferous chondrichthyan fish also reveal that viviparity has been sporadically evolved within the cartilaginous fishes for over 300 million years (Dulvy and Reynolds 1997).

In living actinopterygian bony fishes, viviparity has independently evolved in 14 families, or about 3 percent of living actinopterygian species, still not nearly as prevalent as in the cartilaginous fishes (Wourms and Lombardi 1992). It also has been convergently evolved in the sarcopterygian bony fish, and is present in the living-fossil coelocanths (Wourms and Lombardi 1992).

In the transition to terrestrial habitats, viviparity has convergently evolved within both the protostome panarthropods and the deuterostome amphibians. Living insects exhibit sporadic incidences of the independent evolution of viviparity, as it is present in ten separate groups (Hagan 1951). All three groups of living nonamniote tetrapods, the amphibians, have convergently made the evolutionary transition from oviparity to viviparity (table 2.20): twice each in frog lineages and caecilian lineages, but only once within the salamanders (Wake 1992, 1993). In both groups, however, viviparous species are quite rare compared to the large number of species that still lay eggs.

Of the two lineages of amniote tetrapods, the sauropsids and the synapsids, the sauropsids exhibit an astonishing number of separate acquisitions of viviparity, a trait normally thought of as characteristic of synapsid mammals. Estimates of the independent evolution of viviparity in squamate sauropsids have ranged from 98 times (Blackburn 1992) to 100 times (Blackburn 1993) to 105 times (Blackburn 2000). Rigorous phylogenetic analyses for all sauropsids are currently not available, but Lee and Shine (1998) have conducted the most complete phylogenetic analysis of the incidence of viviparity within the reptiles. Phylogenetic relationships within the speciose clades of scincid lizards, colubrid snakes, and elapids snakes were too inadequate to be included in their study, but phylogenetic analyses of the remaining reptiles (Lee and Shine 1998) reveals the independent evolution of viviparity 39 separate times (table 2.21). In table 2.21, I have listed examples of basal or near-basal species for each independent evolution of viviparity in sauropsids from the cladograms of Lee and Shine (1998); future inclusion of the colubrid and elapid snake and scincid lizard species will certainly increase the length of this list. Evolution of viviparity in the sauropsids is also ancient, as

Table 2.20
Convergent evolution of animal reproductive systems: viviparity in nonamniote animals

Convergent structure and function: VIVIPARITY (development of embryos within the body of the female in order to produce more highly developed offspring at birth to enhance their potential for survival)

Convergent lineages:

1 Oligohaline bristle worm (Bilateria: Protostomia: Lophotrochozoa: Annelida: Polychaeta: Nereidae; *Nereis limnicola*)

2 Peripatid velvet worm (Protostomia: Ecdysozoa: Panarthropoda: Onychophora: Peripatidae; *Peripatus antiguensis*)

3 Pacific beetle-mimic cockroach (Panarthropoda: Arthropoda: Mandibulata: Hexapoda: Blattodea: Blaberidae; *Diploptera punctata*)

4 Megathrips (Arthropoda: Mandibulata: Hexapoda: Thysanoptera: Phlaeothripidae; *Megathrips lativentris*)

5 Body louse (Arthropoda: Mandibulata: Hexapoda: Anoplura: Pediculidae; *Pediculus corporis*)

6 Viviparous caddisfly (Arthropoda: Mandibulata: Hexapoda: Ephemeroptera: Ephemeridae; *Notanatolica vivipara*)

7 Brown soft-scale insect (Arthropoda: Mandibulata: Hexapoda: Homoptera: Coccidae; *Lecanium hesperidum*)

8 Viviparous moth (Arthropoda: Mandibulata: Hexapoda: Lepidoptera: Tineidae; *Tinea vivipara*)

9 Flesh fly (Arthropoda: Mandibulata: Hexapoda: Diptera: Sarcophagidae; *Sarcophaga carnaria*)

10 Leaf beetle (Arthropoda: Mandibulata: Hexapoda: Coleoptera: Chrysomelidae; *Chrysomela varians*)

11 Ichneumon fly (Arthropoda: Mandibulata: Hexapoda: Hymenoptera: Ichneumonidae; *Paniscus testaceus*)

12 Viviparous earwig (Arthropoda: Mandibulata: Hexapoda: Dermaptera: Hemimeridae; *Hemimerus vosseleri*)

13 Sea star (Bilateria: Deuterostomia: Echinodermata: Asteroidea: Asterinidae; *Cryptasterina hystera*)

14 Deep-water feather star (Echinodermata: Crinoidea: Comasteridae; *Comatilia iridometriformis*)

15 Sea cucumber (Echinodermata: Holothuroidea: Apodida: Synaptidae; *Leptosynapta clarki*)

16 Small brittle star (Echinodermata: Ophiuroidea: Amphiuridae; *Amphipholis squamata*)

17 Placoderm "mother fish" (Deuterostomia: Chordata: Vertebrata: Gnathostomata: Placodermi: Ptyctodontida: *Materpiscis attenboroughi* †Devonian)

18 Carboniferous ratfish (Vertebrata: Chondrichthyes: Holocephali; *Delphyodontus dacriformes* †Carboniferous)

19 Blunt-nose cow shark (Chondrichthyes: Elasmobranchii: Squalea: Hexanchiformes: Hexanchidae; *Hexanchus griseus*)

20 Short-tailed nurse shark (Chondrichthyes: Elasmobranchii: Galeomorphii: Orectolobiformes: Ginglymostomatidae; *Pseudoginglymostoma brevicaudatum*)

21 Cobbler wobbegong shark (Chondrichthyes: Elasmobranchii: Galeomorphii: Orectolobiformes: Orectolobidae; *Sutorectus tentaculatus*)

22 Goblin shark (Chondrichthyes: Elasmobranchii: Galeomorphii: Lamniformes: Mitsukurinidae; *Mitsukurina owstoni*)

Table 2.20
(continued)

23 African sawtail catshark (Chondrichthyes: Elasmobranchii: Galeomorphii: Carcharhiniformes: Scyliorhinidae; *Galeus polli*)
24 Speckled catshark (Chondrichthyes: Elasmobranchii: Galeomorphii: Carcharhiniformes: Scyliorhinidae; *Halaelurus boesemani*)
25 Cuban ribbontail catshark (Chondrichthyes: Elasmobranchii: Galeomorphii: Carcharhiniformes: Proscylliidae; *Eridacnis barbouri*)
26 Slender smooth-hound shark (Chondrichthyes: Elasmobranchii: Galeomorphii: Carcharhiniformes: Proscylliidae; *Gollum attenuatus*)
27 Barbeled hound shark (Chondrichthyes: Elasmobranchii: Galeomorphii: Carcharhiniformes: Leptochariidae; *Leptocharius smithii*)
28 Reef ocean perch (Vertebrata: Osteichthyes: Actinoptergyii: Teleostei: Scorpaeniformes: Sebastidae; *Helicolenus percoides*)
29 Large Baikal golomyanka (Actinoptergyii: Teleostei: Scorpaeniformes: Comephoridae; *Comophorus baicalensis*)
30 Four-eyed fish (Actinoptergyii: Teleostei: Cyprinodontiformes: Anablepidae; *Anableps anableps*)
31 Black sailfin goodeid (Actinoptergyii: Teleostei: Cyprinodontiformes: Goodeidae; *Giardinichthys viviparous*)
32 One-sided livebearer (Actinoptergyii: Teleostei: Cyprinodontiformes: Jenynsidae; *Jenynsia lineata*)
33 Live-bearing tooth-carp fish (Actinopterygii: Teleostei: Cyprinodontiformes: Poeciliidae; see Table 2.21)
34 Mousey klipfish (Actinopterygii: Teleostei: Perciformes: Clinidae; *Fucomimus mus*)
35 Pile perch (Actinopterygii: Teleostei: Perciformes: Embiotocidae; *Rhacochilus vacca*)
36 Redrump blenny (Actinopterygii: Teleostei: Perciformes: Labrisomidae; *Xenomedea rhodopyga*)
37 Viviparous eelpout (Actinopterygii: Teleostei: Perciformes: Zoacidae; *Zoarces viviparous*)
38 Viviparous brotula (Actinopterygii: Teleostei: Ophidiiformes: Brotulidae; *Dinematichthys ilucoeteoides*)
39 Viviparous bythithid (Actinopterygii: Teleostei: Ophidiiformes: Bythitidae; *Bythites islandicus*)
40 False cusk (Actinopterygii: Teleostei: Ophidiiformes: Parabrotulidae; *Parabrotula plagiophthalmus*)
41 Viviparous halfbeak (Actinopterygii: Teleostei: Beloniformes: Hemiramphidae; *Zenarchopterus gilli*)
42 Coelacanth (Osteichthyes: Sarcopterygii: Actinistia: Coelacanthidae; *Latimeria chalumnae*)
43 Cayenne caecilian (Sarcopterygii: Tetrapoda: Batrachomorpha: Lissamphibia: Gymnophiona: Typhlonectidae; *Typhlonectes compressicauda*)
44 Banded caecilian (Lissamphibia: Gymnophiona: Scolecomorphidae; *Scolecomorphus vittatus*)
45 African live-bearing toad (Lissamphibia: Batrachia: Anura: Bufonidae; *Nectophynoides occidentalis*)
46 Golden coqui (Lissamphibia: Batrachia: Anura: Leptodactylidae; *Eleutherodactylus jasperi*)
47 Fire salamander (Lissamphibia: Batrachia: Urodela: Salamandridae; *Salamandra salamandra*)

Note: The geological age of extinct species is marked with a †. For data sources, see text.

Table 2.21
Convergent evolution of animal reproductive systems: viviparity in amniote animals

Convergent structure and function: VIVIPARITY (development of embryos within the body of the female in order to produce more highly developed offspring at birth to enhance their potential for survival)

Convergent lineages:

1 Ichthyosaur (Amniota: Sauropsida: Diapsida: Ichthyosauria: Ichthyosauridae; *Icthyosaurus platyodon* †Jurassic)

2 Plesiosaur (Sauropsida: Diapsida: Lepidosauromorpha: Sauropterygia: Pachypleurosauridae; *Keichousaurus hui* †Triassic)

3 Black-headed dwarf chameleon (Lepidosauromorpha: Lepidosauria: Squamata: Iguania: Chameleonidae; *Brachypodion melanocephalum*)

4 Side-striped chameleon (Lepidosauria: Squamata: Iguania: Chameleonidae; *Chamaeleo bitaeniatus*)

5 Pygmy lizard (Squamata: Iguania: Agamidae; *Cophotis ceylanica*)

6 Arabian toad-headed lizard (Squamata: Iguania: Agamidae; *Phrynocephalus arabicus*)

7 Helmeted basilisk (Squamata: Iguania: Agamidae; *Corytophanes cristatus*)

8 Patagonian lizard (Squamata: Iguania: Iguanidae; *Phymaturus patagonicus*)

9 Rock horned lizard (Squamata: Iguania: Iguanidae; *Phrynosoma ditmarsi*)

10 Chihuahua desert horned lizard (Squamata: Iguania: Iguanidae; *Phrynosoma orbiculare*)

11 Santa Cruz Island spiny lizard (Squamata: Iguania: Iguanidae; *Sceloporus angustus*)

12 Trans-volcanic bunchgrass lizard (Squamata: Iguania: Iguanidae; *Sceloporus bicanthalis*)

13 Crevice swift lizard (Squamata: Iguania: Iguanidae; *Sceloporus torquatus*)

14 Mexican emerald spiny lizard (Squamata: Iguania: Iguanidae; *Sceloporus formosus*)

15 Florida worm lizard (Squamata: Scleroglossa: Amphisbaenia: Amphisbaenidae; *Rhineura floridana*)

16 Jewelled gecko (Squamata: Scleroglossa: Gekkonta: Gekkontidae; *Naultinus gemmeus*)

17 Skinks (Squamata: Scleroglossa: Autarchoglossa: Scincomorpha: Scincidae; see Table 2.22)

18 Cape grass lizard (Squamata: Scleroglossa: Autarchoglossa: Scincomorpha: Cordylidae; *Chamaesaura anguina*)

19 Arizona night lizard (Squamata: Scleroglossa: Autarchoglossa: Scincomorpha: Xantusiidae; *Xantusia arizonae*)

20 Common lizard (Squamata: Scleroglossa: Autarchoglossa: Scincomorpha: Lacertidae; *Lacerta vivipara*)

21 Gobi racerunner lizard (Squamata: Scleroglossa: Autarchoglossa: Scincomorpha: Lacertidae; *Eremias przewalskii*)

22 California apod lizard (Squamata: Scleroglossa: Autarchoglossa: Anguimorpha: Anguidae: Anniellinae; *Anniella pulchra*)

23 Peloponnesa slow-worm lizard (Squamata: Scleroglossa: Autarchoglossa: Anguimorpha: Anguidae: Anguinae; *Anguis cephallonica*)

24 Striped grass lizard (Squamata: Scleroglossa: Autarchoglossa: Anguimorpha: Anguidae: Diploglossinae; *Ophiodes striatus*)

25 Imbricate alligator lizard (Squamata: Scleroglossa: Autarchoglossa: Anguimorpha: Anguidae; *Barisia imbricata*)

26 Flathead knob-scaled lizard (Squamata: Scleroglossa: Autarchoglossa: Anguimorpha: Xenosauridae; *Xenosaurus platyceps*)

27 Mosasaur (Squamata: Scleroglossa: Autarchoglossa: Anguimorpha: Mosasauridae; *Platecarpus ictericus* †Cretaceous)

Table 2.21
(continued)

28 Blind snake (Squamata: Scleroglossa: Autarchoglossa: Anguimorpha: Serpentes: Typhlopidae; *Typhlops diardi*)
29 False coral snake (Squamata: Scleroglossa: Autarchoglossa: Anguimorpha: Serpentes: Aniliidae; *Anilius scytale*)
30 Western sand boa (Squamata: Scleroglossa: Autarchoglossa: Anguimorpha: Serpentes: Boidae: Erycinae; *Eryx jaculus*)
31 Banded dwarf boa (Squamata: Scleroglossa: Autarchoglossa: Anguimorpha: Serpentes: Tropidophiidae; *Tropidophis feicki*)
32 Rasp-skinned water snake (Squamata: Scleroglossa: Autarchoglossa: Anguimorpha: Serpentes: Acrochordidae; *Acrochordus granulatus*)
33 Brown snake (Squamata: Scleroglossa: Autarchoglossa: Anguimorpha: Serpentes: Colubridae; *Storeria dekayi*)
34 Northern death adder (Squamata: Scleroglossa: Autarchoglossa: Anguimorpha: Serpentes: Elapidae; *Acanthophis praelongus*)
35 Massasauga rattlesnake (Squamata: Scleroglossa: Autarchoglossa: Anguimorpha: Serpentes: Viperidae: Crotalinae; *Sistrurus catenatus*)
36 Green pit viper (Squamata: Scleroglossa: Autarchoglossa: Anguimorpha: Serpentes: Viperidae: Crotalinae; *Trimerosurus albolabris*)
37 Gabon viper (Squamata: Scleroglossa: Autarchoglossa: Anguimorpha: Serpentes: Viperidae: Viperinae; *Bitis gabonica*)
38 Common European adder (Squamata: Scleroglossa: Autarchoglossa: Anguimorpha: Serpentes: Viperidae: Viperinae; *Vipera berus*)
39 Russell's viper (Squamata: Scleroglossa: Autarchoglossa: Anguimorpha: Serpentes: Viperidae: Viperinae: *Daboia russelli*)
40 Mongolian multituberculate (Amniota: Synapsida: Therapsida: Mammalia: Multituberculata: Djadochtatheriidae; *Kryptobaatar daszevegi* †Cretaceous)
41 Virginia opossum (Mammalia: Theria: Marsupialia: Didelphimorphia: Didelphidae; *Didelphis virginiana*)

Note: The geological age of extinct species is marked with a †. For data sources, see text.

seen in the fossil record (Carter 2008) and in particular in the marine reptiles that convergently returned to the marine realm from the terrestrial in the Mesozoic—the ichthyosaurs, plesiosaurs, and mosasaurs (Lee and Shine 1998; Cheng et al. 2004; Carter 2008).

The numerous incidences of convergent evolution of viviparity within the squamates are evidence of the adaptive value of this reproductive mode, or the role of functional constraint in convergent evolution, but evidence of developmental constraint may be present as well. Thompson and Speake (2006) argue that the unique development of an intravitelline mesoderm and the formation of an isolated yolk sac makes it particularly easy for squamates to make the developmental shift to viviparity, which they have done repeatedly (table 2.21). On the other hand, the highly speciose clades of both the plesiomorphic turtles and highly derived archosaurs have remained oviparous. Lee and Shine (1998) suggest that this also may be due to developmental constraint, in that embryonic diapause is widespread in both turtle and archosaur species, but not in squamates. In embryonic diapause, the egg will suspend development if retained in the uterus of the female, and only resumes development after it is laid. Thus, prolonged retention of the egg within the female will not result in significant development of the embryo in utero, a process that is a precursor to hatching the egg in utero and live birth.

In synapsid amniotes, it was long thought that lecithotrophic viviparity was evolved once by the marsupials (table 2.21) and obligate placentotrophy once by the eutherians (table 2.22). Evidence now suggests that the extinct multituberculate mammals independently evolved lecithotrophic viviparity as well (Kielan-Jaworowska 1979), bringing the total to twice for the mammals as opposed to 39 times for the reptiles.

The evolution of extensive matrotrophy (Types III and IV) with complex placentae and obligate placentotrophy is much rarer than the evolution of viviparity (Thompson and Speake 2006). Obligate placen-

Table 2.22
Convergent evolution of animal reproductive systems: complex placentae

1 Convergent structure and function: COMPLEX PLACENTAE (extensive matrotrophy for the production of highly developed offspring at birth to enhance their potential for survival)

Convergent lineages:

1.1 Dwarf livebearer fish (Osteichthyes: Actinopterygii: Teleostei: Cyprinodontiformes: Poeciliidae; *Heterandria formosa*)

1.2 Live-bearing tooth-carp fish (Actinopterygii: Teleostei: Cyprinodontiformes: Poeciliidae; *Poeciliopsis paucimaculata*)

1.3 Live-bearing tooth-carp fish (Actinopterygii: Teleostei: Cyprinodontiformes: Poeciliidae; *Poeciliopsis prolifica*)

1.4 Live-bearing tooth-carp fish (Actinopterygii: Teleostei: Cyprinodontiformes: Poeciliidae; *Poeciliopsis turneri*)

2 Convergent structure and function: OBLIGATE PLACENTOTROPHY (Type IV full matrotrophy for the production of highly developed offspring at birth to enhance their potential for survival)

Convergent lineages:

2.1 Brazilian scincid lizard (Amniota: Sauropsida: Diapsida: Lepidosauromorpha: Lepidosauria: Squamata: Scleroglossa: Autarchoglossa: Scincomorpha: Scincidae; *Mabuya heathi*)

2.2 Italian three-toed skink (Squamata: Scleroglossa: Autarchoglossa: Scincomorpha: Scincidae; *Chalcides chalcides*)

2.3 Australian grass skink (Squamata: Scleroglossa: Autarchoglossa: Scincomorpha: Scincidae; *Pseudemoia entrecasteauxii*)

2.4 Western serpentiform skink (Squamata: Scleroglossa: Autarchoglossa: Scincomorpha: Scincidae; *Eumecia anchietae*)

2.5 Placental mammal "dawn mother" (Amniota: Synapsida: Therapsida: Mammalia: Theria: Eutheria; *Eomaia scansoria* †Cretaceous)

Note: The geological age of extinct species is marked with a †. For data sources, see text.

totrophy was long considered to be an autapomorphy and defining characteristic of the eutherians, the placental mammals, the first of which evolved 125 million years ago in the Early Cretaceous (Ji et al. 2002). This view was overturned with the discovery of Type IV matrotrophy in the Brazilian scincid lizard *Mabuya heathi* (Blackburn et al. 1984), and subsequent discoveries of obligate placentotrophy in three other skink species (Blackburn 1993, 2000; Flemming and Blackburn 2003). Thus, both the synapsid and sauropsid lineages of amniotes have convergently evolved obligate placentotrophy (table 2.22). The mammalian evolutionary pathway from oviparity (monotremes) to lecithotrophic viviparity (marsupials) to obligate placentotrophy (eutherians) has been independently traveled by the squamate reptiles.

Is complex placentation thus a unique trait of land animals? Perhaps not. Reznick et al. (2002) have developed a "matrotrophy index" (MI) for species comparisons, which they have used to argue that complex placentation has convergently evolved four separate times in fish (table 2.22). They point out that species with lecithotrophic viviparity typically have an MI of 0.6 to 0.7, whereas the fish species *Poeciliopsis retropinna* has an MI of 117. Although the argument for the evolution of complex placentation remains controversial (see Morell 2002), many of the viviparous fish species remain to be analyzed using the methodology of Reznick et al. (2002).

In concluding this section of the chapter on the convergent evolution of reproductive traits in animals, I would like to mention the land plants. Plants are very different organisms from animals; for one thing, they can synthesize their own food. Nevertheless, as we shall see in the next chapter, plants faced the same functional problems that animals faced in their transition from the sea to the land, and thus their evolution is as rife with convergence as we have seen in this chapter among the animals. That convergence extends even to the plants' mode of reproduction, for the early free-spore-reproducing land plants were as dependent on a source of water for their reproduction as were the amphibian animals, and the evolution of the seed by land plants was the equivalent of the evolution of the amniote egg by tetrapods, a key trait in their respective successful invasion of terrestrial habitats (Niklas 1997).

3 Convergent Plants

Phenotypic correspondence among unrelated species provides strong circumstantial evidence for adaptive evolution because it shows that organisms differing in genetic and developmental capabilities can converge on comparable solutions to life's exigencies when confronted with the same or very similar selection pressures. Here we shall consider a much touted example of convergent evolution—arborescence, the tree growth habit.
—Niklas (1997, 316)

Arborescence

The reverse of the old adage "not seeing the forest for the trees"—the condition in which people overlook a larger overall pattern because they are too narrowly focused on individual particulars—is "not seeing the trees for the forest." The trees in a forest do vary—an oak tree is different from a maple tree—but the apparent similarity of tree form that can be seen throughout thousands of trees in a forest may give one the impression that all trees are variants of a single common ancestor. This is not so. Nine separate groups of plants have independently and convergently evolved the tree form (Niklas 1997; Donoghue 2005; Stein et al. 2007).

Trees are an adaptation to life on land. Plant growth in a two-dimensional plane soon leads to crowding and overgrowth, with one plant shading out another plant in the competition for light from the sun, the energy source for the survival of the plant. Just as in human cities, when crowding occurs in the two-dimensional plane of the land surface, the solution is to move into the third dimension above the land surface—to construct towering skyscraper buildings or to evolve trees.

In order to grow into the third dimension of height, a major force must be overcome—gravity. Tree forms all have a central support structure, the trunk, that rises vertically from the land surface. At some distance above the ground, branches extend out from the tree trunk in order to capture as much sunlight as possible for the tree's survival and, in order to ensure the survival of the species, to facilitate fertilization and dispersal of the tree's offspring.

The first trees appeared in the Middle Devonian, over 390 million years ago. Plants began their full-scale invasion of the terrestrial realm in the Late Devonian and Carboniferous, and huge forests spread across the landscapes of the Earth. The trees in those forests were very different from those that constitute the forests of the Earth today. The famous Gilboa fossil forest in New York State, from the Middle Devonian, gives us a glimpse of the earliest forests on Earth. Numerous upright trunks of trees are preserved as sandstone casts, each one broken off about a meter above the ancient ground surface. The trunks range in size from about a half meter to a meter in diameter with a wider, bulbous base with numerous small anchoring roots, revealing that these earliest trees had no taproot system for anchoring. The morphology of the upper portions of the trees was completely unknown until the spectacular discovery in 2007 of fossils of the complete tree, logs that had fallen over and were preserved horizontally, in rock outcrops only 13 kilometers east of the Gilboa forest (Stein et al. 2007). Some of these trees were over 8 meters tall and, most surprisingly, had no laminar leaves or horizontal branches. Instead, they were topped with a crown resembling a spiky pipe brush or shaving brush, consisting of numerous digitately branched offshoots held at an acute angle to the trunk stem. The complete tree *Wattieza (Eospermatopteris) erianus* belongs to an extinct group, the Cladoxylopsida (table 3.1), that is related to modern surviving horsetail rushes (see the appendix). Stein et al. (2007, 906) note that the cladoxylopsid tree's "body plan now stands unequivocally as the oldest known arborescent terrestrial plant form" and that it is "therefore interesting to see how instantly recognizable and, in a significant sense, 'modern' the tree-like architecture of *Wattieza* seems to be."

Both the more plesiomorphic lycophyte club mosses and the cladoxylopsid-related equisetophyte horsetail rushes independently evolved towering trees (table 3.1), although the tallest members of these groups today is around 10 centimeters. In the lycophytes, the ancient lepidodendrid scale trees constructed their trunks with two outer layers of cortex beneath an outermost layer of leaf cushions, whereas in the equisetophytes, the ancient calamite horsetail trees had trunks supported by internal wedges of wood (Niklas 1997).

Table 3.1
Convergent evolution of arborescent plant morphologies

Convergent structure and function: TREE TRUNK (vertical support structure to resist gravity for the elevation and spatial dispersal of photosynthetic and reproductive substructures)

Convergent lineages:

1 Lepidodendrid scale tree (Tracheophyta: Lycophyta: Lepidodendracea; *Lepidodendron rhodumnense* †Carboniferous)

2 Gilboa-forest cladoxylopsid tree (Tracheophyta: Euphyllophyta: Moniliformopses: Cladoxylopsida: Pseudosporochnaceae; *Wattieza (Eospermatopteris) erianus* †Middle Devonian)

3 Calamite horsetail tree (Euphyllophyta: Moniliformopses: Equisetophyta: Calamitaceae; *Calamites cistiiformes* †Carboniferous)

4 Psaronis-type tree ferns (Euphyllophyta: Moniliformopses: Filicophyta: Marattiaceae; *Psaronius schopfii* †Carboniferous)

5 Tempskya-type tree ferns (Euphyllophyta: Moniliformopses: Filicophyta: Tempskyaceae; *Tempskya dernbachii* †Cretaceous)

6 Australian tree fern (Euphyllophyta: Moniliformopses: Filicophyta: Cyatheaceae; *Cyathea cooperi*)

7 Archaeopterid-type woody trees (Euphyllophyta: Lignophyta: Archaeopteridales: Archaeopteridaceae; *Archaeopteris hibernica* †Late Devonian)

8 Cycad palm (Euphyllophyta: Lignophyta: Spermatophyta: Cycadophyta: Cycadaceae; *Cycas media*)

9 Date palm (Euphyllophyta: Lignophyta: Spermatophyta: Angiospermae: Euangiosperms: Monocotyledons: Aracaceae; *Phoenix dactylifera*)

Note: The geological age of extinct species is marked with a †. For data sources, see text.

Three separate groups of the ferns have convergently evolved the tree form through time: the psaronis-type tree ferns in the Carboniferous, the tempskya-type tree ferns in the Cretaceous, and the tree ferns that survive today (table 3.1). Each group invented a different type of trunk in their evolution: the Carboniferous tree ferns had trunks supported by an outer mantle of adventitious roots, the Cretaceous tree ferns had trunks consisting of interweaved stems bound together by adventitious roots, and living tree ferns have a lower columnar base of adventitious roots and an upper trunk supported by an outer layer of cortex and external layer of leaf bases (Niklas 1997).

The first of the lignophyte trees that are so abundant today appeared in the Devonian, particularly the archaeopterid-type woody trees of the Late Devonian, which still reproduced with spores but had trunks supported by innermost concentric layers of heartwood (Niklas 1997). These trees gave rise to the first seed plants in the Late Devonian, a reproductive event in plant evolution equivalent to the evolution of the amniote egg in the tetrapods that we considered in the previous chapter, thus a vital event in the plant conquest of the terrestrial realm. The spermatophyte seed plants are incredibly diverse. Although they descend from the original lignophyte trees, at least two groups of spermatophytes have lost and secondarily reevolved the tree form: the cycad palms and the date palms (table 3.1). The cycad palms have trunks that are supported by an outermost layer of persistent leaf bases, whereas the date palms have trunks supported by a fibrous complex of primary vascular bundles, increasing in density of occurrence from the center to the outer margin of the trunk (Niklas 1997), very different from the heartwood trunks of most lignophyte trees today.

Niklas (1997, 316) argues strongly for the role of functional constraint in the evolution of the tree form: "convergence among so many different plant lineages for many biomechanical traits rather than only one or a few important ones strongly suggests that these traits are the result of adaptive evolution and not fortuitous expressions of diffusive evolution." Would not these same functional constraints produce the same evolutionary result in alien worlds, as they have nine independent times here on Earth? If sessile multicellular photoautotrophic organisms evolve on distant Earth-like worlds, and if those organisms emerge from the seas to invade the land, then they should necessarily—and convergently—evolve the geometry of the tree form. Or, in the spirit of theoretical morphology, can we conceive of an alternative geometry that these organisms might evolve to solve the problems of overcrowding, gravity, and maximizing surface area for photosynthesis? We shall return to this question in the next chapter, when we consider the possible evolution of balloon-like plants, plants that could float in the air, and why such hypothetical plants do not exist on the Earth.

Niklas (1997) points out that both the ancient lepidodendrid scale trees and the calamite horsetail trees not only convergently evolved tree-trunk support structures, but also independently evolved traits such as wood tissue, leaves, cones for sporangia, and heterosporous reproduction. These ancient trees depended upon a readily available and dependable supply of water for their free-spore mode of reproduction, and thus were confined to swampy regions of the terrestrial realm—a developmentally imposed constraint similar to that seen in early tetrapod invaders of the terrestrial realm, the amphibians (as discussed in the previous chapter). And just as the tetrapods escaped that constraint in the evolution of the amniotes, the plants escaped it in the evolution of the seed plants, an event in reproductive evolution that we shall consider in detail later in this chapter.

Because land plants are surrounded by air, not water, they require a system to transport water from the soil below to the plant tissues above, where it is needed. The simplest of the embryophytes, land plants such as liverworts, have no roots and no vascular system. The derived and highly successful vascular land plants are classified as tracheophytes, as they possess tracheids, specialized cells for the transport of water up the stem of the plant against the force of gravity (Lecointre and Le Guyader 2006). Advanced embryophytes, such as *Cooksonia hemisphaerica* from the Late Silurian and *Aglaophyton major* from the Devonian, are now considered to be transitional to the earliest true tracheophytes (Niklas 1997). These horneophytes had no roots or leaves, but did have small stems that stood up against the force of gravity and bifurcated to produce multiple tips that contained the sporangia. The tracheids of the earliest tracheophytes were arranged in a series of circular rings, stacked one atop another up the stem of the plant, or in a continuous climbing helix. The walls of the tracheids were lignified, and thus also provided some support to hold the stem of the plant erect against the force of gravity. The question is still open as to whether tracheids are a synapomorphy, arising only once in plant evolution, and whether the Tracheophyta are monophyletic. A possible alternative is that tracheids are convergent, independently evolved in the lycophytes and the euphyllophytes (Niklas 1997).

The next evolutionary step beyond tracheids for water transport within the plant is the evolution of vessels for rapid and higher-volume water transport. The evolution of vessels is clearly convergent: at least nine separate phylogenetic lineages of tracheophytes have independently evolved vessels (table 3.2). The angiosperms alone have convergently evolved vessels several times (Niklas 1997); for example, the basal angiosperm nymphaealeans have independently evolved vessels at least twice (Schneider and Carlquist 1995). Rates of water and food transport within a plant are size limiting; with the convergent evolution of vessels, plants began to achieve large sizes, and with the convergent evolution of trunks (table 3.1) the invasion of land was in full force.

Photoautotrophs: Light Capture

Plants are photoautotrophs: they produce their own food via photosynthesis. Early land plants, such as *Cooksonia hemisphaerica* some 425 million years ago, had small green stems and no leaves. A stem is a small cylinder, with a fairly large surface area where photosynthesis can take

Table 3.2
Convergent evolution of water-conducting structures in plants

Convergent structure and function: VESSELS (water-conducting cellular structures for rapid water transport within the plant)

Convergent lineages:

1 Resurrection plant (Tracheophyta: Lycophyta: Selaginellaceae; *Selaginella lepidophylla*)

2 Woodland horsetail (Tracheophyta: Euphyllophyta: Moniliformopses: Equisetopyta: Equisetaceae; *Equisetum sylvaticum*)

3 Bracken fern (Euphyllophyta: Moniliformopses: Filicophyta: Dennstaedtiaceae; *Pteridium aquilinum*)

4 Gigantopterid seed fern (Euphyllophyta: Lignophyta: Spermatophyta: Gigantopteridales: Gigantopteridaceae; *Gigantopteris nicotianaefolia* †Permian)

5 Arizona ephedra (Euphyllophyta: Lignophyta: Spermatophyta: Gnetophyta: Ephedraceae; *Ephedra fasciculata*)

6 Malaysian "water lily" (Euphyllophyta: Lignophyta: Spermatophyta: Angiospermae: Nymphaeales: Nymphaeaceae; *Barclaya kunstleri*)

7 Giant "water lily" (Euphyllophyta: Lignophyta: Spermatophyta: Angiospermae: Nymphaeales: Nymphaeaceae; *Victoria amazonica*)

8 Pumpkin (Euphyllophyta: Lignophyta: Spermatophyta: Angiospermae: Euangiosperms: Eudicots: Rosidae: Cucurbitaceae; *Cucurbita maxima*)

Note: The geological age of extinct species is marked with a †. For data sources, see text.

place, and a small volume of internal tissue in need of food. The key word in the last sentence is "small": the larger the cylinder, the smaller its surface area relative to its volume, as surface area increases as a square function of linear dimension whereas volume increases as a cubic function. Thus, the larger the stem of the plant, the smaller its surface area, where photosynthesis occurs, relative to its volume of internal tissue, where food is needed.

The solution to this classic area-volume problem is to somehow increase the surface area of the plant that is capable of photosynthesizing food—that is, to evolve leaves. The evolution of the leaf has occurred independently in at least nine different phylogenetic lineages of land plants (table 3.3). Interestingly, eight of these lineages convergently developed leaves in the Devonian, from the simple liverworts to the first of the derived lignophyte trees. Thus, the many leaves we see around us each day, from ferns to maple tree leaves, can be traced back in time to that explosion of convergent evolution in the Devonian.

Botanists divide leaves into two types: microphylls, which are small with only one vein and no leaf gaps, and macrophylls, which are large with complex venation and leaf gaps (Tomescu 2008). Not surprisingly, microphylls were convergently evolved by Devonian liverworts, mosses,

Table 3.3
Convergent evolution of leaves in plants

Convergent structure and function: LEAVES (expanded-surface-area phototrophic structures for light capture and photosynthesis)

Convergent lineages:

1 Devonian liverwort leaf (Embryophyta: Marchantiophyta: Pallavincinaceae; *Pallavicinites devonicus* †Late Devonian)

2 Devonian moss leaf (Embryophyta: Stomatophyta: Hemitracheophyta: Bryophyta *sensu stricto*: Sporogitales; *Sporogonites exuberans* †Early Devonian)

3 Devonian lycopod leaf (Hemitracheophyta: Polysporangiophyta: Tracheophyta: Lycophyta: Protolepidodendraceae; *Leclercqia andrewsii* †Middle Devonian)

4 Devonian horsetail leaf (Tracheophyta: Euphyllophyta: Moniliformopses: Equisetophyta: Sphenopsidaceae; *Sphenophyllum involutum* †Late Devonian)

5 Whisk fern leaf (Euphyllophyta: Moniliformopses: Filicophyta: Psilotaceae; *Psilotum nudum*)

6 Devonian fern leaf (Euphyllophyta: Moniliformopses: Filicophyta: Zygopteridaceae; *Ellesmeris sphenopteroides* †Late Devonian)

7 Devonian aneurophyte leaf (Euphyllophyta: Lignophyta: Aneurophytales: Pseudosporochnaceae; *Rellimia thomsonii* †Middle Devonian)

8 Devonian archaeopterid leaf (Euphyllophyta: Lignophyta: Archaeopteridales: Archaeopteridaceae; *Archaeopteris hibernica* †Late Devonian)

9 Carboniferous seed-fern leaf (Euphyllophyta: Lignophyta: Spermatophyta: Lyginopteridales: Lyginopteridaceae; *Telangium bifidum* †Early Carboniferous)

Note: The geological age of extinct species is marked with a †. For data sources, see text.

and lycopods (Gensel and Kasper 2005); that is, in land plants that are small and thus less in need of increased surface area for photosynthesis. The analyses of Boyce and Knoll (2002) demonstrate that macrophylls evolved convergently at least four separate times in the equisetophytes, filicophytes, "progymnosperms" (nonspermatophyte lignophytes), and spermatophytes. In table 3.3 I have listed a leaf species that is one of the most ancient for each of these groups, taken from the analyses of Boyce and Knoll (2002). Tomescu (2008) agrees that macrophylls originated independently a minimum of four times, but presents phylogenetic analyses that suggest the actual number of convergences may be nine. He demonstrates that leaves were independently developed at least twice in the ferns (separately in the Psilotaceae and the Zygopteridaceae), and perhaps as many as five times. Likewise, he argues that leaves arose independently in both the aneurophyte and archaeopterid nonspermatophyte lignophytes (table 3.3).

Before leaving this section on photosynthetic light-capture structures in plants, I would like to touch briefly on the convergent evolution of a different type of light-capture structure: eyes. Eyes are usually considered to be unique to animals, and we examined their convergent evolution in detail in the previous chapter (see table 2.15). However, not only have eyes evolved in the unikontan branch of the eukaryote phylogenetic tree—the branch leading to animals—but they have convergently evolved in the bikontan branch as well, the branch that leads to the plants (see the appendix). And not just once—simple eyespots have independently evolved in a wide variety of phototactic algae and algal phytoflagellates (table 3.4). Eyespots have independently evolved no less than 11 separate times in groups as diverse as green algae, brown algae, yellow-green algae, and many algal flagellates (Foster and Smyth 1980; Kivic and Walne 1983). Kivic and Walne (1983) document the convergent evolution of different types of eyespots in the algae (table 3.4), and Foster and Smyth (1980, 619) also demonstrate that the phototaxis

Table 3.4
Convergent evolution of eyes in algae and algal phytoflagellates

Convergent structure and function: EYESPOTS (photoreceptor cells aggregated into distinct spot regions for light-detection abilities)

1 EYESPOT TYPE A:

Convergent lineages:

1.1 Chroomonas cryptophyte (Eukarya: Bikonta: Chromoalveolata: Cryptophyta; *Chroomonas salina*)

1.2 Perdinium dinoflagellate (Eukarya: Bikonta: Chromoalveolata: Dinophyta; *Peridinium westii*)

1.3 Chlamydomonas green alga (Eukarya: Bikonta: Green eukaryotes: Chlorobionta: Ulvophyta: Chlorophyceae; *Chlamydomonas reinhardtii*)

1.4 Platymonas flagellated alga (Eukarya: Bikonta: Green eukaryotes: Chlorobionta: Prasinophyta; *Platymonas subcordiformis*)

2 EYESPOT TYPE B:

Convergent lineages:

2.1 Pavlova coccolithophorid (Eukarya: Bikonta: Chromoalveolata: Haptophyta; *Pavlova helicata*)

2.2 Ochromonas golden alga (Eukarya: Bikonta: Chromoalveolata: Stramenopiles: Chrysophyceae; *Ochromonas villosa*)

2.3 Fucus brown alga (Eukarya: Bikonta: Chromoalveolata: Stramenopiles: Phaeophyceae; *Fucus serratus*)

2.4 Heterococcus yellow-green alga (Eukarya: Bikonta: Chromoalveolata: Stramenopiles; Xanthophyceae; *Heterococcus pleurococcoides*)

3 EYESPOT TYPE C:

Convergent lineages:

3.1 Euglena euglenophyte (Eukarya: Bikonta: Excavobionta: Euglenobionta; *Euglena gracilis*)

3.2 Wolozynskia dinoflagellate (Eukarya: Bikonta: Chromoalveolata: Dinophyta; *Wolozynskia coronata*)

3.3 Polydriella eustigmatophyte alga (Eukarya: Bikonta: Chromoalveolata: Stramenopiles: Eustigmatophyceae; *Polydriella helvetica*)

Note: For data sources, see text.

control system for the motile species often has "dramatically different machinery" and has been independently derived in each group.

While present in some of the most plesiomorphic Chlorobionta, like the green algae, eyespots are lost in higher plants, particularly sessile land plants. Land plants have evolved a series of phototropic mechanisms instead, as seen in plant species that slowly turn in the light, tracking the path of the sun across the sky, or opening and closing their flowers in response to the rising and setting of the sun. Charles Darwin was fascinated with heliotropic movement in plants, and discovered in his experiments that his plants stopped tracking the sun during the day if a filter removing blue wavelength light was placed between them and the sun. We now know that the genes encoding for heliotropic movement in plants have been convergently evolved in animals, where they are used in maintaining circadian rhythms (this example of molecular convergence will be considered further in chapter 5).

Photoautotrophs: Carbon Processing

Life has evolved a variety of autotrophic pathways: chemoautotrophic eubacteria obtain energy for organic synthesis from oxidizing methane, ammonia, nitrites, sulfites, or other inorganic mineral compounds present on the Earth. Photoautotrophic eubacteria obtain the energy they need from a source not on the Earth, namely by capturing photons from the sun. The oxygenic type of photosynthesis was evolved by the cyanobacteria over 3.5 billion years ago, as witnessed by the presence of stromatolite bacterial mounds in the Earth's fossil record; the subsequent production of free oxygen by these bacteria has transformed the very atmosphere of the planet. In the Cenozoic, some 3.47 billion years later, oxygen levels in the atmosphere have risen to such high levels, and carbon dioxide content fallen to such low levels, that they actually interfere with the photosynthetic process the cyanobacteria evolved such a long time ago.

The chloroplasts in the cells of all photosynthetic eukaryotes, and hence all plants, are symbiotic cyanobacteria. Carbon fixation for the production of hydrocarbon molecules, or food, by plants is catalyzed by *Rubisco* (ribulose-1,5-bisphosphate carboxylase/oxygenase) in an ancient process known as C_3 photosynthesis. In the normal C_3 photosynthetic pathway, carbon is fixed directly by Rubisco carboxylation in mesophyll-tissue chloroplasts. However, if carbon dioxide partial pressures in the atmosphere are low and oxygen partial pressures are high, the presence of oxygen can interfere with Rubisco carboxylation and produces oxygenation instead; fixed carbon is then wastefully lost in photorespiration.

About 30 to 50 million years ago this threshold was reached in the evolution of the Earth's atmosphere. This event was partially due to the activity of land plants themselves, as they have been extracting carbon dioxide from the atmosphere and releasing free oxygen back into the atmosphere for the past 425 million years, ever since the rapid spread of land plants in the Late Silurian. It was also partially due to the formation of the gigantic Himalayan mountain chain in the Cenozoic, produced by the plate-tectonic-driven collision of the Indian subcontinent with Asia, and subsequent weathering processes that have removed a substantial amount of carbon dioxide from the Earth's atmosphere. In any event, plants began to experience difficulty in fixing carbon using the C_3 photosynthetic pathway.

To deal with the problem of Rubisco oxygenation, plants have evolved a more efficient C_4 photosynthetic pathway, one that builds upon the original C_3 pathway but adds an additional carbon dioxide–concentrating pathway. In the C_4 photosynthetic pathway, carbon is primarily fixed by PEPC (phosphoenol pyruvate carboxylase) carboxylation in the mesophyll tissue, and then four carbon acids are shuttled into bundle-sheath tissue regions of the leaf where Rubisco and the chloroplasts have been concentrated (Sage 1999, 2001a). This more efficient C_4 photosynthetic pathway of carbon dioxide concentration can overcome photorespiration that would normally be triggered by high oxygen levels in the atmosphere.

Evolution of the C_4 photosynthetic pathway is not trivial. Land plants were developmentally constrained by the original ancient Rubisco-based biochemistry and cellular systems evolved over 425 million years ago (Sage 1999). Thus, it might appear that only a few types of plants (or perhaps only one) could have made the biochemical and cellular changes necessary to evolve a new photosynthetic pathway. But this is not the case: the C_4 photosynthetic pathway has been independently evolved, over and over, by different plant groups in the past 30 million years (table 3.5). The phylogenetic analyses of Kellogg (1999) have revealed that the C_4 photosynthetic pathway was independently evolved by 18 separate angiosperm families, and that it has convergently arisen at least four independent times in the grass family Poaceae alone (Sinha and Kellogg 1996). In table 3.5 I have listed basal or near basal species for each clade of plants that have independently evolved C_4 photosynthesis from the cladograms of Kellogg (1999) and Sinha and Kellogg (1996). At present, the known number of convergences of C_4 photosynthesis stands at 33, but future analyses of previously unstudied angiosperm groups are almost certainly going to lengthen the list of convergences in the future (Sage 2001a, 2001b).

Is C_4 photosynthesis developmentally constrained to the flowering plants? Sage (2001b) notes that it is surprising that C_4 photosynthesis has not been evolved by the nonangiosperm tracheophytes, particularly the ferns, and suggests that only the flowing plants have the developmental flexibility to evolve the biochemical and cellular changes necessary for C_4 photosynthesis. In addition, some geochemical models suggest that atmospheric oxygen and carbon dioxide levels may have approached today's levels back in the Carboniferous, largely due to the effects of the explosive spread of the Earth's first forests. The fact that tracheal-breathing size-limited hexapods, such as the dragonflies, could evolve species with 75-centimeter wingspans supports the model predictions of higher oxygen levels in the Carboniferous. Yet no evidence exists that any of the Carboniferous plants groups (all nonangiosperm, for the angiosperms evolved much later in the Cretaceous) evolved the C_4 photosynthetic pathway (Sage 1999).

Although at present it is known that only about 3 percent of angiosperm species are C_4 photosynthesizers, that 3 percent accounts for about 30 percent of total terrestrial primary productivity on a global scale, and up to as much as 80 percent of the primary productivity in certain ecosystems, such as tropical grasslands (Sage 2001b). Following the expectations of the theory of natural selection, one would predict the further convergent evolution of C_4 photosynthesis in new angiosperm groups, and continued ecological expansion of existing C_4 photosynthesizers (table 3.5) into the future. However, Sage (2001a) has noted that human burning of fossil fuels in the past several hundred years has led to progressively increasing levels of carbon dioxide in the Earth's atmosphere, a process that favors the original C_3 photosynthesizers, while other human activities, such as slash-and-burn agriculture and deforestation in the tropics, has favored the spread of C_4 grasslands and weeds. Thus, predicting the future direction of plant evolution appears to be more a function of predicting the future activity of humans than that of plants.

In the past two sections of the chapter we have considered some of the functional problems associated with a photoautrophic mode of life, and the convergencies that necessarily result from those functional

Table 3.5
Convergent evolution of the C_4 photosynthetic pathway in plants in the past 30 million years

---

Convergent pathway and function: C_4 PHOTOSYNTHESIS (more efficient photosynthetic pathway to counteract Rubisco oxidation and photorespiration)

Convergent lineages:

1 Water thyme (Euangiosperms: Monocotyledons: Alismatales: Hydrocharitaceae; *Hydrilla verticillata*)

2 Switchgrass (Euangiosperms: Monocotyledons: Eumonocotyledons: Commelinidae: Poaceae; *Panicum virgatum*)

3 Maize (Euangiosperms: Monocotyledons: Eumonocotyledons: Commelinidae: Poaceae; *Zea mays*)

4 Australian eriachne grass (Euangiosperms: Monocotyledons: Eumonocotyledons: Commelinidae: Poaceae; *Eriachne triodioides*)

5 Sand dropseed (Euangiosperms: Monocotyledons: Eumonocotyledons: Commelinidae: Poaceae; *Sporobolus cryptandrus*)

6 Flatspike sedge (Euangiosperms: Monocotyledons: Eumonocotyledons: Commelinidae: Cyperaceae; *Abildgaardia ovata*)

7 Egyptian papyrus (Euangiosperms: Monocotyledons: Eumonocotyledons: Commelinidae: Cyperaceae; *Cyperus papyrus*)

8 Amphibious sedge (Euangiosperms: Monocotyledons: Eumonocotyledons: Commelinidae: Cyperaceae; *Eleocharis vivipara*)

9 Rhynchospora (Euangiosperms: Monocotyledons: Eumonocotyledons: Commelinidae: Cyperaceae; *Rhynchospora rubra*)

10 Mongolian calligonum (Euangiosperms: Eudicots: Core eudicots: Caryophyllales: Polygonaceae; *Calligonum mongolicum*)

11 Pindan pink (Euangiosperms: Eudicots: Core eudicots: Caryophyllales: Caryophyllaceae; *Polycarpaea longiflora*)

12 Tassel flower (Euangiosperms: Eudicots: Core eudicots: Caryophyllales: Amaranthaceae; *Amaranthus edulis*)

13 Black saxual (Euangiosperms: Eudicots: Core eudicots: Caryophyllales: Chenopodiaceae; *Haloxylon aphyllum*)

14 Suaeda halophyte (Euangiosperms: Eudicots: Core eudicots: Caryophyllales: Chenopodiaceae; *Suaeda aralocaspica*)

15 Four-wing saltbush (Euangiosperms: Eudicots: Core eudicots: Caryophyllales: Chenopodiaceae; *Atriplex canescens*)

16 Slender snakecotton (Euangiosperms: Eudicots: Core eudicots: Caryophyllales: Chenopodiaceae; *Froelichia gracilis*)

17 Bienertia halophyte (Euangiosperms: Eudicots: Core eudicots: Caryophyllales: Chenopodiaceae; *Bienertia sinuspersici*)

18 Little hogweed (Euangiosperms: Eudicots: Core eudicots: Caryophyllales: Portulacaceae; *Portulaca oleracea*)

19 Australian anacampseros (Euangiosperms: Eudicots: Core eudicots: Caryophyllales: Portulacaceae; *Anacampseros australiana*)

20 Shoreline purslane (Euangiosperms: Eudicots: Core eudicots: Caryophyllales: Aizoaceae; *Sesuvium portulacastrum*)

21 Scarlet spiderling (Euangiosperms: Eudicots: Core eudicots: Caryophyllales: Nyctaginaceae; *Boerhavia coccinea*)

22 Naked-stem carpetweed (Euangiosperms: Eudicots: Core eudicots: Caryophyllales: Molluginaceae; *Mollugo nudicaulis*)

23 Hawaiian spurge (Euangiosperms: Eudicots: Rosidae: Eurosids I: Euphorbiaceae; *Chamaecyce forbesii*)

Table 3.5
(continued)

24 Desert forb (Euangiosperms: Eudicots: Rosidae: Eurosids I: Zygophyllales: Zygophyllaceae; *Zygophyllum simplex*)
25 Arizona poppy (Euangiosperms: Eudicots: Rosidae: Eurosids I: Zygophyllales: Zygophyllaceae; *Kallstroemia grandiflora*)
26 Purple mistress (Euangiosperms: Eudicots: Rosidae: Eurosids II: Brassicaceae; *Moricandia arvensis*)
27 Spider wisp (Euangiosperms: Eudicots: Rosidae: Eurosids II: Brassicaceae; *Cleome gynandra*)
28 Pineland heliotrope (Euangiosperms: Eudicots: Asteridae: Euasterids I: Boraginaceae; *Heliotropium polyphyllum*)
29 Blepharis (Euangiosperms: Eudicots: Asteridae: Euasterids I: Acanthaceae; *Blepharis ciliaris*)
30 Anticharis bush (Euangiosperms: Eudicots: Asteridae: Euasterids I: Scrophulariaceae; *Anticharis linearis*)
31 Australian yellow weed (Euangiosperms: Eudicots: Asteridae: Euasterids II: Asteraceae; *Flaveria australasica*)
32 Narrow-leaf lemon weed (Euangiosperms: Eudicots: Asteridae: Euasterids II: Asteraceae; *Pectis linifolia*)
33 Rock anethum (Euangiosperms: Eudicots: Asteridae: Euasterids II: Asteraceae; *Glossocardia bosvallea*)

Note: For data sources, see text.

constraints on possible evolutionary pathways. One way around these constraints would be to switch from an autotrophic mode of life to a heterotrophic one; that is, to evolve a mode of life similar to that seen in the animals and the fungi. In the next chapter we shall consider plants that have done just that: the convergent evolution of carnivorous plants (plants that eat animals) and herbivorous plants (plants that eat other plants).

Defense: Antiherbivore Adaptations

Herbivores are to plants what carnivores are to animals. A typical animal anticarnivore defense is to run away, as discussed in the previous chapter (see table 2.12), but this option is not open to plants. Being unable to flee, plants have evolved both physical and chemical defensive systems, in many ways convergent on the defensive spines, stingers, and poison defenses evolved by the animals (see tables 2.13, 2.14).

The leaves of a plant are its food factories, and one could predict that there is a selective advantage for plants that defend those structures. It is not surprising that a wide range of distantly related plants all have convergently evolved leaf spines, from the nonflowering cycads to highly derived asterid angiosperms (table 3.6). The similarity of some of these

Table 3.6
Convergent evolution of physical defenses in plants

---

1 Convergent structure and function: LEAF SPINES (spines on apices of leaf lobes or around leaf margins to discourage herbivore biting)

Convergent lineages:

1.1 Cycad (Euphyllophyta: Lignophyta: Spermatophyta: Cycadophyta: Cycadales: Zamiaceae; *Encephalartos ferox*)

1.2 Aloe yucca (Spermatophyta: Angiospermae: Euangiosperms: Monocotyledons: Eumonocotyledons: Liliales: Liliaceae; *Yucca aloifolia*)

1.3 Cogon grass (Euangiosperms: Monocotyledons: Eumonocotyledons: Commelinidae: Poaceae; *Imperata cylindrical*)

1.4 Black oak (Euangiosperms: Eudicots: Rosidae: Eurosids I: Fagaceae; *Quercus velutina*)

1.5 Scorpion weed (Euangiosperms: Eudicots: Asteridae: Euasterids I: Boraginaceae; *Phacelia purshii*)

1.6 European holly (Euangiosperms: Eudicots: Asteridae: Euasterids II: Aquifoliaceae; *Ilex aquifolium*)

1.7 Bull thistle (Euangiosperms: Eudicots: Asteridae: Euasterids II: Asteraceae; *Cirsium vulgare*)

2 Convergent structure and function: TWIGS/BARK WITH THORNS/SPINES (thorns or spines on twigs and/or trunk bark to discourage herbivores)

Convergent lineages:

2.1 Candle cactus (Euangiosperms: Eudicots: Core eudicots: Caryophyllales: Cactaceae; *Pilocereus lanuginosus*)

2.2 Knotweed (Euangiosperms: Eudicots: Core eudicots: Caryophyllales: Polygonaceae; *Polygonum pennsylvanicum*)

2.3 Madagascar spiny plant (Euangiosperms: Eudicots: Core eudicots: Caryophyllales: Didiereaceae; *Alluaudia montagnacii*)

2.4 Canary Island spurge (Euangiosperms: Eudicots: Rosidae: Eurosids I: Euphorbiaceae; *Euphorbia canariensis*)

2.5 French rose (Euangiosperms: Eudicots: Rosidae: Eurosids I: Rosaceae; *Rosa gallica*)

2.6 European buckthorn (Euangiosperms: Eudicots: Rosidae: Eurosids I: Rhamnaceae; *Rhamnus cathartica*)

2.7 Honey locust (Euangiosperms: Eudicots: Rosidae: Eurosids I: Fabaceae; *Gleditsia triacanthos*)

2.8 Common prickly-ash (Euangiosperms: Eudicots: Rosidae: Eurosids II: Rutaceae; *Zanthoxylum americanum*)

2.9 Bottle tree (Euangiosperms: Eudicots: Asteridae: Euasterids I: Apocynaceae; *Pachypodium lealii*)

3 Convergent structure and function: STINGERS (stinging poison hairs and needles to discourage herbivores)

Convergent lineages:

3.1 Stinging nettle (Euangiosperms: Eudicots: Rosidae: Eurosids I: Urticaceae; *Urtica dioica*)

3.2 Spurge stinging nettle (Euangiosperms: Eudicots: Rosidae: Eurosids I: Euphorbiaceae; *Cnidoscollus stimulosus*)

3.3 Cowhage vine (Euangiosperms: Eudicots: Rosidae: Eurosids I: Fabaceae; *Mucuna pruriens*)

Table 3.6
(continued)

4 Convergent structure and function: HOSTED INSECT DEFENDERS (mutualistic relationship with animals who also serve to defend the host plant)
Convergent lineages:
4.1 Ant rattan (Euangiosperms: Monocotyledons: Eumonocotyledons: Commelinidae: Aracaceae; *Daemonorops formicaria*)
4.2 Whistling-thorn acacia (Euangiosperms: Eudicots: Rosidae: Eurosids I: Fabaceae; *Acacia drepanolobium*)
4.3 Valley oak (Euangiosperms: Eudicots: Rosidae: Eurosids I: Fagaceae; *Quercus lobata*)
4.4 Common figs (Euangiosperms: Eudicots: Rosidae: Eurosids I: Moraceae; *Ficus carica*)
4.5 Ant plant (Euangiosperms: Eudicots: Asteridae: Euasterids I: Rubiaceae; *Hydnophytum formicarum*)

Note: For data sources, see text.

biological structures to human deadly devices is often apparent; for example, the aloe yucca is also known as the "Spanish bayonet" because of the shape of its sharp-edged leaves. Cogon-grass blades contain silica crystals that render the margins of the leaf as sharp as the blade of a serrated saw, and every child quickly learns not to run barefoot around holly trees. Scotland is said to have chosen the thistle as its national emblem because ancient Scots were warned of approaching invaders in the night when one of the invaders inadvertently stepped on a thistle and let out a loud yelp of pain.

Other plants have convergently evolved defensive structures for the twigs and bark of the plant as well as its leaves (table 3.6). Cactus and spurge spines are familiar convergent examples of armor covering the entire plant. Rose thorns are a hazard every gardener learns to be wary of, and this same family of plants produces the cockspur hawthorn, *Crataegus crus-galli*, a name illustrating the convergence of a plant structure to one seen in animals. Another animal-form convergent example in the family of the thorny honey locusts is the catclaw acacia, *Acacia greggii*, with sharp sickle-shaped spines that can shred the clothing and skin, ruining many a hiking experience in southwestern North America.

Still other plants add poison to their defensive structures (table 3.6), producing stingers convergent to those found in animals (see tables 2.7, 2.14). The poisonous tiny hairs of the stinging nettle produce painful burns if brushed against skin tissue; this same family produces the Australian stinging tree, *Dendrocnide moroides*, with tiny spines containing neurotoxin. The "cnide" in the plant's generic name is apt, as the sting of this tree is like that of a cnidarian jelly fish and can actually be fatal (Stewart 2009). The Euphorbiaceae family is famous for its eerie convergence on numerous cactus forms, which we shall consider later in this chapter, but it has also produced a convergent stinging nettle that is virtually identical to a true nettle (table 3.6).

Last, some plants have recruited animals to defend them (table 3.6). The defender species are sometimes ants, like those hosted by the whistling-thorn acacia, or wasps, hosted by the valley oak. This convergent mutualism is not common; much more abundant is the convergent assistance of animals in plant reproduction, a topic that will be taken up later in the chapter.

Convergent with animals (see table 2.13), plants have evolved chemical defenses as well. Unlike animals, however, plants have independently acquired a staggering number of chemical defense systems (table 3.7). This is not surprising, as the number of defensive strategies available to organisms that cannot flee or actively fight is more limited. In table 3.7 I have listed numerous examples of the convergent evolution of poison in different plant structures, from leaves to seeds to roots, taken largely from the works of Stewart (2009) and Agosta (1996) and phylogenetically cross-checked using the APG II (2003). The most poisonous plant families are the nightshades (Solanaceae), cashews (Anacardiaceae), spurges (Euphorbiaceae), and parsleys (Apiaceae), groups that Stewart (2009, 71) collectively terms "botanical crime families." But this designation is only from our perspective; from the plant's perspective, it is simply defending itself. Although these families contain the most numerous, and most infamous, species of poisonous plants, poison tissues are found scattered across the plant phylogenetic spectrum, from nonflowering pines and cycads, to more basal core angiosperms like magnoliids and monocotyledons, to highly derived eudicots like rosids and asterids. For each separate clade of plants, I have listed a single species example in table 3.7, regardless of whether the clade contains many poisonous species or only a few, in order to illustrate the extent of convergent evolution of chemical defenses across all clades.

The poison produced in convergent chemical defense is not always the same, but even here biochemical convergence is widespread (Stewart 2009). For example, cyanide has been independently developed in monocotyledons (Johnsongrass leaf), eurosids (rubber tree seeds), and asterids (black elderberry), to name only a few examples. Poisonous alkaloid neurotoxins like nicotine and cocaine have been independently evolved by monocotyledons (betel nut), basal eudicots (monkshood plant), rosids

Table 3.7
Convergent evolution of chemical defenses in plants

1 Convergent structure and function: LEAF/STEM POISON (poison concentrated in leaves and/or stems to protect phototrophic structures from herbivores)

Convergent lineages:

1.1 Jack-in-the-pulpit leaf (Euangiosperms: Monocotyledons: Alismatales: Araceae; *Arisaema triphyllum*)
1.2 False hellebore leaf (Euangiosperms: Monocotyledons: Eumonocotyledons: Liliales: Liliaceae; *Veratrum viride*)
1.3 Yellow lady's slipper leaf (Euangiosperms: Monocotyledons: Eumonocotyledons: Asparagales: Orchidaceae; *Cypripedium calceolus*)
1.4 Blue agave stem (Euangiosperms: Monocotyledons: Eumonocotyledons: Asparagales: Agavaceae; *Agave tequilana*)
1.5 Johnsongrass leaf (Euangiosperms: Monocotyledons: Eumonocotyledons: Commelinidae: Poaceae; *Sorghum halepense*)
1.6 Swallowwort leaf (Euangiosperms: Eudicots: Ranunculales: Papaveraceae; *Chelidonium majus*)
1.7 Mayapple leaf (Euangiosperms: Eudicots: Ranunculales: Berberidaceae; *Podophyllum peltatum*)
1.8 Curare vine leaf (Euangiosperms: Eudicots: Ranunculales: Menispermaceae; *Chondrodendron tomentosum*)
1.9 Kratom leaf (Euangiosperms: Eudicots: Core eudicots: Caryophyllales: Cactaceae; *Trichocereus pachanoi*)
1.10 Soapwort leaf (Euangiosperms: Eudicots: Core eudicots: Caryophyllales: Caryophyllaceae; *Saponaria officinalis*)
1.11 Marijuana leaf (Euangiosperms: Eudicots: Rosidae: Eurosids I: Cannabinaceae; *Cannabis sativa*)
1.12 Khat shrub leaf (Euangiosperms: Eudicots: Rosidae: Eurosids I: Celastraceae; *Catha edulis*)
1.13 Coca shrub leaf (Euangiosperms: Eudicots: Rosidae: Eurosids I: Erythroxylaceae; *Erythroxylum coca*)
1.14 Ratbane tree leaf (Euangiosperms: Eudicots: Rosidae: Eurosids I: Dichapetalaceae; *Dichapetalum toxicarium*)
1.15 Pencil cactus stem (Euangiosperms: Eudicots: Rosidae: Eurosids I: Euphorbiaceae; *Euphorbia tirucalli*)
1.16 Upas tree leaf (Euangiosperms: Eudicots: Rosidae: Eurosids I: Moraceae; *Antiaris toxicaria*)
1.17 Black locust bark (Euangiosperms: Eudicots: Rosidae: Eurosids I: Fabaceae; *Robinia pseudoacacia*)
1.18 Tansy mustard leaf (Euangiosperms: Eudicots: Rosidae: Eurosids II: Brassicaceae; *Descurainia pinnata*)
1.19 Poison sumac leaf (Euangiosperms: Eudicots: Rosidae: Eurosids II: Anacardiaceae; *Toxicodendron vernix*)
1.20 Tea leaf (Euangiosperms: Eudicots: Asteridae: Ericales: Thaeaceae; *Thea sinensis*)
1.21 Common rhododendron leaf (Euangiosperms: Asteridae: Ericales: Ericaceae; *Rhododendron ponticum*)
1.22 Strychnine vine leaf (Euangiosperms: Eudicots: Asteridae: Euasterids I: Loganiaceae; *Strychnos toxifera*)
1.23 Tobacco leaf (Euangiosperms: Eudicots: Asteridae: Euasterids I: Solanaceae; *Nicotiana tabacum*)
1.24 Diviner's sage leaf (Euangiosperms: Eudicots: Asteridae: Euasterids I: Lamiaceae; *Salvia divinorum*)

Table 3.7
(continued)

1.25 Kombe vine leaf (Euangiosperms: Eudicots: Asteridae: Euasterids I: Apocynaceae; *Strophanthus kombe*)
1.26 Sweet woodruff leaf (Euangiosperms: Eudicots: Asteridae: Euasterids I: Rubiaceae; *Galium odoratum*)
1.27 Indian tobacco leaf (Euangiosperms: Eudicots: Asteridae: Euasterids II: Campanulaceae; *Lobelia inflata*)
1.28 White snakeroot leaf (Euangiosperms: Eudicots: Asteridae: Euasterids II: Asteraceae; *Eupatorium rugosum*)
1.29 English ivy leaf (Euangiosperms: Eudicots: Asteridae: Eurasterids II: Araliaceae; *Hedera helix*)
1.30 Blister bush leaf (Euangiosperms: Eudicots: Asteridae: Euasterids II: Apiaceae; *Peucedanum galbanum*)

2 Convergent structure and function: SEED/FRUIT POISON (poison concentrated in seeds and/or fruit to protect reproductive structures from herbivores)
Convergent lineages:
2.1 Sweet shrub seed (Euangiosperms: Magnoliidae: Laurales: Calycanthaceae; *Calycanthus floridus*)
2.2 Betel nut (Euangiosperms: Monocotyledons: Eumonocotyledons: Commelinidae: Arecaceae; *Areca catechu*)
2.3 Opium poppy seedpod (Euangiosperms: Eudicots: Ranunculales: Papaveraceae; *Papaver somniferum*)
2.4 Blue cohosh seed (Euangiosperms: Eudicots: Ranunculales: Berberidaceae; *Caulophyllum thalictroides*)
2.5 Castor bean (Euangiosperms: Eudicots: Rosidae: Eurosids I: Euphorbiacea; *Ricinus communis*)
2.6 Rosary pea seed (Euangiosperms: Eudicots: Rosidae: Eurosids I: Fabaceae; *Abrus precatorius*)
2.7 Habanero chile pod (Euangiosperms: Eudicots: Rosidae: Eurosids I: Solanaceae; *Capsicum chinense*)
2.8 Squirting cucumber (Euangiosperms: Eudicots: Rosidae: Eurosids I: Cucurbitaceae; *Ecballium elaterium*)
2.9 Coyotillo berries (Euangiosperms: Eudicots: Rosidae: Eurosids I: Rhamnaceae; *Karwinskia humboldtiana*)
2.10 Cacao bean (Euangiosperms: Eudicots: Rosidae: Eurosids II: Malvaceae; *Theobroma cacao*)
2.11 Fetid buckeye (Euangiosperms: Eudicots: Rosidae: Eurosids II: Sapindaceae; *Aesculus glabra*)
2.12 Lime fruit (Euangiosperms: Eudicots: Rosidae: Eurosids II: Rutaceae; *Citrus aurantifolia*)
2.13 Coffee bean (Euangiosperms: Eudicots: Asteridae: Euasterids I: Rubiaceae; *Coffea arabica*)
2.14 Suicide-tree seed (Euangiosperms: Eudicots: Asteridae: Euasterids I: Apocynaceae; *Cerbera odollam*)
2.15 Lantana berries (Euangiosperms: Eudicots: Asteridae: Euasterids I: Verbenaceae; *Lantana camara*)
2.16 Morning glory seed (Euangiosperms: Eudicots: Asteridae: Euasterids I: Convolvulaceae; *Ipomoea tricolor*)
2.17 Black elderberry (Euangiosperms: Eudicots: Asteridae: Euasterids II: Adoxaceae; *Sambucus nigra*)

Table 3.7
(continued)

3 Convergent structure and function: ROOT/BULB POISON (poison concentrated in roots and/or bulbs to protect nutrient uptake and reproductive structures from herbivores)

Convergent lineages:

3.1 Fly-poison bulb (Euangiosperms: Monocotyledons: Eumonocotyledons: Liliaceae; *Amianthium muscaetoxicum*)

3.2 Bloodroot root (Euangiosperms: Eudicots: Ranunculales: Papaveraceae; *Sanguinaria canadensis*)

3.3 Pokeberry root (Euangiosperms: Eudicots: Core eudicots: Caryophyllales: Phytolaccaceae; *Phytolacca americana*)

3.4 Cassava root (Euangiosperms: Eudicots: Rosidae: Eurosids I: Euphorbiaceae; *Manihot esculenta*)

3.5 Potato tuber (Euangiosperms: Eudicots: Asteridae: Euasterids I: Solanaceae; *Solanum tuberosum*)

3.6 Iboga root (Euangiosperms: Eudicots: Asteridae: Euasterids I: Apocynaceae; *Tabernanthe iboga*)

3.7 Water hemlock root (Euangiosperms: Eudicots: Asteridae: Eurasterids II: Apiaceae; *Cicuta maculata*)

4 Convergent structure and function: ENTIRE PLANT POISON (poison distributed throughout entire plant to repel herbivores)

Convergent lineages:

4.1 Sago palm (Euphyllophyta: Lignophyta: Spermatophyta: Cycadophyta: Cycadaceae; *Cycas revoluta*)

4.2 European yew tree (Euphyllophyta: Lignophyta: Spermatophyta: Pinophyta: Taxaceae; *Taxus baccata*)

4.3 Split-leaf philodendron (Spermatophyta: Angiospermae: Euangiosperms: Monocotyledons: Alismatales: Araceae; *Philodendron selloum*)

4.4 Autumn crocus (Euangiosperms: Monocotyledons: Eumonocotyledons: Liliaceae; *Colchicum autumnale*)

4.5 Monkshood (Euangiosperms: Eudicots: Ranunculales: Ranunculaceae; *Aconitum napellus*)

4.6 Peyote cactus (Euangiosperms: Eudicots: Core eudicots: Caryophyllales: Cactaceae; *Lophophora williamsii*)

4.7 Black snakeroot (Euangiosperms: Eudicots: Rosidae: Geraniales: Melanthiaceae; *Zigadenus venenosus*)

4.8 Daphne shrub (Euangiosperms: Eudicots: Rosidae: Eurosids II: Thymelaeaceae; *Daphne mezereum*)

4.9 Mountain laurel (Euangiosperms: Eudicots: Asteridae: Ericales: Ericaceae; *Kalmia latifolia*)

4.10 Oakleaf hydrangea shrub (Euangiosperms: Eudicots: Asteridae: Cornales: Hydrangeaceae; *Hydrangea quercifolia*)

4.11 Deadly nightshade (Euangiosperms: Eudicots: Asteridae: Euasterids I: Solanaceae; *Atropa belladona*)

4.12 Buttonbush (Euangiosperms: Eudicots: Asteridae: Euasterids I: Rubiaceae; *Cephalanthus occidentalis*)

4.13 Common foxglove (Euangiosperms: Eudicots: Asteridae: Euasterids I: Plantaginaceae; *Digitalis purpurea*)

4.14 Butterfly weed (Euangiosperms: Eudicots: Asteridae: Euasterids I: Asclepiadaceae; *Asclepias tuberosa*)

4.15 Yellow jessamine vine (Euangiosperms: Eudicots: Asteridae: Euasterids I: Gelsemiaceae; *Gelsemium sempervirens*)

Note: For data sources, see text.

(coca leaf), and asterids (tobacco leaf). Deadly paralysis-inducing muscle toxins have been independently developed in plants ranging from basal eudicots (curare vine leaf) to highly derived asterids (strychnine vine leaf), and heart-muscle poisons have independently evolved in rosids (upas tree leaf) and asterids (kombe vine leaf). Psychoactive toxins, targeting the brain, have been independently developed by basal core eudicots (kratom leaf), rosids (marijuana leaf), and asterids (diviner's sage leaf). Powerful skin-destructive irritants have been independently developed by monocotyledons (blue agave stem), rosids (poison sumac leaf), and asterids (blister bush leaf).

Some of these plants are not really dangerous to humans; morning glory seeds, for example, contain only tiny amounts of the psychoactive toxin LSD, and hydrangea bushes only small amounts of cyanide. But then, morning glories and hydrangea bushes are not defending themselves against large animals like humans; they have smaller animals in target. Others are surprising: cassava roots and potato tubers are poisonous (the latter less so than the former), but are rendered harmless if they are cooked properly, and they are a widely used food source for humans. Humans even deliberately enjoy the effects of the neurotoxin caffeine in tea leaves and coffee beans, and some enjoy the neurotoxin nicotine in tobacco leaves (although the delivery system, burning cigarettes, eventually kills them).

Defense: Antidehydration Adaptations

Dehydration is a very serious problem for land-dwelling organisms. Many of the morphological and physiological changes that took place in the evolution of tetrapods from fish were adaptations for preventing water loss in the dry, hostile terrestrial environment. (Other changes were linked to two other serious problems on land: the crushing force of gravity and rapid temperature fluctuations, as compared to the sea, where neutrally buoyant fish feel no gravity and temperatures vary very little, even on seasonal scales.) Even so, if an animal feels thirsty, it is able to go hunting for water and to replenish its water loss by drinking as much water as it can find.

In contrast, plants cannot go hunting for water. Plants are entirely dependent on the vagaries of terrestrial weather patterns in bringing rain and groundwater, weather patterns that are usually unreliable. A wet spring this year might be followed by a drought next year. One way to deal with an unreliable source of water is to store up as much water as possible when it is available, so the plant can survive periods of time when water is scarce. One such water-storage adaptation among plants is the evolution of succulent stems and leaves, structures that store large amounts of water in fleshy pith and cortex cells (Niklas 1997), resulting in a plant that possesses stems and leaves that appear swollen or fat.

Plants of widely differing phylogenetic lineages have repeatedly and convergently evolved succulent stems and leaves (table 3.8); interestingly, even some animals have converged on a succulent water-storage adaptation, such as the desert-adapted dromedary camel, *Camelus dromedarius*, which can store enormous amounts of water in its tissues. In table 3.8 I have listed examples from families containing the largest numbers of succulent species known to me, such as the Cactaceae (cactuses), Euphorbiaceae (spurges), Aizoaceae (stone plants), and Crassulaceae (jade plants). Many, many other families of plants have a few species, or sometimes only one, that have also convergently evolved succulent stems or leaves while living in semiarid environments; thus the list given in table 3.8 cannot be taken in any sense to be exhaustive.

Table 3.8
Convergent evolution of water-retention structures in plants

Convergent structure and function: SUCCULENT STEMS/LEAVES (water-storage structures to ensure sufficient water for plant survival in semiarid environments or environments in which the water source is unreliable)

Convergent lineages:

1 Quiver tree (Euangiosperms: Monocotyledons: Eumonocotyledons: Asparagales: Asphodelaceae; *Aloe dichotoma*)

2 Agave (Euangiosperms: Monocotyledons: Eumonocotyledons: Asparagales: Agavaceae; *Agave flexispina*)

3 Rosette plant (Euangiosperms: Monocotyledons: Eumonocotyledons: Commelinidae: Bromeliaceae; *Abromeitiella lorentziana*)

4 Candle cactus (Euangiosperms: Eudicots: Core eudicots: Caryophyllales: Cactaceae; *Pilocereus lanuginosus*)

5 Stone plant (Euangiosperms: Eudicots: Core eudicots: Caryophyllales: Aizoaceae; *Carpobrotus edulis*)

6 Madagascar spiny plant (Euangiosperms: Eudicots: Core eudicots: Caryophyllales: Didiereaceae; *Alluaudia montagnacii*)

7 Jade plant (Euangiosperms: Eudicots: Core eudicots: Saxifragales; Crassulaceae; *Crassula ovata*)

8 Canary Island spurge (Euangiosperms: Eudicots: Rosidae: Eurosids I: Euphorbiaceae; *Euphorbia canariensis*)

9 Bottle tree (Euangiosperms: Eudicots: Asteridae: Euasterids I: Apocynaceae; *Pachypodium lealii*)

10 Pickle plant (Euangiosperms: Eudicots: Asteridae: Euasterids II: Asteracea; *Senecio stapeliiformis*)

Note: For data sources, see text.

It can, however, illustrate how phylogenetically widespread the phenomenon is—independently arising in numerous monocotyledons, core eudicots, and highly derived rosids and asterids—and how geographically widespread—from Africa (quiver tree) to North America (candle cactus) to South America (rosette plant) to Madagascar (bottle tree; Niklas 1997).

The most spectacular examples of the convergent evolution of identical morphological structures to prevent dehydration are the desert-adapted cactuses of the Western Hemisphere and spurges of the Eastern Hemisphere (table 3.9). Niklas (1997) documents that both groups started out from morphologically similar beginnings, from tropical broad-leafed trees, such as the leaf cactus and the honey spurge (table 3.9). In subsequent evolution in more arid environments, both groups of plants reduced the amount of their woody stem tissues, the number of their branches, and the sizes of their leaves—progressing from tall, treelike cactus forms to shorter, sparsely branched cactus forms, to squat, barrel-shaped stems, and finally to rounded sea-urchin-shaped domes (table 3.9). Both groups of plants lost their leaves and evolved thorns (cactuses evolved thorns from modified leaves and spurges from modified branches) that not only deter herbivores (table 3.6) but also dissipate heat and break up wind currents from the stem surface, both of which can increase water loss (Niklas 1997).

Table 3.9
Convergent evolution of cactus forms in arid-adapted desert plants of the Western Hemisphere (Cactaceae) and the Eastern Hemisphere (Euphorbiaceae)

Ancestral morphological condition: tropical broad-leafed tree species
Cactaceae: Leaf cactus (*Pereskia sacharosa*)
Euphorbiaceae: Honey spurge (*Euphorbia mellifera*)
Convergent lineages:
1 Tall tree-forms with multiple succulent branches from a central trunk
Cactaceae: Candle cactus (*Pilocereus lanuginosus*)
Euphorbiaceae: Canary Island spurge (*Euphorbia canariensus*)
2 Shorter, sparsely branched succulent stems with zigzag vertical ridges and furrows
Cactaceae: Snake cactus (*Peniocereus serpintinus*)
Euphorbiaceae: Devil's backbone spurge (*Euphorbia cryptospinosa*)
3 Squat, barrel-shaped succulent stem with flat ridges separating recessed furrows
Cactaceae: Compass-barrel cactus (*Ferocactus cylindraceus*)
Euphorbiaceae: Velvet spurge (*Euphorbia valida*)
4 Rounded, sea-urchin-shaped succulent dome
Cactaceae: Sea-urchin cactus (*Astrophytum asterias*)
Euphorbiaceae: Baseball spurge (*Euphorbia obesa*)

Note: For data sources, see text.

Reproduction

In the last chapter we saw that animals have repeatedly and independently evolved the same reproductive systems, particularly viviparity. The same is true of plants, except the convergent reproductive system for these organisms is heterospory. The repeated evolution of heterospory in independent phylogenetic lineages has been called "the most iterative key innovation in the evolutionary history of the plant kingdom" (Bateman and DiMichele 1994, 345).

The original Devonian tracheophytes had sporophytes that shed their spores freely into the air; the spores, which were all the same size, then produced free-living gametophytes that produced the next generation of sporophytes. The free-living gametophytes were bisexual, and required moist conditions not only for their vegetative growth but also for sperm transfer and fertilization of their eggs, as well as the survival of the developing new sporophyte embryo (Niklas 1997). The production of bisexual gametophytes is an ancient reproductive condition known as homospory, which still exists in many living species of club mosses (lycophytes), horsetails (equisetophytes), and ferns (filicophytes). In many ways, homosporous plants "may be thought of as the amphibians of the plant kingdom in the sense that the completion of their sexual life cycle requires a 'return to water'" (Niklas 1997, 190).

A major evolutionary shift in plants' mode of reproduction occurred independently in many separate plant lineages during the Late Devonian (table 3.10). This was the convergent evolution of heterospory, where plants now produced two distinctly different types of spores: numerous, very small microspores and relatively few but much larger megaspores. Specialized plants producing the new microspores developed into sperm-producing gametophytes, and plants producing the larger megaspores developed into egg-bearing gametophytes; thus, heterosporous plants had now evolved gametophytes that were unisexual, not bisexual like the homosporous plants (Niklas 1997). The evolution of heterospory and unisexual gametophytes occurred independently at least eleven separate times in disparate groups of plants (Bateman and DiMichele 1994). Two separate groups of lycopods, one major group of equisetophytes, and three separate groups of filicophytes convergently became heterosporous (table 3.10), although homosporous species of these types of plants exist to the present day. The filicophytes are particularly interesting in that they have convergently evolved heterospory repeatedly throughout geologic time: at least once in the Late Devonian, at least once in the

Table 3.10
Convergent evolution of heterosporous reproduction, and ultimately of the seed

Convergent structure and function: HETEROSPORES (differentiated microspores and megaspores for the production of unisexual gametophytes)

Convergent lineages:

1 Devonian barinophyte spores (Tracheophyta: Lycophyta: Barinophytales: Barinophytaceae; *Protobarinophyton pennsylvanicum* †Late Devonian)

2 Devonian selaginellalean spores (Tracheophyta: Lycophyta: Selaginellales: Protolepidodendraceae; *Barsostrobus famennensis* †Late Devonian)

3 Carboniferous calamite spores (Tracheophyta: Euphyllophyta: Moniliformopses: Equisetophyta: Equisetales: Archaeocalamitaceae; *Protocalamostachys farringtonii* †Early Carboniferous)

4 Devonian stauropterid spores (Euphyllophyta: Moniliformopses: Filicophyta: Stauropteridales: Gillespieaceae; *Gillespiea randolphensis* †Late Devonian)

5 Giant water-fern spores (Euphyllophyta: Moniliformopses: Filicophyta: Salviniales: Salviniaceae; *Salvinia molesta*)

6 European water-clover fern spores (Euphyllophyta: Moniliformopses: Filicophyta: Marsileales: Marsileaceae; *Marsilea quadrifolia*)

7 Australian platyzoma-fern spores (Euphyllophyta: Moniliformopses: Filicophyta: Filicales: Platyzomataceae; *Platyzoma microphylla*)

8 Devonian aneurophyte spores (Euphyllophyta: Lignophyta: Aneurophytales: Chaleuriaceae; *Chaleuria cirrosa* †Middle Devonian)

9 Devonian archaeopterid spores (Euphyllophyta: Lignophyta: Archaeopteridales: Archaeopteridaceae; *Archaeopteris latifolia* †Late Devonian)

10 Carboniferous noeggerathialean spores (Euphyllophyta: Lignophyta: Noeggerathiales: Noeggerathiaceae; *Noeggerathiostrobus vicinalis* †Late Carboniferous)

11 Carboniferous cecropsidalean spores (Euphyllophyta: Lignophyta: Cecropsidales: Cecropiaceae; *Cecropsis luculentum* †Late Carboniferous)

12 Devonian elkinsia seeds (Euphyllophyta: Lignophyta: Spermatophyta: Lagenstomales: Elkinsiaceae; *Elkinsia polymorpha* †Late Devonian)

Note: The geological age of extinct species is marked with a †. For data sources, see text.

Late Cretaceous (the ancestors of living giant water ferns), and separately in the modern European water-clover ferns and Australian platyzoma ferns (table 3.10). Four separate groups of lignophytes independently evolved heterospory in the span of time from the Middle Devonian to the Late Carboniferous: the major plant lineages of the aneurophytales and archaeopteridales, and the more enigmatic noeggerathiales and cecropsidales, all of which are now extinct. In the midst of this rampant convergent evolution of heterospory in plant lineages, one plant group took the next major evolutionary step in plant reproduction: the evolution of the seed (table 3.10).

The evolution of heterospory, the partitioning of resources between numerous small microspores and a few large megaspores, set the stage for the evolution of the seed (Niklas 1997). Production of microspores required minimal resources, and because they were so very small they were easily and widely distributed by wind, ensuring an enlarged geographic dispersal of genetic information and the reduced probability of inbreeding in plants. The concentration of more resources in a few megaspores, or in only one megaspore, increased the probability of successful production of new sporophytes by these megagametophytes. The next step in plant reproductive evolution would be for the sporophyte to retain the megaspores rather than releasing them to produce free-living megagametophytes. Hence the evolution of the seed: "The seed is an adaptive solution to life on land functionally analogous to the way amniotic animals reproduce. . . . The evolution of the seed unquestionably released vascular plant reproduction from the ecological requirement for an external source of water for the growth and survival of the megagametophyte, the fertilization of its eggs, and the early development of the sporophyte embryo. Thus the seed opened the door for the adaptive radiation of vascular plants into habitats drier than those favoring the survival and reproductive success of free-sporing plants" (Niklas 1997, 343–344).

The first known spermatophytes, the seed plant *Elkinsia polymorpha*, are found as fossils in Late Devonian strata of the Appalachian Mountains of eastern North America. Just as multiple plant lineages converged on the evolution of heterospory, the fossil record suggests that several plant lineages were converging on the evolution of the seed. Spermatophytes are heterosporous plants that retain the megagametophyte, and its developing embryos, within sporophyte tissues. Niklas (1997) argues that heterosporous species of the ancient lepidodendrid lycophyte *Lepidocarpon* and the ancient calamite equisetophyte *Calamocarpon* were well on the way to developing the seed habit. However, as far as we yet know from the fossil record, only one lineage of plants successfully developed the seed, and that group, the spermatophytes, dominates the terrestrial plant biosphere just as the amniotes dominate the terrestrial animal biosphere.

The spermatophytes evolved in the Late Devonian and the amniotes soon after in the Early Carboniferous. The amniotes quickly diverged into two distinct phylogenetic lineages—the sauropsids and the synapsids (see appendix)—and the subsequent evolution of members of these lineages has been discussed in chapter 2. The spermatophytes were more divergent in their evolution: they produced three distinct lineages that are now extinct (the lyginopteridales, medullosanales, and gigantopteridales) and the five living ginkgophyte, cycadophyte, pinophyte, gnetophyte, and angiosperm lineages (see appendix). Just as the spermatophytes, the seed plants, are the most successful of all the land plants, the angiosperms, the flowering plants, are the most successful of all the spermatophytes.

Why flowers? The plant seed, like the animal amniote egg, is a reproductive structure clearly linked to antidehydration adaptations in the hostile dry air of terrestrial habitats. Much of the success of flowering plants is linked to solving not dehydration problems but a different reproductive problem: the problem of fertilization in organisms that cannot move about in search for a mate.

The problem of fertilization in sessile organisms is an ancient one that extends back to life in the sea. Many sessile marine animals simply release their gametes directly into the sea water and depend upon water currents to transport them to other animals for fertilization. These animals are often astonishingly convergent to plants in their morphologies: crinoids are deuterostomous bilaterians, closely related to chordate animals such as ourselves, yet many ancient crinoids had root systems that looked like those of land plants, stems like plant trunks, and branching arms like plant branches. Plant similarities are often seen in both the colloquial and scientific names of sessile marine animals: stalked echinoderms are known as "sea lilies" because their long stems and calyxes look like the stems and flowers of lilies, the bryozoa are named "moss animals" because their colonies look like moss, the gorgonian cnidarian "sea fans" whose central colonial stem and lateral branches look like palm leaves, the colorful polyps of an anthozoan cnidarian colony appear similar to a garden of "sea flowers," and so on.

The apparent convergence in form between many sessile marine animals and land plants is in fact reversed: it is the land plants that have converged on the sessile marine animals, not the other way around, as the marine animals evolved these forms first and the plants converged on them secondarily. Rather than currents in the dense water of marine habitats, land plants must depend upon wind in the thin atmosphere of terrestrial habitats to disperse their gametes. Just as many sessile marine animals have elevated themselves above the sea floor in order to catch stronger marine currents, plants have evolved stems and trunks to elevate them up in the air in order to catch the wind with their spores and pollen.

Many land plants are thus not only dependent upon the vagaries of terrestrial weather patterns in bringing rain; they are also dependent upon these same weather patterns in bringing wind for successful fertilization. The spermatophytes largely depended upon wind for fertilization for over 240 million years, from the Late Devonian until the evolution of the first angiosperms some 115 to 125 million years ago (Thien et al. 2009). At that time, the angiosperm spermatophytes evolved an interesting new fertilization strategy: they began to involve a sexual partner who was mobile. Although a few nonangiosperm spermatophytes use pollen-eating insects as pollination assistants, the really elaborate plant-animal reproductive *pas de deux* began with the evolution of the first flowering plants in the Early Cretaceous (and with the evolution of the dioecious angiosperms, which have separate male and female plants, the plant-animal interaction becomes a *ménage à trois*).

The basal angiosperms consist of three sequential cladogram branches, in what has been termed the ANITA grade of angiosperm evolution (Thien et al. 2009). These branches are: first, the Amborellales, second, the Nymphaeales, and third, the Austrobaileyales (APG II 2003; Lecointre and Le Guyader 2006), based upon molecular phylogenies. Although molecular evidence places the Amborellales at the very base of the angiosperm clade, the oldest fossil flowers found thus far belong to the Nymphaeales, one branch up from the base of the clade.

The oldest flower yet discovered is quite small, only about 1 centimeter in diameter. Somewhat younger nymphaealean flowers, around 90 million years old, have fossil morphologies very similar to modern-day flowers that are pollinated by beetles, reinforcing long-held views that beetle pollination was a major form of plant-animal reproductive interaction in the earliest of the flowering plants (Bernhardt 2000; Thien et al. 2009). Examination of living members of the basal angiosperms reveal that flies are the primary pollinators of six families, beetles the primary pollinators of five families, and bees are major pollinators in only one family (Thien et al. 2009). Plant-animal interactions surely must have evolved in the 115 to 125 million years that have passed since the evolution of the first flowering plant, and so the reproductive pattern seen in living basal angiosperms cannot be taken as proof of identical relationships in the Early Cretaceous. Still, it is interesting that living basal angiosperms are pollinated preferentially by flies and beetles, and only rarely by bees. More derived angiosperms not only added bees to the list of sexual partners but added butterflies, moths, wasps, birds, and mammals as well.

Four different types of flower morphologies have been evolved by angiosperms that are pollinated exclusively by beetles: bilobate, brush, chamber-blossom, and painted-bowl flowers (Bernhardt 2000). These cantharophilous flower types have been convergently evolved over and over again in even highly derived angiosperms (table 3.11); that is, beetle pollination is not a reproductive mode confined to the basal angiosperms or plesiomorphic magnoliids. Nine independent groups of monocotyledons, eight of them derived eumonocotyledons, have independently evolved cantharophilous flowers (table 3.11). In the derived eudicot angiosperms, cantharophilous flowers have convergently appeared 12 separate times, ranging from basal eudicots like the Ranunculalaes to the highly derived Rosidae and Asteridae (Bernhardt 2000).

Giant flowers are rarer and more specialized than cantharophilous flowers, yet they too have convergently evolved multiple times in the angiosperms (table 3.12). Floral gigantism is defined as the possession of blossoms—either flowers or flowerlike inflorescences—that exceed 30 centimeters in diameter (Davis et al. 2008). Very large flowers can be up to a meter in diameter, such as those found in rafflesia, or up to three meters in height, like as those found in the corpse lilies (table 3.12). These giant flowers attract a wide range of pollinators: beetles, flies, moths, birds, fruit bats, and even lemur primates (Davis et al. 2008). However, the great majority of pollinators of giant flowers are flies and small-bodied beetles. It appears that the large flower size facilitates the temporary trapping of these smaller animals by the flower, and the animals subsequently pollinate the flower as they seek an escape route. The phylogenetic distribution of the convergent evolution of floral gigantism is across the entire spectrum of the angiosperm clade: from the most basal angiosperms, like the Nymphaeales, to the highly derived Eurasterids (table 3.12).

Even more specialized than giant flowers are flowers that smell like rotting meat. One would think that such a specialized type of flower, specifically adapted to the restricted role of attracting carrion-eating animals, would be unique in the evolution of flowering plants, but this is not the case: sapromyiophilous, or carrion-mimic, flowers have convergently evolved in the angiosperms no less than seven separate times (table 3.13). Some carrion-mimic flowers are also giant, like the corpse lilies and rafflesia (table 3.12), but others are normal sized or even small (Davis et al. 2008; Stewart 2009). Sapromyiophilous flowers have independently evolved in plesiomorphic angiosperms like the magnoliid pawpaws; in monocotyledons like the corpse lilies and carrion flowers; in basal eudicots like the stinking hellebore; and in highly derived euasterids like the Zulu giant carrion plant (table 3.13).

Derived flowers have evolved shapes, colors, and odors to attract the attention of potential pollinators. Flowers also offer food rewards to

Table 3.11
Convergent evolution of cantharophilous flowers for plant pollination by beetles in angiosperms

---

Convergent structure and function: CANTHAROPHILOUS FLOWERS (evolution of specialized flower morphologies that facilitate pollination by beetles)

Convergent lineages:

1 Corpse lily (Euangiosperms: Monocotyledons: Alismatales: Araceae; *Amorphophallus titanum*)
2 Split-leaf cyclanthus (Euangiosperms: Monocotyledons: Eumonocotyledons: Pandanales: Cyclanthaceae; *Cyclanthus bipartitus*)
3 Sun's eye tulip (Euangiosperms: Monocotyledons: Eumonocotyledons: Liliales: Liliaceae; *Tulipa agenensis*)
4 Peacock iris (Euangiosperms: Monocotyledons: Eumonocotyledons: Asparagales: Iridaceae; *Moraea glaucopis*)
5 Giant orchid (Euangiosperms: Monocotyledons: Eumonocotyledons: Asparagales: Orchidaceae; *Pteroglossaspis ecristata*)
6 Star of Bethlehem (Euangiosperms: Monocotyledons: Eumonocotyledons: Asparagales: Hyacinthaceae; *Ornithogalum umbellatum*)
7 Cape star (Euangiosperms: Monocotyledons: Eumonocotyledons: Asparagales: Hypoxidaceae; *Spiloxene capensis*)
8 Orchidantha (Euangiosperms: Monocotyledons: Eumonocotyledons: Commelinidae: Lowiaceae; *Orchidantha maxillarioides*)
9 Root-spine palm (Euangiosperms: Monocotyledons: Eumonocotyledons: Commelinidae: Arecaceae; *Crysophila warscewiczii*)
10 Persian buttercup (Euangiosperms: Eudicots: Ranunculales: Ranunculaceae; *Ranunculus asiaticus*)
11 Corn poppy (Euangiosperms: Eudicots: Ranunculales: Papaveraceae; *Papaver rhoeas*)
12 Sundew (Euangiosperms: Eudicots: Core eudicots: Caryophyllales: Droseraceae; *Drosera pauciflora*)
13 Autograph tree (Euangiosperms: Eudicots: Rosidae: Eurosids I: Clusiaceae; *Clusia rosea*)
14 Tropical chestnut (Euangiosperms: Eudicots: Rosidae: Eurosids II: Malvaceae; *Sterculia stipulata*)
15 Malasian vatica (Euangiosperms: Eudicots: Rosidae: Eurosids II: Dipterocarpaceae; *Vatica parvifolia*)
16 Indian mastixia (Euangiosperms: Eudicots: Asteridae: Cornales: Cornaceae; *Mastixia arborea*)
17 Indian persimmon (Euangiosperms: Eudicots: Asteridae: Ericales: Ebenaceae; *Diospyros malabarica*)
18 Ballhead ipomopsis (Euangiosperms: Eudicots: Asteridae: Ericales: Polemoniaceae; *Ipomopsis congesta*)
19 Unamkodi (Euangiosperms: Eudicots: Asteridae: Euasterids I: Convolvulaceae; *Erycibe paniculata*)
20 African iodes (Euangiosperms: Eudicots: Asteridae: Euasterids I: Icacinaceae; *Iodes africana*)
21 Naked bluebell (Euangiosperms: Eudicots: Asteridae: Euasterids II: Campanulaceae; *Wahlenbergia gymnoclada*)

Note: For data sources, see text.

Table 3.12
Convergent evolution of floral gigantism in plants

Convergent structure and function: FLORAL GIGANTISM (blossoms with a diameter of 30 centimeters or greater that facilitate pollination by flies and small-bodied beetles)

Convergent lineages:

1 Giant "water lily" (Euphyllophyta: Lignophyta: Spermatophyta: Angiospermae: Nymphaeales: Nymphaeaceae; *Victoria amazonica*)

2 Big-leaf magnolia (Angiospermae: Euangiosperms: Magnoliidae: Magnoliales: Magnoliaceae; *Magnolia macrophylla*)

3 Pelican flower (Euangiosperms: Magnoliidae: Piperales: Aristolochiaceae; *Aristolochia grandiflora*)

4 Corpse lily (Euangiosperms: Monocotyledons: Alismatales: Araceae; *Amorphophallus titanum*)

5 Lady slipper orchid (Euangiosperms: Monocotyledons: Eumonocotyledons: Asparagales: Orchidaceae; *Phragmipedium grande*)

6 Sacred lotus (Euangiosperms: Eudicots: Proteales: Nelumbonaceae; *Nelumbo nucifera*)

7 Rafflesia (Euangiosperms: Eudicots: Rosidae: Eurosids I: Rafflesiaceae; *Rafflesia arnoldii*)

8 Beobab (Euangiosperms: Eudicots: Rosidae: Eurosids II: Malvaceae; *Adansonia digitata*)

9 Zulu giant carrion plant (Euangiosperms: Eudicots: Asteridae: Euasterids I: Apocynaceae; *Stapelia gigantea*)

Note: For data sources, see text.

Table 3.13
Convergent evolution of sapromyiophilous flowers that smell like rotting meat

Convergent structure and function: CARRION-MIMIC FLOWERS (production of rotting-meat smell to attract carrion-eating animals for pollination purposes)

Convergent lineages:

1 Pawpaw (Euangiosperms: Magnoliidae: Magnoliales: Annonaceae; *Asimina triloba*)

2 Corpse lily (Euangiosperms: Monocotyledons: Alismatales: Araceae; *Amorphophallus titanium*)

3 Carrion flower (Euangiosperms: Monocotyledons: Eumonocotyledons: Liliales: Liliaceae; *Smilax herbacea*)

4 Stinking iris (Euangiosperms: Monocotyledons: Eumonocotyledons: Asparagales: Iridaceae; *Iris foetidissima*)

5 Stinking hellebore (Euangiosperms: Eudicots: Ranunculalaes: Ranunculaceae; *Helleborus foetidus*)

6 Rafflesia (Euangiosperms: Eudicots: Rosidae: Eurosids I: Rafflesiaceae; *Rafflesia arnoldii*)

7 Zulu giant carrion plant (Euangiosperms: Eudicots: Asteridae: Euasterids I: Apocynaceae; *Stapelia gigantea*)

Note: For data sources, see text.

animal pollinators: the ancient flowers offered only excess pollen for pollen-eating animals, a trait that is actually older than the angiosperms, as it is found in some nonangiosperm spermatophytes like cycads and conifers as well (Labandeira et al. 2007). Derived flowers offer not only excess pollen but also sweet nectar, starchy food bodies, stigmatic secretions, perianth segments, and edible stamenodia (Bernhardt 2000; Thien et al. 2009).

One particularly specialized type of food reward system, known as ornithophily, is adapted specifically to attract nectar-feeding birds. Ornithophilous, or hummingbird, flowers have no smell; they are of a brick-red monocolor, are trumpet-shaped, and produce copious amounts of nectar. The evolution of hummingbird flowers in western North America has been extensively studied by Grant and Grant (1968), who document the independent evolution of ornithophilous flowers in 18 separate families of angiosperms in this region of the world alone (table 3.14). For each family, I have listed a single species representative, but some families have more hummingbird-flowered species than others. For example, the Scrophulariaceae contain eight genera and 74 species with hummingbird flowers, and a more detailed phylogenetic analysis of the family would be needed to determine how many of these hummingbird-flowered species are independently derived and how many are synapomorphic. Other families, like the Cactaceae, Nyctaginaceae, Saxifragaceae, Fouquieriaceae, Convolvulaceae, and Rubiaceae contain a single species with hummingbird flowers, and it is clear that each of these is an independent convergence on ornithophily (Grant and Grant 1968). Last, of the 18 convergently evolved hummingbird flowers listed in table 3.14, seven belong to species that are the sole red-flowered representative of all of the other variously hued flowers belonging to other species within each particular genus: *Silene laciniata*, *Ribes speciosum*, *Astragalus coccineus*, *Gilia aggregata*, *Ipomoea coccinea*, *Monardella macrantha*, *Mimulus cardinalis*; and only two red-flowered species are found in the generally blue- and white-flowered species of the genus *Delphinium*, *D. cardinale* and *D. nudicaule* (Went 1971).

Worldwide, some 65 families of angiosperms have evolved ornithophilous flowers (Cronk and Ojeda 2008). It will take a detailed analysis of the phylogenetic distribution of these flowers to determine how many are independent originations of ornithophily. For example, species in all eight families of the monocotyledon order Zingiberales are pollinated by birds (Cronk and Ojeda 2008). Do these eight families represent eight independent convergencies on ornithophily within the Zingiberales, or

Table 3.14
Convergent evolution of ornithophilous hummingbird flowers of western North America

Convergent structure and function: HUMMINGBIRD FLOWERS (odorless, brick-red, trumpet-shaped flowers rich in nectar that attract nectar-feeding birds for pollination purposes)

Convergent lineages:

1 Firecracker plant (Euangiosperms: Monocotyledons: Eumonocotyledons: Liliales: Liliaceae; *Brodiaea ida-maia*)

2 Schott's century plant (Euangiosperms: Monocotyledons: Eumonocotyledons: Asparagales: Agavaceae; *Agave schottii*)

3 Scarlet larkspur (Euangiosperms: Eudicots: Ranunculalaes: Ranunculaceae; *Delphinium cardinale*)

4 Indian pink (Euangiosperms: Eudicots: Core eudicots: Caryophyllales: Caryophyllaceae; *Silene laciniata*)

5 Hedgehog cactus (Euangiosperms: Eudicots: Core eudicots: Caryophyllales: Cactaceae; *Echinocereus triglochidiatus*)

6 Scarlet four o'clock (Euangiosperms: Eudicots: Core eudicots: Caryophyllales: Nyctaginaceae; *Allionia coccinea*)

7 Fuchsia-flowered gooseberry (Euangiosperms: Eudicots: Core eudicots: Saxifragales: Saxifragaceae; *Ribes speciosum*)

8 California fuchsia (Euangiosperms: Eudicots: Rosidae: Myrtales: Onagraceae; *Zauschneria californica*)

9 Locoweed (Euangiosperms: Eudicots: Rosidae: Eurosids I: Fabaceae; *Astragalus coccineus*)

10 Ocotillo (Euangiosperms: Eudicots: Asteridae: Ericales: Fouquieriaceae; *Fouquieria splendens*)

11 Scarlet skyrocket (Euangiosperms: Eudicots: Asteridae: Ericales: Polemoniaceae; *Gilia aggregata*)

12 Star-glory (Euangiosperms: Eudicots: Asteridae: Euasterids I: Convolvulaceae; *Ipomoea coccinea*)

13 Scarlet monardella (Euangiosperms: Eudicots: Asteridae: Euasterids I: Lamiaceae; *Monardella macrantha*)

14 Crimson monkey-flower (Euangiosperms: Eudicots: Asteridae: Euasterids I: Scrophlariaceae; *Mimulus cardinalis*)

15 Desert honeysuckle (Euangiosperms: Eudicots: Asteridae: Euasterids I: Acanthaceae; *Anisacanthus thurberi*)

16 Scarlet bouvardia (Euangiosperms: Eudicots: Asteridae: Euasterids I: Rubiaceae; *Bouvardia glaberrima*)

17 Cardinal flower (Euangiosperms: Eudicots: Asteridae: Euasterids II: Campanulaceae; *Lobelia cardinalis*)

18 Arizona honeysuckle (Euangiosperms: Eudicots: Asteridae: Euasterids II: Caprifoliaceae; *Lonicera arizonica*)

Note: For data sources, see text.

is ornithophily a synapomorphy for the entire order, with nonornithophilous species within the order having secondarily lost the bird-pollination mode of reproduction? In other cases convergent evolution has clearly occurred, as with the species-rich eudicot family Asteraceae that possesses a single South American genus, *Mutisia*, that is bird-pollinated (Cronk and Ojeda 2008).

In the Western Hemisphere, ornithophilous flowers are pollinated by species of hummingbirds (family Trochilidae), in Africa and Asia by species of sunbirds (family Nectariniidae), and in Australia and New Zealand by species of honeyeaters (family Meliphagidae)—a striking ecological convergence that we shall examine in more detail in the next chapter. Regardless of the particular pollinating bird species, ornithophilous flowers around the world are remarkably convergent in form and color. The nectar-producing capabilities of the flowers are easily understandable, as this is the food reward system that the plants use to attract nectar-feeding birds. But why are the flowers all red? Curiously, the red color is thought to be as much for bee deterrence as much as it is for bird attraction (Cronk and Ojeda 2008). Bees see color up to light wavelengths of 550 nanometers, whereas birds can see colors that extend to wavelengths of 660 nanometers, much further into the red end of the electromagnetic spectrum. The red color of many ornithophilous flowers has a wavelength of 585 nanometers, outside the visual range of bees but clearly visible to birds (Cronk and Ojeda 2008).

Other aspects of ornithophilous flower morphology also act to deter bees. Hummingbirds with long, curved bills can easily drink nectar from trumpet- or tubular-shaped flowers, whereas bees cannot. Since hummingbirds feed while hovering, hummingbird flowers typically have no landing platforms that bees might use. In addition, most birds have a very poor sense of smell (excepting carrion hunters, such as vultures), and so ornithophilous flowers have no scent that might attract unwanted insects.

In summary, convergent fertilization systems are rife in the seed plants. Even such highly specialized systems as cantharophily, sapromyiophily, and ornithophily have been evolved over and over again in independent plant lineages. There appear to be a limited number of successful fertilization systems, and plant evolution has been constrained to reusing these systems repeatedly in independent plant lineages.

While there are a limited number of ways to achieve fertilization in spermatophytes, there also appear to be a limited number of ways to disperse the product of fertilization: the seed. Species whose members are geographically widespread are more resistant to extinction than geographically endemic, locally restricted species. Whereas animal species can expand their geographic ranges by simply walking or flying to new territories, plants cannot. Thus, the sessile nature of plants presents two reproductive problems, not just one: the problem of fertilization in organisms that cannot move about in search for a mate, which we have just considered, and the problem of attaining geographic dispersal for the plant's offspring. Spermatophytes have evolved two pathways to try to obtain geographic dispersal for their fertile seeds: physical seed-dispersal systems and the evolution of fruit, in which the plant once again uses an attract-and-reward system to elicit the help of a mobile partner in carrying the plant's offspring away to other territories.

Just as many plants use wind for pollination, they also depend upon wind for dispersal of their seeds. Some of these plants have evolved seeds with fine, filamentous tufts with much larger surface areas than those of the seed; these tufts can catch breezes which lift the seed up into the air to be carried away by the wind (Loewer 1995; Stewart 2009). Tufted seeds have independently appeared in both the rosid and asterid eudicot clades, and are perhaps most familiar in the cottonwoods and common dandelion (table 3.15). Other plants have seeds that develop wings—enlarged surface-area structures that extend away from the seed center—that can not only catch breezes to be lifted up into the air, but can also maintain themselves in the air for considerable distances due to the helicopter-lift effect of their spinning wings even if the initial breeze has died. Winged seeds have been independently developed throughout the clade of seed plants, from nonangiosperm spermatophytes like pinophytes to plesiomorphic angiosperm spermatophytes like magnoliids and to highly derived asterid eudicots (table 3.15). The spinning seeds of maple trees are perhaps the most familiar example of this type of seed-dispersal system in eastern North America. However, the East Indian cucurbit vine, which aptly bears the species name *macrocarpa*, has seeds with wings that are 15 centimeters in total length, and the falling seeds descend in a spiral path with a 7-meter diameter (Loewer 1995), a feat our most massive maples cannot duplicate.

An alternative physical-dispersal system that does not use wind is the development of seeds that are capable of hitching a ride on mobile animals. In this case the animal is not rewarded for carrying the seed away, as often the animal is not even aware that a seed has attached itself to its body. I have termed this type of seed-dispersal system the "Velcro system," as it is a historical fact that the idea for the creation of Velcro came from the observation of the adhesive capabilities of the thistle-shaped burrs of the greater burdock, *Arctium lappa* (Stewart 2009). Many different plant lineages, from the monocotyledon foxtails and needle grasses to highly derived asterid devil's claws, beggar's lice, and cocklebururs have independently evolved seeds with tiny barbs or hooks that attached themselves to moving animals (table 3.15).

A last physical-dispersal system is to produce seed containers that actually explode, sending the seeds of the plant rocketing off into the distance. The eastern dwarf mistletoe is capable of blasting its seeds away at the velocity of 90 kilometers an hour, and gorse seedpods explode with a noise like that of a gunshot (Stewart 2009). The wild geranium can throw its seeds some 3 meters away; squirting cucumbers and Indian sword beans can propel their seeds up to 6 meters away; exploding witch hazel seedpods send their seeds up to 10 meters away; and the exploding fruit of the sandbox tree is able to propel its seeds up to 100 meters into the distance (Loewer 1995; Stewart 2009). The exploding-seed-container dispersal system has been developed multiple times independently throughout the eudicot clade (table 3.15). The list of convergent species given in table 3.15 is not exhaustive, but was chosen to illustrate the widespread phylogenetic distribution of independent origins of the same dispersal system across the spermatophyte clade. A rigorous analysis of the phylogenetic distribution of all physical-dispersal systems in spermatophytes has yet to be conducted. However, such analyses have been conducted, at least in part, for those plants that have evolved an alternative dispersal system: fleshy fruit.

By far the most familiar angiosperm seed-dispersal system is the evolution of fruit, in which the mobile animal partner is enticed to eat the sweet fruit and swallow the enclosed seeds, which are defecated by the animal much later, hopefully (from the plant's point of view) to then sprout and grow at some distance removed from the original fruit-bearing plant. Fleshy fruit and berry structures have convergently and independently evolved in so many plant lineages that it will take many detailed systematic analyses to sort out the many separate phylogenetic originations of the fruit habit. The fruit habit is also not an exclusively angiosperm trait: the odiferous fruit of the nonangiosperm spermatophyte *Ginkgo biloba* is familiar to many, as are the aromatic "berries" of the common juniper, a nonangiosperm pinophyte spermatophyte.

Commercially produced fruit that are popular with human consumers have convergently evolved in many branches of the eudicot clade. The highly derived Eurosid-I family Rosaceae in particular contains many fruit-bearing species, giving us apples, plums, peaches, pears, and cherries.

Table 3.15
Convergent evolution of physical seed-dispersal systems in plants

1 Convergent structure and function: TUFTED SEEDS (fine, filamentous, parachute-like tufts for geographic dispersal of seeds by wind)

Convergent lineages:

1.1 Fireweed (Euangiosperms: Eudicots: Rosidae: Myrtales: Onagraceae; *Epilobium angustifolium*)

1.2 Eastern cottonwood (Euangiosperms: Eudicots: Rosidae: Eurosids I: Salicaceae; *Populus deltoides*)

1.3 Common milkweed (Euangiosperms: Eudicots: Asteridae: Euasterids I: Asclepiadaceae; *Asclepias syriaca*)

1.4 Dogbane (Euangiosperms: Eudicots: Asteridae: Euasterids I: Apocynaceae; *Apocynum cannabinum*)

1.5 Common dandelion (Euangiosperms: Eudicots: Asteridae: Euasterids II: Asteraceae; *Taraxacum officinale*)

2 Convergent structure and function: WINGED SEEDS (helicopter-like wings for geographic dispersal of seeds by wind)

Convergent lineages:

2.1 Eastern white pine (Euphyllophyta: Lignophyta: Spermatophyta: Pinophyta: Pinaceae; *Pinus strobus*)

2.2 Yellow poplar (Spermatophyta: Angiospermae: Euangiosperms: Magnoliidae: Magnoliales: Magnoliaceae; *Liriodendron tulipifera*)

2.3 Curly dock weed (Euangiosperms: Eudicots: Core eudicots: Caryophyllales: Polygonaceae; *Rumex crispus*)

2.4 Winged elm (Euangiosperms: Eudicots: Rosidae: Eurosids I: Ulmaceae; *Ulmus alata*)

2.5 East Indian cucurbit vine (Euangiosperms: Eudicots: Rosidae: Eurosids I: Cucurbitaceae; *Macrozanonia macrocarpa*)

2.6 Silver maple (Euangiosperms: Eudicots: Rosidae: Eurosids II: Sapindaceae; *Acer saccharinum*)

2.7 Alianthus (Euangiosperms: Eudicots: Rosidae: Eurosids II: Simaroubaceae; *Alianthus altissima*)

2.8 Common hoptree (Euangiosperms: Eudicots: Rosidae: Eurosids II: Rutaceae; *Ptelea trifoliata*)

2.9 Alpine rockcress (Euangiosperms: Eudicots: Rosidae: Eurosids II: Brassicaceae; *Arabis alpina*)

2.10 Horseradish tree (Euangiosperms: Eudicots: Rosidae: Eurosids II: Moringaceae; *Moringa pterygosperma*)

2.11 White ash (Euangiosperms: Eudicots: Asteridae: Euasterids I: Oleaceae; *Fraxinus americana*)

2.12 Princess tree (Euangiosperms: Eudicots: Asteridae: Euasterids I: Scrophulariaceae; *Paulownia tomentosa*)

3 Convergent structure and function: VELCRO SEEDS (geographic dispersal of seeds by adhesion to mobile animals)

Convergent lineages:

3.1 Foxtail (Euangiosperms: Monocotyledons: Eumonocotyledons: Commelinidae: Poaceae; *Hordeum murinum*)

3.2 Sand-tick trefoil (Euangiosperms: Eudicots: Rosidae: Eurosids I: Fabaceae; *Desmodium lineatum*)

3.3 Devil's claw (Euangiosperms: Eudicots: Asteridae: Euasterids I: Pedaliaceae; *Harpagophytum procumbens*)

3.4 Beggar's lice (Euangiosperms: Eudicots: Asteridae: Euasterids I: Boraginaceae; *Cynoglossum officinale*)

Table 3.15
(continued)

3.5 Unicorn plant (Euangiosperms: Eudicots: Asteridae: Euasterids I: Martyniaceae; *Proboscidea louisianaca*)
3.6 Queen Anne's lace (Euangiosperms: Eudicots: Asteridae: Euasterids II: Apiaceae; *Daucus carota*)
3.7 Cocklebur (Euangiosperms: Eudicots: Asteridae: Euasterids II: Asteraceae; *Xanthium strumarium*)

4 Convergent structure and function: EXPLODING SEED CONTAINERS (exploding seed pods or fruit for geographic dispersal of seeds)
Convergent lineages:
4.1 Eastern dwarf mistletoe (Euangiosperms: Eudicots: Core eudicots: Santalales: Santalaceae; *Arceuthobium pusillum*)
4.2 Witch hazel (Euangiosperms: Eudicots: Core eudicots: Saxifragales: Hamamelidaceae; *Hamamelis virginiana*)
4.3 Wild geranium (Euangiosperms: Eudicots: Rosidae: Geraniales: Geraniaceae; *Geranium maculatum*)
4.4 Sandbox tree (Euangiosperms: Eudicots: Rosidae: Eurosids I: Euphorbiacea; *Hura crepitans*)
4.5 Gorse (Euangiosperms: Eudicots: Rosidae: Eurosids I: Fabaceae; *Ulex europaeus*)
4.6 Squirting cucumber (Euangiosperms: Eudicots: Rosidae: Eurosids I: Cucurbitaceae; *Ecballium elaterium*)
4.7 Creeping woodsorrel (Euangiosperms: Eudicots: Rosidae: Eurosids I: Oxalidaceae; *Oxalis corniculata*)
4.8 Pale touch-me-not (Euangiosperms: Eudicots: Asteridae: Ericales: Balsaminaceae; *Impatiens pallida*)

Note: For data sources, see text.

But melons are produced by a different Eurosid-I family, Cucurbitaceae, and grapes are even more distantly placed, in the basal rosid family Vitaceae. Tart persimmons come from a much more distantly related family in the asterid clade (Ebenaceae), as do our popular tomatoes and eggplants (Solanaceae). A modern molecular phylogeny for the family Brassicaceae in the eudicot Eurosid-II clade has revealed that fruit has evolved three separate times independently in this single family alone (Mummenhoff et al. 2008).

While sorting out the many convergent occurrences of the fruit habit in the huge clade of the eudicots is going to take some time, rigorous phylogenetic analyses of convergent fruit evolution have been completed for the clade of the monocotyledons (Givnish et al. 2005). Fleshy fruits have independently evolved in 22 separate lineages within the monocolyledons (table 3.16). In table 3.16 I have listed basal or near-basal species for each independent evolution of fruit in the cladogram of Givnish et al. (2005); subsequent evolution in some of these clades have produced numerous fruit-bearing species (such as the clade of

Table 3.16
Convergent evolution of fleshy fruit in monocotyledon angiosperms

Convergent structure and function: FLESHY FRUIT (geographic dispersal of seeds by fruit-eating animals)

Convergent lineages:

1 Settler's flax (Euangiosperms: Monocotyledons: Alismatales: Araceae; *Gymnostachys anceps*)
2 White bat flower (Euangiosperms: Monocotyledons: Eumonocotyledons: Dioscoreales: Dioscoreaceae; *Tacca integrifolia*)
3 Screwpine (Euangiosperms: Monocotyledons: Eumonocotyledons: Pandanales: Pandanaceae; *Pandanus tectorius*)
4 Giant lily (Euangiosperms: Monocotyledons: Eumonocotyledons: Liliales: Liliaceae; *Cardiocrinum giganteum*)
5 Supplejack vine fruit (Euangiosperms: Monocotyledons: Eumonocotyledons: Liliales: Ripogonaceae; *Ripogonum scandens*)
6 Night heron (Euangiosperms: Monocotyledons: Eumonocotyledons: Liliales: Colchicaceae; *Disporum cantoniense*)
7 Large-flowered trillium (Euangiosperms: Monocotyledons: Eumonocotyledons: Liliales: Melanthiaceae; *Trillium grandiflorum*)
8 Forest smilax (Euangiosperms: Monocotyledons: Eumonocotyledons: Asparagales: Agavaceae; *Behnia reticulata*)
9 Smooth Solomon's seal (Euangiosperms: Monocotyledons: Eumonocotyledons: Asparagales: Ruscaceae; *Polygonatum biflorum*)
10 Hyacinthina (Euangiosperms: Monocotyledons: Eumonocotyledons: Asparagales: Amaryllidaceae; *Griffinia hyacinthina*)
11 Cardwell lily (Euangiosperms: Monocotyledons: Eumonocotyledons: Asparagales: Amaryllidaceae; *Proiphys amboinensis*)
12 Scrambling lily (Euangiosperms: Monocotyledons: Eumonocotyledons: Asparagales: Hemerocalidaceae; *Geitonoplesium cymosum*)
13 Palm grass (Euangiosperms: Monocotyledons: Eumonocotyledons: Asparagales: Hyperoxidaceae; *Curculigo capitulata*)
14 Tasmanian Christmas bells (Euangiosperms: Monocotyledons: Eumonocotyledons: Asparagales: Blandfordiaceae; *Blandfordia punicea*)
15 Neuwiedia orchid (Euangiosperms: Monocotyledons: Eumonocotyledons: Asparagales: Orchidaceae; *Neuwiedia veratrifolia*)
16 'Ohe (Euangiosperms: Monocotyledons: Eumonocotyledons: Commelinidae: Poales: Joinvilleaceae; *Joinvillea ascendens*)
17 Whip vine (Euangiosperms: Monocotyledons: Eumonocotyledons: Commelinidae: Poales: Flagelliariaceae; *Flagellaria indica*)
18 Brocchinia bromeliad (Euangiosperms: Monocotyledons: Eumonocotyledons: Commelinidae: Poales: Bromeliaceae; *Brocchinia reducta*)
19 Rattan palm (Euangiosperms: Monocotyledons: Eumonocotyledons: Commelinidae: Arecales: Arecaceae; *Calamus laoensis*)
20 Malaysian hanguana (Euangiosperms: Monocotyledons: Eumonocotyledons: Commelinidae: Commelinales: Hanguanaceae; *Hanguana malayana*)
21 Amischolotype commelinid (Euangiosperms: Monocotyledons: Eumonocotyledons: Commelinidae: Commelinales: Commelinaceae; *Amischolotype hispida*)
22 Wild ginger (Euangiosperms: Monocotyledons: Eumonocotyledons: Commelinidae: Zingiberales: Zingiberaceae; *Siphonochilus aethiopicus*)

Note: For data sources, see text.

the Arecaceae), while others have not (such as the clade of the Ruscaceae).

Interestingly, Givnish et al. (2005) also have demonstrated that former fruit-bearing species have secondarily lost the fruit habit 11 separate times within the clade of the monocotyledons. Their careful ecological analyses of the phylogenetic distribution of fruit occurrence within the monocotyledons reveals that species that have invaded shaded habitats have often subsequently evolved fleshy fruit, whereas species possessing fruit have often lost the fruit habit if those species returned to open habitats. Apparently animal dispersal of seeds is needed for plants growing under closed canopies, whereas in open habitats less costly mechanisms of wind dispersal seem to suffice. Up to 95 percent of woody understory species in neotropical rain forests have evolved fleshy fruits, a phenomenon that Givnish et al. (2005, 1481) have termed "concerted convergence"; they point out that this pattern is among "the strongest ever demonstrated for evolutionary convergence in individual traits and the predictability of evolution."

Even more specialized than the evolution of fruit and subsequent seed dispersal by vertebrate animals is the evolution of seed dispersal by ants, or myrmecochory (Dunn et al. 2007). Myrmecochorous plants have evolved elaiosome, a specialized lipid-rich structure that ants use as a handle in carrying the seed from the plant back to their nest. At the nest, the ants remove the elaiosome (which is the plant's reward to the ants), and the seed is discarded in a midden or outside of the nest itself. The elaiosome has clearly been convergently evolved by numerous different plants, as the structure has multiple developmental origins (Dunn et al. 2007).

As phylogenetic relationships within the clade of the monocotyledons have now been resolved (Givnish et al. 2005), Dunn et al. (2007) have used this phylogeny to study the distribution of elaiosome origins. Myrmecochory has independently evolved 19 separate times within the clade of the monocotyledons (table 3.17). In table 3.17 I have listed basal or near-basal species for each independent evolution of elaiosomes in the cladogram of Dunn et al. (2007). Unlike monocotyledons that have convergently evolved fruit (table 3.16), Dunn et al. (2007) have demonstrated that the evolution of myrmecochory in monocotyledons is independent of whether habitats are shaded or open, and hence independent of the evolution of fleshy fruits themselves. Rather, the evolution of myrmecochory in plants appears to be directly linked to the evolutionary diversification and increase in population sizes of ants

Table 3.17
Convergent evolution of myrmecochory in monocotyledon flowering plants

Convergent structure and function: MYRMECOCHORY (geographic dispersal of seeds by ants)

Convergent lineages:

1 Large-flower bellwort (Euangiosperms: Monocotyledons: Eumonocotyledons: Liliales: Colchicaceae; *Uvularia grandiflora*)

2 Men-in-a-boat (Euangiosperms: Monocotyledons: Eumonocotyledons: Liliales: Colchicaceae; *Androcymbium striatum*)

3 Dogtooth violet (Euangiosperms: Monocotyledons: Eumonocotyledons: Liliales: Liliaceae; *Erythronium americanum*)

4 Yellow Star-of-Bethlehem (Euangiosperms: Monocotyledons: Eumonocotyledons: Liliales: Liliaceae; *Gagea lutea*)

5 Bigelow's adder's tongue (Euangiosperms: Monocotyledons: Eumonocotyledons: Liliales: Liliaceae; *Scoliopus bigelovii*)

6 Large-flowered trillium (Euangiosperms: Monocotyledons: Eumonocotyledons: Liliales: Melanthiaceae; *Trillium grandiflorum*)

7 Palm grass (Euangiosperms: Monocotyledons: Eumonocotyledons: Asparagales: Hypoxidaceae; *Curculigo capitulata*)

8 Cyanastrum (Euangiosperms: Monocotyledons: Eumonocotyledons: Asparagales: Tecophilaeaceae; *Cyanastrum cordifolium*)

9 Turbinate hensmannia (Euangiosperms: Monocotyledons: Eumonocotyledons: Asparagales: Hemerocallidaceae; *Hensmannia turbinata*)

10 Summer snowflake (Euangiosperms: Monocotyledons: Eumonocotyledons: Asparagales: Amaryllidaceae; *Leucojum aestivum*)

11 Yellow iris (Euangiosperms: Monocotyledons: Eumonocotyledons: Asparagales: Iridaceae; *Iris pseudacorus*)

12 Hooker's dasypogon (Euangiosperms: Monocotyledons: Eumonocotyledons: Commelinidae: Dasypogonaceae; *Dasypogon hookeri*)

13 Pennsylvania sedge (Euangiosperms: Monocotyledons: Eumonocotyledons: Commelinidae: Poales: Cyperaceae; *Carex pensylvanica*)

14 Switchgrass (Euangiosperms: Monocotyledons: Eumonocotyledons: Commelinidae: Poales: Poaceae; *Panicum virgatum*)

15 Silver vase bromeliad (Euangiosperms: Monocotyledons: Eumonocotyledons: Commelinidae: Poales: Bromeliaceae; *Aechmea fasciata*)

16 Blushing bromeliad (Euangiosperms: Monocotyledons: Eumonocotyledons: Commelinidae: Poales: Bromeliaceae; *Nidularium fulgens*)

17 Spiral ginger (Euangiosperms: Monocotyledons: Eumonocotyledons: Commelinidae: Zingiberales: Costaceae; *Costus barbatus*)

18 Zebra plant (Euangiosperms: Monocotyledons: Eumonocotyledons: Commelinidae: Zingiberales: Marantaceae; *Calathea zebrina*)

19 Dancing girl ginger (Euangiosperms: Monocotyledons: Eumonocotyledons: Commelinidae: Zingiberales: Zingiberaceae; *Globba winitii*)

Note: For data sources, see text.

beginning in the Late Eocene, some 37 million years ago. The peak in the evolution of fleshy fruit in the monocotyledons began earlier, some 85 million years ago, and was triggered by the spread of Late Cretaceous forests and closed-canopy habitats (Givnish et al. 2005).

Although myrmecochorous plants are equally likely to be found in either open habitats or closed-canopy habitats, Dunn et al. (2007) do demonstrate that monocotyledons in open habitats that are fire-prone, such as in Mediterranean ecosystems, have preferentially evolved myrmecochory. They suggest that transport of seeds to ant nests confers the selective advantage of reducing the risk of seed mortality due to fire, in addition to the selective advantage of geographic dispersal of the seeds. This conclusion is supported by the documented convergent evolution of myrmecochory in the fire-prone shrub lands of both South African and Australian ecosystems (Milewski and Bond 1982), which provides an example of ecosystem convergence as well, the subject of the next chapter of this book.

In summary, the incidences of reproductive convergence in plants, just as in the animals we considered in the previous chapter, vastly overshadow the classic examples of convergent evolution that are usually offered for both groups. The spectacular convergent evolution of wings in animals pales in comparison with the astonishing number of independent origins of viviparity, and the convergent evolution of towering trees in plants pales in comparison with the astonishing number of independent origins of the fruit habit. I predict the 22 separate convergences on the fruit habit discovered in the monocotyledons alone (table 3.16) are going to be a small number indeed when rigorous phylogenetic analyses are eventually applied to mapping the number of independent origins of fleshy fruit in the huge clade of the eudicots.

In many ways, plant evolution is more complex and more interesting than animal evolution because of its overtly multispecies aspect. Plants coevolve with the numerous animal partners that they have involved in their reproduction and defense, and these beetles, flies, bees, wasps, butterflies, moths, ants, birds, and mammals have themselves evolved in response to plant evolution. In the next chapter we shall focus on the phenomenon of convergent evolution at the multispecies level: that is, the convergent evolution of entire ecosystems.

4 Convergent Ecosystems

In chemistry, the urge of scientists to order and classify natural phenomena resulted in the well-known periodic table of the elements, which allowed chemists to predict new elements and their chemical properties. . . . Some ecologists wonder whether something like a "periodic table of niches" might be possible.
—Pianka (1978, 267)

One Ecological Role, Many Convergent Players

Imagine a universe in which there are an unlimited number of ways to make a living. In such a universe, each species would have its own unique way of making a living, different from all other species. When we examine the ecological structure of living organisms on Earth, we can clearly see that we do not inhabit such a universe.

In our universe the number of ways of making a living, of ecological roles or niches, is demonstrably limited. Multiple species are constrained in their evolution to playing the same ecological role, filling the same ecological niche, as best they can given their own phylogenetic and developmental backgrounds. The ecological role is the same, but the species players are from many different evolutionary pathways that have independently converged on filling that niche.

Ecological niche convergence is best recognized by first considering truly bizarre ways of making a living—ecological roles that are so strange that, at first glance, it would seem probable that only one species would have evolved such a restricted pathway. And then one realizes that, astonishingly, the ecological evolution of life on Earth is so constrained that multiple species have converged on that pathway.

One such bizarre pathway is that of plants that eat animals—an ecological role so strange that it fascinated Darwin (1875). Carnivorous plants have evolved adaptations to actively trap, kill, and digest animals.

Although such adaptations seem so improbable that one might think they could have evolved only once in the plants, Darwin's (1875) own researches convinced him that the carnivorous plants were polyphyletic, and led him to believe that plants had independently evolved carnivory in three separate lineages. Over a century and a quarter later, modern phylogenetic analyses reveal that carnivory has arisen independently in at least six separate plant clades: within the commelinid monocotyledons, the caryophyllalean core eudicots, the eurosids-I eudicots, the ericalean-asterid eudicots, and at least twice in the eurasterids-I eudicots (Albert et al. 1992; Cameron et al. 2002; Ellison and Gotelli 2009).

Not only is the trait of carnivory convergent in plants, but the means of carnivory itself is also convergent, in that independent plant lineages have evolved the same mechanism for it. Although six separate plant clades have evolved carnivory, carnivorous plants use only three different types of animal traps: flypaper traps, pitcher traps, and mechanical traps (table 4.1). Carnivorous plants with flypaper traps have leaves or hairs that ooze sticky droplets to capture animals that touch them. Flypaper-type traps have convergently evolved at least five separate times (table 4.1), and it is possible that they have independently arisen more than once in the clade of the Caryophyllales. Within the carnivorous Caryophyllales, the sundews (Droseraceae) are basal; thus, at first one might conclude that flypaper traps are a synapomorphy for all plants in this clade with that trait, and so I have listed its evolution only once in table 4.1. However, higher in the clade, the dewy pines (Drosophyllaceae) and the African lianas (*Triphyophyllum peltatum*, Dioncophyllaceae) may have reevolved flypaper traps a second time, in that the Nepenthaceae are below these two families in the clade of the Caryophyllales, and the Nepenthaceae do not have flypaper-type traps (Ellison and Gotelli 2009).

Carnivorous plants with pitcher traps have tubular leaves or flowers with a pool of digestive fluids at the base of the tube. Animals that enter the plant's tubes fall into the cup at its base to drown and be consumed. In some cases the plant's tubes have downward-pointing hairs or spines that prevent the prey from trying to climb back out of the tube structure to escape. Pitcher traps have convergently evolved at least four separate times (table 4.1). In some plant lineages, pitcher traps are developmental modifications of preexisting flypaper traps (Asian pitcher plants in the Caryophyllales), whereas in other lineages pitcher traps have evolved independently of any lineage possessing flypaper traps (Australian pitcher plants, Brocchinia bromeliads). While insects are the prey of

Table 4.1
Convergent evolution of carnivory in plants

Convergent ecological role: CARNIVOROUS PLANTS ("Venus flytrap niche")

1 Convergent lineages with "flypaper" traps:
- 1.1 Sundew (Euangiosperms: Eudicots: Core eudicots: Caryophyllales: Droseraceae; *Drosera intermedia*)
- 1.2 Roridula (Euangiosperms: Eudicots: Asteridae: Ericales: Roridulaceae; *Roridula gorgonias*)
- 1.3 Devil's claw (Euangiosperms: Eudicots: Asteridae: Euasterids I: Martyniaceae; *Ibicella lutea*)
- 1.4 Rainbow plant (Euangiosperms: Eudicots: Asteridae: Euasterids I: Byblidaceae; *Byblis gigantea*)
- 1.5 Butterwort (Euangiosperms: Eudicots: Asteridae: Euasterids I: Lentibulariaceae; *Pinguicula vulgaris*)

2 Convergent lineages with "pitcher" traps:
- 2.1 Brocchinia bromeliad (Euangiosperms: Monocotyledons: Eumonocotyledons: Commelinidae: Bromeliaceae; *Brocchinia reducta*)
- 2.2 Asian pitcher plant (Euangiosperms: Eudicots: Core eudicots: Caryophyllales: Nepenthaceae; *Nepenthes truncata*)
- 2.3 Australian pitcher plant (Euangiosperms: Eudicots: Rosidae: Eurosids I: Cephalotaceae; *Cephalotus follicularis*)
- 2.4 American pitcher plant (Euangiosperms: Eudicots: Asteridae: Ericales: Sarraceniaceae; *Sarracenia purpurea*)

3 Convergent lineages with "mechanical" traps:
- 3.1 Venus flytrap (Euangiosperms: Eudicots: Core eudicots: Caryophyllales: Droseraceae; *Dionaea muscipula*)
- 3.2 Bladderwort (Euangiosperms: Eudicots: Asteridae: Euasterids I: Lentibulariaceae; *Utricularia inflata*)
- 3.3 Corkscrew plant (Euangiosperms: Eudicots: Asteridae: Euasterids I: Lentibulariaceae; *Genlisea margaretae*)

Note: For data sources, see text.

choice for most carnivorous plants, species of the pitcher-type carnivores are known to trap and eat animals as large as mice (Stewart 2009).

The most complicated of carnivorous-plant capture mechanisms are the mechanical traps, the most famous of which is the snap trap of the Venus flytrap (table 4.1). These snap traps are highly modified leaves with projecting antennae-like spikes around their perimeters. Animals alighting on the leaf, or touching the spikes in approaching the leaf, trigger the snap and the leaf folds shut, trapping the animal inside, with a rapidity that is the fastest movement known to exist in plants (Cameron et al. 2002). Darwin (1875) argued that the snap traps of the Venus flytrap, *Dionaea muscipula*, and the waterwheel, *Aldrovanda vesiculosa*, were independently evolved. Ecologically the two species are very different: the Venus flytrap is a terrestrial plant whereas the waterwheel is fully aquatic and does not possess roots; in addition, the Venus flytrap captures flying animals whereas the waterwheel captures underwater swimming animals. However, Darwin's (1875) hypothesis of the convergent evolution of snap traps has been disproved by modern phylogenetic molecular analyses, which reveal that the Venus flytrap and the waterwheel are sister taxa, and that the snap trap is a synapomorphy, thus not convergent (Cameron et al. 2002; Ellison and Gotelli 2009).

The bladderworts are also aquatic, but rather than having snap-trap leaves like the waterwheel, they have modified their leaves into bladder-like structures with a snap-trigger lid. When triggered, these bladders develop an internal negative pressure that is capable of sucking swimming animals into the bladder and closing the lid, trapping the animal inside to be digested. The Venus flytrap usually eats only insects, but the bladderworts can trap and eat animals as large as tadpoles, and they can reset their traps every thirty minutes, making them one of the most voracious of the carnivorous plants (Stewart 2009).

The corkscrew plants have modified some of their leaves into geometric traps, twin corkscrews that meet at a feeding pit containing digestive fluid. These corkscrewed leaves are actually developed underground, in the place of roots, and primarily capture bacteria. Upon entering the corkscrew, the prey organism is unable to leave because inward-pointing hairs and the geometry of the screw prevent backward movement, and the prey can only move forward to its death in the feeding pit.

The bladderworts and the corkscrew plants both belong to the family of the Lentibulariaceae (table 4.1), yet the nature of their mechanical leaf traps is so different that it is believed that the two types of trap were independently developed; that is, neither plant is in an ancestor-descendant relationship to the other, and their different traps represent separate modifications of the flypaper-type trap of the basal-lentibulariacean butterworts (Ellison and Gotelli 2009).

What is convergent in all of these plants is the ecological role of carnivory, rather than the overall morphology of the plants (as in chapter 3). There is an underlying ecological trigger, a selective advantage, that is found in most carnivorous plants: they are able to invade nitrogen-poor habitats preferentially over other plants. Most carnivorous plants are found in boggy or swampy environments where nutrients are in short supply, and therefore these plants have evolved a new way to acquire additional nutrients, namely by eating animals. Nitrogen, in particular, is preferentially allocated to reproductive structures in many carnivorous plants (Ellison and Gotelli 2009), triggering the differential reproductive process of natural selection in their favor.

Plants by definition are multicellular photoautotrophs—that is, they produce their own food by the process of photosynthesis. Even the carnivorous plants still have green leaves and photosynthesis, so they have not fully abandoned photoautotrophy. Still other plants have taken another bizarre evolutionary pathway: they have become heterotrophic, like animals. Such plants have drastically reduced their reliance on photosynthesis, and some have eventually lost it entirely, becoming parasitic like many animal species or saprophytic like many fungal species, who are close relatives to the animals. These are plants that, in essence, eat other plants.

Perhaps the most notorious of the parasitic plants is the dodder vine (table 4.2), a parasite that one botanist describes as resembling "an alien life form that has come to suck the life out of [E]arth's vegetation" (Stewart 2009, 147). Dodder vines lack the green color of chlorophyll, and instead are eerie shades of orange, red, and yellow. Dodder leaves have been reduced to near-microscopic scalelike structures, such that the vines appear to be naked, brightly colored strings smothering the plant that is being parasitized. The dodder vine is a highly derived asterid eudicot, yet the very distantly related, near-basal-Euangiosperms clade of the magnoliids has convergently evolved its ecological equivalent in the macabre-named "love vine," although its embrace is just as unloving as that of the dodder, at least from the point of view of the parasitized plant (table 4.2). All in all, at least seven independent phylogenetic

Table 4.2
Convergent evolution of heterotrophy in plants

Convergent ecological role: PARASITIC AND SAPROPHYTIC PLANTS ("dodder niche")

Convergent lineages:

1 Love vine (Euangiosperms: Magnoliidae: Laurales: Lauraceae; *Cassytha filiformis*)
2 Coral root orchid (Euangiosperms: Monocotyledons: Eumonocotyledons: Asparagales: Orchidaceae; *Corallorhiza maculata*)
3 Eastern dwarf mistletoe (Euangiosperms: Eudicots: Core eudicots: Santalales: Santalaceae; *Arceuthobium pusillum*)
4 Rafflesia (Euangiosperms: Eudicots: Rosidae: Eurosids I: Rafflesiaceae; *Rafflesia arnoldii*)
5 Indian pipe (Euangiosperms: Eudicots: Asteridae: Ericales: Ericaceae; *Monotropa uniflora*)
6 Witch weed (Euangiosperms: Eudicots: Asteridae: Euasterids I: Scrophulariaceae; *Striga asiatica*)
7 Dodder (Euangiosperms: Eudicots: Asteridae: Euasterids I: Convolvulaceae; *Cuscuta epithymum*)

Note: For data sources, see text.

lineages of plants have converged upon the ecological role of heterotrophy, the most aptly named of which is perhaps the "witch weed" (table 4.2).

Striking and seemingly improbable examples of ecological niche convergence are not limited to the plants. Consider the ecological role of insectivory. Insects are an abundant source of food on land, and so it is not surprising that many animals have evolved insectivorous modes of life. Ecological convergence exists at all levels and types of insectivory, such as the fairly restricted ecological role of eating-insects-on-the-wing, in which the insectivore captures and eats insects in the air without pausing from insect to insect. Dragonflies, which are themselves insects (Odonata), avian swifts (Apodidae), and mammalian bats (Microchiroptera), to name only a few separate phylogenetic lineages, are all species that have converged on filling this same ecological niche.

Let us consider an extremely specialized form of insectivory: eating insects found only in tree bark. There exists a highly specialized group of species that have evolved adaptations to fill this ecological niche: the avian woodpeckers (Picidae). These birds have evolved long, sharp, pointed beaks for boring and digging into tree bark, and very long tongues for extracting exposed insects. The jaws and cranium of the woodpecker are structured in a fashion that cushions the brain of the bird as it hammers away, preventing concussion. In the foot of the woodpecker, the fourth digit has been rotated backward, joining the first digit, such that the bird has feet with two toes pointing forward and two backward. These clawed feet, combined with the stiffened tail feathers of the woodpecker, allow the bird to firmly anchor itself against a tree and to forcefully hammer the bark of the tree without becoming dislodged from its triangular base of two grasping feet and supporting prop of tail feathers.

Thus, the woodpeckers are unique, highly specialized animals that have evolved to fill an extremely specialized and restricted ecological role, yes? Not so. They are not unique, in that no less than six additional phylogenetic lineages of animals have converged on filling the ecological niche of eating wood-boring insects (table 4.3). Three of these lineages are avian, but they are very different types of birds from woodpeckers, with very different adaptations. Both the akiapola'au honeycreepers of Hawaii and the sickle-billed vangas of Madagascar extract insects from tree bark with their bills, but their bills are not like those of woodpeckers. The akiapola'au honeycreeper has a thick lower beak that is only half as long as its curved upper beak. It drills and digs into tree bark with its

Table 4.3
Convergent evolution of woodpecker ecological equivalents in animals

Convergent ecological role: EATERS OF WOOD-BORING INSECTS ("woodpecker niche")

Convergent lineages:

1 Red-headed woodpecker (Amniota: Sauropsida: Archosauromorpha: Dinosauria: Saurischia: Theropoda: Maniraptora: Aves: Neognathae: Neoaves: Piciformes; Picidae; *Melanerpes erythrocephalus*)

2 Akiapola'au honeycreeper (Aves: Neognathae: Neoaves: Passeriformes: Drepanididae; *Hemignathus munroi*)

3 Sickle-billed vanga (Aves: Neognathae: Neoaves: Passeriformes: Vangidae; *Falculea palliata*)

4 Woodpecker finch (Aves: Neognathae: Neoaves: Passeriformes: Emberizidae; *Cactospiza pallida*)

5 Striped possum (Amniota: Synapsida: Therapsida: Mammalia: Marsupialia: Diprotodontia: Petauridae; *Dactylopsila trivirgata*)

6 Apatemyid (Mammalia: Eutheria: Boreoeutheria: Apatemyidae; *Apatemys chardini* †Paleogene)

7 Aye-aye (Eutheria: Euarchontoglires: Primates: Lemuriformes: Daubentoniidae; *Daubentonia madagascariensis*)

Note: For data sources, see text.

lower beak, and then extracts insects from the bark with the hooked upper beak. As their name implies, the sickle-billed vangas have sickle-shaped beaks, which they insert into bark cracks and openings; they then use a lever action to pull out the insect.

The woodpecker finches of the Galapagos have evolved a totally different approach to extracting wood-boring insects—they use tools. These birds use their beaks, not to drill with, but to hold a long, sharp tool, usually a cactus spine. They insert the cactus spine into bark cracks and openings, and also use a lever action to pull out insects, but using the tool rather than a specialized beak morphology. We shall consider the convergent evolution of tool usage by animals in detail in chapter 6.

Not only did tool usage and three distinctly different types of beak morphologies evolve in independent bird lineages to fill the same ecological niche, but no less than three separate lineages of mammals have also converged on the same restricted ecological role of eating wood-boring insects (table 4.3). All three of these mammal groups use very elongated fingers, tipped with claws, to insert into tree bark and to pull out enclosed insects. The particular fingers used, however, are different in each group. The striped possum of New Guinea, a marsupial mammal, has evolved a very elongated fourth digit (the ring finger in humans) for this function, whereas the aye-aye of Madagascar, a placental mammal, has evolved an elongated third digit (the middle finger in humans) for the same function. These two groups of mammals are alive today, but there existed a third group, the extinct apatemyids, that also converged on this same ecological role (table 4.3). The apatemyids evolved two elongated fingers rather than one, the second digit (the pointing finger in humans) and the third digit, for use in extracting insects from tree bark (Koenigswald and Schierning 1987).

Again, what is convergent in all of these animals is the ecological role of eating wood-boring insects, not the morphologies of the animals as previously considered in detail in chapter 2. Birds, bats, pterosaurs, and butterflies have all convergently evolved a morphologically similar structure, namely wings, to accomplish the same function: powered flight (see chapter 2). Here, greatly different morphological structures and behaviors have evolved in independent phylogenetic lineages in order to convergently fill the same ecological niche: eating wood-boring insects. A bat superficially looks like a bird, but a red-headed woodpecker looks very different from a striped possum. Morphologically, these animals are very different, but ecologically they are equivalents.

Consider yet another highly specialized form of insectivory: attacking and eating ants in their nests. Huge numbers of ants can inhabit an ant colony, and they are highly aggressive in defending that colony. Thus, attacking an ant nest is not a trivial task, and one might conclude that only one type of highly specialized animal would fill the niche of such a restrictive mode of food acquisition. But within the mammals alone, five separate groups have converged on myrmecophagy, three times independently in the placental mammals, once in the marsupial mammals, and once in the monotreme mammals (table 4.4). Interestingly, in the placental mammals both morphological convergence and ecological convergence have occurred, in that the South American giant anteater, the Asian pangolin, and the African aardvark have all evolved long claws and strong forelimbs for tearing into ant nests. All three groups of animals have evolved tubular-shaped mouths that contain a very long, sticky tongue for capturing ants. The anteaters and pangolins have totally lost their teeth, and the aardvark retains only a few teeth. All three groups have highly developed salivary glands, which assist in swallowing copious amounts of ants, and all three have convergently evolved strongly muscular pyloric regions in their stomachs to assist in digesting ants (Lecointre and Le Guyader 2006). Yet these animal groups are only very distantly related: the anteaters are xenarthrans, basal eutherians; the aardvarks are afrotherians; and the pangolins are laurasiatherians (table 4.4).

Table 4.4
Convergent evolution of myrmecophagous ecological equivalents in animals

Convergent ecological role: ANT-EATING ANIMALS ("anteater niche")
Convergent lineages:
1 Thorny devil lizard (Amniota: Sauropsida: Diapsida: Lepidosauromorpha: Squamata: Iguania: Agamidae; *Moloch horridus*)
2 Desert horned lizard (Lepidosauromorpha: Squamata: Iguania: Phrynosomatidae; *Phrynosoma platyrhinos*)
3 "Desert bird" alvarezsaurid (Diapsida: Archosauromorpha: Dinosauria: Saurischia: Theropoda: Coelurosauria: Maniraptora: Alvarezsauridae; *Shuvuuia deserti* †Cretaceous)
4 Long-beaked echidna (Amniota: Synapsida: Therapsida: Mammalia: Monotremata: Tachyglossidae; *Zaglossus bruijni*)
5 Numbat (Mammalia: Marsupialia: Myrmecobiidae; *Myrmecobius fasciatus*)
6 Giant anteater (Mammalia: Eutheria: Xenarthra: Myrmecophagidae; *Myrmecophaga tridactyla*)
7 Aardvark (Eutheria: Afrotheria: Tubulidentata: Orycteropodidae; *Orycteropus afer*)
8 Giant pangolin (Eutheria: Laurasiatheria: Pholidota: Manidae; *Manis gigantea*)

Note: The geological age of extinct species is marked with a †. For data sources, see text.

Although today these animal groups are geographically separated ecological equivalents, the enigmatic fossil species *Eurotamandua joresi*, from the Eocene Messel strata in Germany, indicates that the xenarthrans once extended their range into that of the laurasiatherian pangolins (Gaudin and Branham 1998).

The marsupial mammals and the plesiomorphic, egg-laying monotreme mammals are even more distantly related, but they too have converged on the ecological role of myrmecophagy with the numbat of Australia and the long-beaked echidna of New Guinea (table 4.4). Like the placental mammals, the myrmecophagous echidnas have evolved toothless, tubular-shaped mouths with long, sticky tongues for capturing ants (Lecointre and Le Guyader 2006).

Last, the ecological niche of myrmecophagy has also been convergently filled by three independent groups of animals in the clade of sauropsid amniotes, the reptiles. The desert horned lizards of North American, phrynosomatid lepidosaurs, and the thorny devil lizards of Australia, agamid lepidosaurs, have independently evolved myrmecophagy (table 4.4). Both of these lizard groups have reduced their dentition, a morphological trait convergent on the myrmecophagous mammals, have expanded their stomachs to assist in digesting large numbers of ants, and have strikingly similar dorsoventrally flattened bodies covered with sharp spines (Pianka and Parker 1975). These two lizard groups have also convergently evolved much quicker prey capture and processing behaviors than any other iguanian lizards; with their particularly rapid tongue protrusion and retraction, the thorny desert lizards can consume up to 2,000 ants per day (Meyers and Herrel 2005).

Myrmecophagy also appears to have evolved within the clade of the dinosaurs, as evidenced by the peculiar alvarezsaurids of the Cretaceous (table 4.4). One of the best preserved of these is the "desert bird," *Shuvuuia deserti*, which is not a true bird but a related maniraptoran theropod. It has powerful forearms and hands with a single finger, tipped with a very large, hooked claw that is hypothesized to have been used to rip open ant or termite nests, as the living mammalian anteaters do. Its snout is also elongated and roughly tubular, but the morphology of the tongue contained inside is unknown. These Cretaceous animals are enigmatic in that ants did not become a significant part of the terrestrial ecosystem until the Eocene, as discussed in chapter 3 in the discussion of the convergent evolution of myrmecochory in plants (see table 3.17). Thus, they may not have been exclusively specialized for eating only ants, but may have also preyed on wood-nesting termites (Longrich and Currie 2008).

In general for the major heterotrophic organisms, animals are different types of carnivores or herbivores, ultimately using living organisms as a source of food, whereas the closely related fungi are saprophytes, feeding on dead and decaying organisms. However, there exist some animals, the carrion-eaters, who have converged on the saprophytic, necrophagous mode of life. The corpse-seeking carrion beetles and hyenas are very different-looking types of animals, one an arthropod and the other a mammal, yet they are ecological equivalents (table 4.5). However, the most spectacular convergence on the ecological role of necrophagy is exhibited by the birds, where morphological convergence has paralleled ecological convergence. The Eurasian black vulture looks very similar to the North American turkey vulture, but the accipitrid Eurasian black vulture is a member of the clade of the Falconiformes, closely related to hawks and eagles, whereas the cathartid North American turkey vulture is a member of the clade of the Ciconiiformes, closely related to the storks and flamingos (Lecointre and Le Guyader 2006). Nevertheless, both bird groups have independently evolved featherless, bald heads and necks that allow them to push their heads into rotting carcasses without fouling their body feathers. Both bird groups have also evolved extremely acidic stomachs to protect them from bacterial poisoning in eating carrion. Even more amazing, both bird groups have independently evolved similar behaviors: both Eastern Hemisphere and Western Hemisphere vultures aggregate in social flocks in trees, and both have evolved

Table 4.5
Convergent evolution of necrophagous ecological equivalents in animals

Convergent ecological role: CARRION-EATING ANIMALS ("vulture niche")
Convergent lineages:
1 American carrion beetle (Bilateria: Protostomia: Ecdysozoa: Arthropoda: Hexapoda: Coleoptera: Staphylinoidea: Silphidae; *Necrophila americana*)
2 Eurasian black vulture (Bilateria: Deuterostomia: Chordata: Osteichthyes: Sarcopterygii: Reptiliomorpha: Amniota: Sauropsida: Archosauromorpha: Dinosauria: Saurischia: Theropoda: Maniraptora: Aves: Neognathae: Neoaves: Falconiformes: Accipitridae; *Aegypius monachus*)
3 Turkey vulture (Aves: Neognathae: Neoaves: Ciconiiformes: Cathartidae; *Cathartes aura*)
4 Teratorn (Aves: Neognathae: Neoaves: Ciconiiformes: Teratornithidae; *Argentavis magnificens* †Miocene)
5 Striped hyena (Amniota: Synapsida: Therapsida: Mammalia: Eutheria: Laurasiatheria: Carnivora: Feliformia: Hyaenidae; *Hyaena hyaena*)

Note: The geological age of extinct species is marked with a †. For data sources, see text.

the soaring-in-circular-formation flight pattern in searching for rotting corpses (we shall consider the convergent evolution of behavior in animals in detail in chapter 6).

The North American ciconiiform cathartid vultures also contain the largest flying bird species alive on the Earth today, the California condor, *Gymnogyps californianus*, with a wingspan of three meters. Yet the ciconiiform birds apparently converged on the ecological role of necrophagy at least twice in their evolutionary history (table 4.5): in the Miocene, some 10 million years ago, they produced another family of gigantic vultures, the Teratornithidae, with wingspans of 6 meters, the largest flying birds known in Earth's history (Benton 2005).

Compared to necrophages, convergent animals who survive by eating rotting corpses, the ecological role of eating nectar would seem to be a sweet alternative. Nectarivory is a very highly specialized ecological niche, one that (at first glance) would not seem to be easy to fill. Nectar is very sugar-rich, but it is very difficult to survive on a diet of sugar. The proteins, amino acids, vitamins, and trace minerals necessary for animal growth and reproduction are present in nectar in extremely low levels, if at all (Gartrell 2000). The best-known nectarivores, the hummingbirds, have evolved a suite of adaptations directly related to the difficulty of surviving on nectar: small body sizes, lowered metabolic rates, lowered protein requirements, and substantial changes in both their digestive and renal physiologies (Gartrell 2000). All this is in addition to their evolution of extremely long beaks and tubular tongues in order to feed from flowers, and glossy feathers packed tightly to their bodies, an adaptation believed to prevent their feathers from being soiled by sticky nectar (Gartrell 2000).

Thus, the hummingbird would seem to be an exquisite but unique animal highly adapted to a very restrictive ecological role. But it is not unique—no less than eleven other phylogenetic lineages of animals have independently converged on the ecological role of nectarivory (table 4.6). The hummingbirds, avian family Trochilidae, are found only in the Western Hemisphere, in both North and South America (Cronk and Ojeda 2008). In the tropics of the Eastern Hemisphere, from Africa to Asia and Australia, a separate group of birds have independently evolved to fill the nectarivore ecological role: the sunbirds, avian family Nectariniidae (table 4.6). In Australia, New Zealand, and Pacific islands as far away as Hawaii, a third group of birds have convergently filled the nectarivore niche: the honeyeaters, avian family Meliphagidae.

The trochilids, nectariniids, and meliphagids constitute the three main phylogenetic lineages of birds that have converged on filling the

Table 4.6
Convergent evolution of flying nectarivorous ecological equivalents in animals

Convergent ecological role: FLYING NECTAR FEEDERS ("hummingbird niche")
Convergent lineages:
1 Hummingbird clearwing moth (Bilateria: Protostomia: Ecdysozoa: Arthropoda: Hexapoda: Lepidoptera: Sphingidae; *Hemaris thysbe*)
2 Ruby-throated hummingbird (Bilateria: Deuterostomia: Chordata: Osteichthyes: Sarcopterygii: Reptiliomorpha: Amniota: Sauropsida: Archosauromorpha: Dinosauria: Saurischia: Theropoda: Maniraptora: Aves: Neognathae: Neoaves: Apodiformes: Trochilidae; *Archilochus colubris*)
3 Ruby-cheeked sunbird (Aves: Neognathae: Neoaves: Passeriformes: Nectariniidae; *Chalcoparia singalensis*)
4 Red-throated honeyeater (Aves: Neognathae: Neoaves: Passeriformes: Meliphagidae; *Myzomela eques*)
5 Red-legged American honeycreeper (Aves: Neognathae: Neoaves: Passeriformes: Thraupidae; *Cyanerpes cyaneus*)
6 'I'iwi Hawaiian honeycreeper (Aves: Neognathae: Neoaves: Passeriformes: Drepanididae; *Vestiaria coccinea*)
7 Japanese white-eyes (Aves: Neognathae: Neoaves: Passeriformes: Zosteropidae; *Zosterops japonicus*)
8 South African sugarbird (Aves: Neognathae: Neoaves: Passeriformes: Promeropidae; *Promerops cafer*)
9 Tennessee warbler (Aves: Neognathae: Neoaves: Passeriformes: Parulidae: *Vermiuora peregrina*)
10 Swift parrot (Aves: Neognathae: Neoaves: Psittaciformes: Psittacidae; *Lathamus discolor*)
11 Mexican long-tongued bat (Amniota: Synapsida: Therapsida: Mammalia: Eutheria: Laurasiatheria: Chiroptera: Microchiroptera: Phyllostomidae; *Choeronycteris mexicana*)

Note: For data sources, see text.

nectarivore ecological niche (Cronk and Ojeda 2008). But they are not alone, for no less than five other families of passeriform birds and one family of psittaciform birds have convergently evolved nectarivory (table 4.6). In the Western Hemisphere the red-legged American honeycreepers (Thraupidae) and Tennessee warblers (Parulidae) have convergently evolved nectarivory, and coexist with hummingbirds. In the Eastern Hemisphere, the South African sugarbirds (Promeropidae) and Japanese white-eyes (Zosteropidae) have convergently evolved nectarivory, and coexist with the sunbirds. In Australia the swift parrots (Psittacidae) and in Hawaii the 'I'iwi honeycreepers (Drepanididae) have convergently evolved nectarivory, and coexist with honeyeaters. All of these birds have independently evolved many of the nectarivorous adaptations seen in the hummingbirds: lowered metabolic rates and protein requirements in honeyeaters; glossy feathers and slender, thin beaks in swift parrots; extensible brush tongues independently in both honeyeaters and swift parrots, and so on (Gartrell 2000).

Convergence on this extremely specialized way of surviving is not limited to the birds. In North America, the hummingbird clearwing moth is an insect that has not merely converged on the ecological role of nectarivory, it is also a hummingbird morphological mimic. This moth has converged so closely on the shape and color of a hummingbird that it is often mistaken for one—only when one notices a pair of antennae on the head of the animal does one suddenly realizes that it is not a bird at all. And last, even the mammals have evolved nectarivory (table 4.6). The Mexican long-tongued bat, aptly named as its tongue can be as long as one-quarter of its body length, has convergently filled the nectarivore ecological niche even though it looks nothing like a hummingbird.

I will use one last—and major—example of ecological niche convergence to conclude this section of the chapter, and to set the stage for the next section, in which we will consider the ecological convergence of entire ecosystems. This is the convergent evolution of ground-dwelling, flightless birds. Many different phylogenetic lineages of birds have independently devolved the ability to fly, losing the characteristic set of traits that sets them apart from the other theropod dinosaurs (Roff 1994; Roots 2006). Their wings become reduced and vestigial, with the shortening and weakening of the humerus, ulna, and radius within the wing. Their sterna become flat, losing the prominent keel for anchoring powerful flight muscles. Flight feathers in the wing and tail are shortened until they become useless for flight. Finally, the feathers lose their aerodynamic asymmetry and their smooth aerodynamic contours, which were maintained by barbules and microscopic hooklets, and become loosely constructed and hairlike, giving a shaggy appearance to many flightless birds (Feduccia 1996).

Not all changes in the evolution of flightlessness in birds are trait losses. Because they no longer fly and no longer need to be lightweight, constraints on body size are lifted and flightless birds can become quite large. To adapt to their new life as bipedal ground dwellers, they develop larger legs and powerful thigh muscles, not only to outrun predators or to run down prey but also to defend themselves with powerful kicks of their hind limbs (Roots 2006). As summarized by Feduccia (1996, 289), "whatever the group of birds, when flightlessness occurs, the same general morphological features appear time and time again, convergently, providing a veritable lesson in the rules of evolutionary design."

From an ecological perspective, what is particularly striking is the convergent evolution of multiple mammalian ecological equivalents in flightless birds—these birds do not converge on a single mammalian ecological role, but rather have evolved to fill a suite of niches usually occupied by the ground-dwelling mammals (table 4.7). Ground-dwelling birds have evolved carnivores, insectivores, omnivores, herbivores, and frugivores that convergently play the mammalian ecological roles of cats, moles, badgers, ungulates, and tapirs. Other flightless birds have become swimmers, occupying both marine and freshwater habitats, converging on the ecological role of pinniped mammals (table 4.7). Ecologically considered, the penguin is a bird that has convergently evolved into a seal.

Not long after the destruction of the dinosaurian ecosystem in the end-Cretaceous mass extinction, a major group of ground-dwelling predatory birds, the phorusrhacids or "terror birds," evolved in the Paleocene (Benton 2005). Anatomically, some of these birds are eerily similar to the Mesozoic dromaeosaurs and troodontids, as discussed in chapter 2 (see table 2.7). Therefore, rather than saying that the birds converged on the mammalian role of a ground-dwelling felid carnivore in the Cenozoic, it would be more accurate to argue that the felids convergently filled the ecological niche of the extinct coelurosaurs. Some of the extinct phorusrhacids were quite large; the typical species of *Phorusrhacos* were 1.5 meters tall, but the Eocene *Phorusrhacos longissimus* reached almost 3 meters and the titan phorusrhacid, *Titanus walleri*, was slightly taller than 3 meters (Feduccia 1996; Benton 2005). The red-legged seriema of South America is a surviving cousin of the extinct phorusrhacids, but this ground-dwelling predator is only about 40 centimeters in height (Feduccia 1996).

The phorusrhacids survived from the Paleocene to the Pliocene and, although early members of the group are found in Europe, the family diversified and flourished in isolation in South America. Another group of ground-dwelling avian carnivores, the diatrymids, were more successful in the Northern Hemisphere, but persisted only from the Paleocene through the Eocene. Some of these birds were also quite large: the Eocene *Diatryma gigantea* stood at a height of 2 meters (table 4.7). Their diet may have been more diversified than the felid-like phorusrhacid carnivores in that some may have been primarily scavengers and some more omnivorous, eating some plant material as well as animal prey (Feduccia 1996; Benton 2005).

No other group of birds has produced as many convergent flightless species as the rails (family Rallidae; table 4.7). At least 24 living species of rails have independently evolved flightlessness (Roff 1994; Roots 2006), and the number of extinct species of flightless rails that existed on tropical Pacific islands before the arrival of humans (and the mammalian predators they brought with them) is estimated to have been between 500 to 1,600 (Steadman 2006). However, the New Zealand weka rails are formidable predators that have added human-introduced rats and mice to their own diet, rather than the other way around (Feduccia 1996). I have listed some 13 species of rails whose diet primarily consists of animal prey (although they also consume some plant material) in table 4.7. Rather than felids, these 13 rail species are probably more convergent on omnivorous predators like the canids, and may be considered as ecological equivalents of foxes. Other species of rails clearly are more in the category of foraging omnivores, and a few are herbivores (table 4.7). The South Island takahe of New Zealand is the largest living rail, yet its diet is exclusively vegetarian (Roots 2006).

The living ostriches, rheas, and the extinct palaeotids are plesiomorphic paleognath flightless birds that have converged on filling the ecological niche of foraging omnivores. Their diet consists primarily of plant material (for which they convergently evolved gizzards to process, as discussed for the ostrich in chapter 2), but they consume a considerable number of animals as well, chiefly insects, small mammals, and small reptiles. Captive rheas in zoos are even known to have developed a taste for sparrows (Roots 2006). As such, these ground-dwelling birds have converged on mammalian ecological roles ranging from badgers' to wild

Table 4.7
Convergent evolution of mammalian ecological equivalents in birds

1 Convergent ecological role: FLIGHTLESS, GROUND-DWELLING AVIAN CARNIVORES ("cat niche, fox niche")

Convergent lineages:

1.1 Giant diatrymid (Aves: Neognathae: Neoaves: Gastornithiformes: Diatrymidae; *Diatryma gigantea* †Eocene)

1.2 Titan phorusrhacid (Aves: Neognathae: Neoaves: Gruiformes: Cariamae: Phorusrhacidae; *Titanus walleri* †Pliocene)

1.3 Red-legged seriema (Aves: Neognathae: Neoaves: Gruiformes: Cariamae: Cariamidae; *Cariama cristata*)

1.4 North Island weka rail (Aves: Neognathae: Neoaves: Gruiformes: Rallidae; *Gallirallus australis grayi*)

1.5 Black weka rail (Aves: Neognathae: Neoaves: Gruiformes: Rallidae; *Gallirallus australis australis*)

1.6 Stewart Island weka rail (Aves: Neognathae: Neoaves: Gruiformes: Rallidae; *Gallirallus australis scotti*)

1.7 New Britain rail (Aves: Neognathae: Neoaves: Gruiformes: Rallidae; *Gallirallus insignis*)

1.8 New Caledonia wood rail (Aves: Neognathae: Neoaves: Guiformes: Rallidae; *Gallirallus lafresnayus*)

1.9 Calayan rail (Aves: Neognathae: Neoaves: Guiformes: Rallidae; *Gallirallus calayanensis*)

1.10 Okinawa rail (Aves: Neognathae: Neoaves: Gruiformes: Rallidae; *Rallus okinawa*)

1.11 Henderson Island rail (Aves: Neognathae: Neoaves: Gruiformes: Rallidae; *Porzana atra*)

1.12 Aldabra rail (Aves: Neognathae: Neoaves: Gruiformes: Rallidae; *Canirallus cuvieri aldabranus*)

1.13 Snoring rail (Aves: Neognathae: Neoaves: Gruiformes: Rallidae; *Aramidopsis plateni*)

1.14 Guam rail (Aves: Neognathae: Neoaves: Gruiformes: Rallidae; *Rallus owstoni*)

1.15 Lord Howe Island rail (Aves: Neognathae: Neoaves: Gruiformes: Rallidae; *Tricholimnas sylvestris*)

1.16 Auckland Island rail (Aves: Neognathae: Neoaves: Gruiformes: Rallidae; *Rallus pectoralis muelleri*)

2 Convergent ecological role: FLIGHTLESS, GROUND-DWELLING AVIAN INSECTIVORES ("mole niche")

Convergent lineages:

2.1 Kiwi (Aves: Paleognathae: Struthioniformes 3: Apterygidae; *Apteryx haasti*)

2.2 Kagu (Aves: Neognathae: Neoaves: Gruiformes: Rhynochetidae; *Rhynchetos jubatus*)

3 Convergent ecological role: FLIGHTLESS, GROUND-DWELLING AVIAN FORAGING OMNIVORES ("badger niche, wild pig niche")

Convergent lineages:

3.1 Palaeotis (Aves: Paleognathae: Lithornithiformes: Palaeotidae; *Palaeotis weigelti* †Eocene)

3.2 Ostrich (Aves: Paleognathae: Struthioniformes 1: Struthionidae; *Struthio camelus*)

3.3 Rhea (Aves: Paleognathae: Struthioniformes 2: Rheidae; *Rhea americana*)

3.4 New Guinea flightless rail (Aves: Neognathae: Neoaves: Gruiformes: Rallidae; *Megacrex inepta*)

3.5 Inaccessible Island rail (Aves: Neognathae: Neoaves: Gruiformes: Rallidae; *Atlantisia rogersi*)

Table 4.7
(continued)

3.6 Woodford's rail (Aves: Neognathae: Neoaves: Gruiformes: Rallidae; *Nesoclopeus woodfordi*)
3.7 Samoan moorhen (Aves: Neognathae: Neoaves: Gruiformes: Rallidae; *Gallinula pacificus*)
3.8 San Cristobal moorhen (Aves: Neognathae: Neoaves: Gruiformes: Rallidae; *Gallinula silvestris*)
3.9 Gough Island moorhen (Aves: Neognathae: Neoaves: Gruiformes: Rallidae; *Gallinula nesiotis comeri*)
3.10 Giant coot (Aves: Neognathae: Neoaves: Gruiformes: Rallidae; *Fulica gigantea*)
3.11 Zapata rail (Aves: Neognathae: Neoaves: Gruiformes: Rallidae; *Cyanolimnus cerverai*)

4 Convergent ecological role: FLIGHTLESS, GROUND-DWELLING AVIAN HERBIVORES ("ungulate niche")
Convergent lineages:
4.1 Emu (Aves: Paleognathae: Struthioniformes 3: Dromaiidae; *Dromaius novaehollandiae*)
4.2 Moa-nalo (Aves: Neognathae: Galloanserae: Anseriformes: Anatidae; *Chelychelynechen quassus* †Holocene)
4.3 Kakapo parrot (Aves: Neognathae: Neoaves: Psittaciformes: Psittacidae; *Strigops habroptilus*)
4.4 Broad-billed parrot (Aves: Neognathae: Neoaves: Psittaciformes: Psittacidae; *Lophopssittacus mauritianus* †Holocene)
4.5 South Island takahe (Aves: Neognathae: Neoaves: Gruiformes: Rallidae; *Notornis mantelli hochsteteri*)
4.6 Tasmanian native hen (Aves: Neognathae: Neoaves: Gruiformes: Rallidae; *Tribonyx mortieri*)
4.7 Invisible rail (Aves: Neognathae: Neoaves: Gruiformes: Rallidae; *Habroptila wallaci*)

5 Convergent ecological role: FLIGHTLESS, GROUND-DWELLING AVIAN FRUGIVORES ("tapir niche")
Convergent lineages:
5.1 Eremopezus (Aves: Paleognathae: Aepyornithiformes: Eremopezidae; *Eremopezus eocaenus* †Oligocene)
5.2 Elephant bird (Aves: Paleognathae: Aepyornithiformes: Aepyornithidae; *Mullerornis agilis* †Holocene)
5.3 Cassowary (Aves: Paleognathae: Struthioniformes 3: Casuariidae; *Casuarius casuarius*)
5.4 Dodo (Aves: Neognathae: Neoaves: Columbiformes: Raphidae; *Raphus cucullatus* †Holocene)
5.5 Solitaire (Aves: Neognathae: Neoaves: Columbiformes: Raphidae; *Pezophaps solitaria* †Holocene)

6 Convergent ecological role: FLIGHTLESS, SWIMMING AVIAN PISCIVORES ("pinniped niche")
Convergent lineages:
6.1 Patagopteryx (Aves: Ornithothoraces: Patagopterygiformes: Patagopterygidae; *Patagopteryx deferrariisi* †Cretaceous)
6.2 Hesperornis (Aves: Ornithurae: Hesperornithiformes: Hesperornithidae; *Hesperornis regalis* †Cretaceous)
6.3 Steamer duck (Aves: Neognathae: Galloanserae: Anseriformes: Anatidae; *Tachyeres brachypterus*)

Table 4.7
(continued)

6.4 Auckland Islands teal (Aves: Neognathae: Galloanserae: Anseriformes: Anatidae; *Anas aucklandica*)
6.5 Law's diving goose (Aves: Neognathae: Galloanserae: Anseriformes: Anatidae; *Chendytes lawi* †Holocene)
6.6 King penguin (Aves: Neognathae: Neoaves: Sphenisciformes: Spheniscidae; *Aptenodytes patagonica*)
6.7 Flightless cormorant (Aves: Neognathae: Neoaves: Pelecaniformes: Phalacrocoracidae; *Nannopterum harrisi*)
6.8 Murres (Aves: Neognathae: Neoaves: Charadriiformes: Alcidae; *Uria aalge*)
6.9 Titicaca grebe (Aves: Neognathae: Neoaves: Podicipediformes: Podicipedidae; *Rollandia microptera*)
6.10 Junín grebe (Aves: Neognathae: Neoaves: Podicipediformes: Podicipedidae; *Podiceps taczanowskii*)
6.11 Atitlán grebe (Aves: Neognathae: Neoaves: Podicipediformes: Podicipedidae; *Podilymbus gigas* †Holocene)

Note: The geological age of extinct species is marked with a †. For data sources, see text.

pigs', depending upon the proportions of plant and animal content in their diet.

It was long thought that the living "ratite" ground-dwelling birds—ostriches, rheas, emus, cassowaries, and kiwis—all descended from a single common ancestor. Today they exist geographically scattered across the Earth: ostriches in Africa, rheas in South America, emus and cassowaries in Australia, and kiwis in New Zealand. The classic explanation for this puzzling geographic dispersal was that the flightless ancestors of these birds evolved on the Mesozoic supercontinent of Gondwana, which consisted of the present-day landmasses of South America, Africa, Madagascar, India, Australia, the New Zealand islands, and Antarctica all combined. It is known that Gondwana began to break up in the Cretaceous, and thus it was hypothesized that the present-day ratites evolved separately, each on its own fragment of Gondwana, a process of vicariant evolution driven by plate-tectonic continental drift.

However, the monophyletic origin of the ratites came into question when molecular phylogenies for these birds began to be constructed. Recent molecular analyses reveal that the ratites are polyphyletic, and that the close morphological similarities between these paleognath birds are not synapomorphies, but rather are due to convergence (Harshman et al. 2008). In retrospect, this is not surprising, given the incredible number of independent morphological convergences seen in other flightless birds (table 4.7). Molecular evidence now indicates that the ostriches (Struthionidae) and the rheas (Rheidae) independently evolved flightlessness. It also indicates that the living emus (Dromaiidae), cassowaries (Casuariidae), and kiwis (Apterygidae), as well as the extinct moas (Dinornithidae), are a monophyletic group that evolved flightlessness independently of both the ostriches and the rheas (Harshman et al. 2008). Thus, in table 4.7, I have divided the Struthioniformes—the "ostrich-like" birds—into the Struthioniformes 1, 2, and 3 to indicate that they are polyphyletic and that they represent three separate convergent lineages.

The end result is not only two independent morphological convergences on flightlessness in the ostrichs and the rheas, but also two independent ecological convergences on foraging omnivory (table 4.7). In contrast, the Australian emus and their New Zealand cousins the moas (*Dinornis giganteus*, now extinct due to human activities) are considered to be ecologically equivalent to browsing ungulate mammals (Daugherty et al., 1993). In addition to some rail species, the ecological role of browsing herbivory is convergently filled by two species of flightless parrots and one species of a flightless browsing "goose," the moa-nalo. These ground-dwelling birds convergently filled the ecological niche of ungulate mammals in widely separated regions of the Earth: the kakapo parrot still survives in New Zealand, whereas the recently extinct broad-billed parrot was a native of Mauritius, and the moa-nalo lived in Hawaii (Roots 2006).

Rather than these ground-dwelling birds converging on the ecological role of browsing herbivorous mammals, it may well be that the Cenozoic browsing herbivorous mammals were themselves convergent on the ecological role of the Mesozoic browsing herbivorous dinosaurs. Barrett (2005) argues that the ornithomimosaurs were herbivores, and points out that many traits now associated with herbivory in living birds, such as gastric mills and large intestines, were also a feature of these extinct theropods. Thus, rather than being convergent on mammals, the living herbivorous ground-dwelling birds may have reevolved a more ancient ecological role, that of the ornithomimosaur niche.

Even though the emu-cassowary-kiwi clade is now proved to be monophyletic (Harshman et al. 2008), the members of the clade are ecologically divergent (table 4.7). The peculiar New Zealand kiwis are not herbivores, but rather are ecologically convergent on the mammalian mole ecological role, specialized in hunting down and eating animals that dwell in the soil. While they do not tunnel through the soil like moles, they do live in burrows and hunt for insects at night in the dark, using their highly developed sense of smell and long beaks to locate earthworms and other burrowing prey underground, as moles do subterraneally (Daugherty et al. 1993). Far to the north of New Zealand, in New Caledonia, another group of ground-dwelling birds have convergently filled the mole ecological niche. These are the kagu, which specialize in locating underground prey in a manner similar to the kiwis (Roots 2006).

The Australian cassowaries are also not herbivores, but frugivores (table 4.7). They are the last surviving large ground-dwelling birds that have convergently filled the mammalian frugivore niche. The large mammalian frugivores themselves are in danger of extinction—the largest remaining South American frugivore is the 300-kilogram Brazilian tapir, compared to the extinct 7,580-kilogram gomphothere and the extinct 6,300-kilogram giant ground sloth (Hansen and Galetti 2009). The recently extinct elephant bird of Madagascar was a 450-kilogram frugivore (Hansen and Galetti 2009) that may have been related to the more ancient giant flightless eremopezus, a bird of Egypt. Eremopezus is argued not to have been related to the African ostrich (Rasmussen et al. 2001), and molecular evidence indicates that the elephant bird also was "clearly not the result of a recent divergence from the ostrich, or any other ratite lineage" (Cooper et al. 2001). Yet another group of birds, the columbiform pigeons, independently filled the ground-dwelling frugivore ecological niche on isolated islands further to the east from Madagascar and Africa. The recently extinct dodo was a large, ground-dwelling frugivore on Mauritius Island, and the recently extinct solitaire independently filled the ground-dwelling frugivore niche on Rodriguez Island (Feduccia 1996). In all cases, the extinction of these large frugivores—whether mammal or flightless bird—has led to serious disruption of seed-dispersal interactions in the affected ecosystems (Hansen and Galetti 2009).

Ground-dwelling birds are not the only birds to have converged on mammalian ecological roles. Multiple lineages of birds have become flightless swimming and diving piscivores, converging on the pinniped mammal niche (table 4.7). Just as the seals and sea lions have devolved their walking legs into flippers for swimming (as discussed in chapter 2; see table 2.1), so have the penguins devolved their wings. Unlike the ground-dwelling flightless birds, the penguins have not lost the strong keel on their sternum, nor have they lost their powerful pectoral muscles. Instead of using these traits for flight, however, they are used for swimming in water. As water is much denser than air, swimming through this dense medium is even more energetic than flying (Roots 2006). One has only to observe the graceful, seemingly effortless, swimming of seals in any zoo large enough to have underwater viewing stations in the seal exhibit area, and to repeat those observations in the penguin exhibit area, to appreciate the power of convergent evolution (I recommend the Schönbrunn zoo in Vienna, my favorite, where the seal and penguin exhibition areas are side by side).

Other swimming birds did not devolve their wings into flippers, and thus have convergently lost their keeled sternum and powerful pectoral muscles, as the ground-dwelling birds have (Roff 1994; Roots 2006). Instead of using their wings as flippers, they have evolved powerful leg muscles and large, webbed feet. They hold their vestigial wings tight against their bodies and smoothly propel themselves through the water with their feet. The large flightless cormorant of the Galapagos, the large steamer ducks of South America, and the murres of the North Atlantic swim in this fashion (table 4.7). Two other groups of the duck family have independently converged on the swimming piscivore niche: the Auckland Island teal and the recently extinct Law's diving goose of western North America.

The convergent evolution of flightless, swimming piscivorous birds is not confined to the marine realm. Three species of grebes are known to have independently filled this ecological niche in freshwater lakes in South America, two of which survive today (table 4.7). How many other instances there may be of the convergent evolution of swimming birds in freshwater habitats is unknown, as the fossil record of these environments is very poor.

However, the fossil record does reveal that the convergent evolution of flightless, swimming birds is an ancient phenomenon that predates the destruction of dinosaurian ecosystems in the Cretaceous and the subsequent evolution of the mammalian ecosystems of the Cenozoic (Benton 2005). Two independent lineages of birds are known to have devolved their wings and converged on the swimming piscivore niche in the Cretaceous (table 4.7), long before the evolution of pinniped mammals. Thus, the pinniped mammals may themselves be viewed as having converged on a more ancient ecological role, the hesperornithid niche.

One Ecosystem Play, Many Convergent Casts of Actors

In the previous section of the chapter, we have seen that the number of ecological roles or niches available for Earth organisms is demonstrably limited, in that species from many different phylogenetic lineages have been constrained in their evolution to filling the same ecological niche, even if that ecological role is extremely specialized. Ecological niche convergence is the rule, rather than the exception, in evolution.

In this section we shall consider the ecological level of communities and ecosystems, which are assemblages of multiple ecological roles. Comparisons of ecosystems from different regions of the Earth, both in space and in time, reveal a remarkable similarity in the assemblages of ecological roles within those ecosystems. One might think of an ecosystem as being analogous to a theatrical performance: in any given region of the Earth, an ecosystem play is being performed with a script specifying the interaction of the various ecological roles in that play. What is unexpected is one's observation that the ecosystem play—the script of ecological roles—seems to be very similar in many parts of the Earth, even though the casts of animal and plant actors in those ecological roles may be very different.

The flightless, ground-dwelling birds of New Zealand reveal a convergent phenomenon that is larger than that of the "one ecological role, many convergent players" thesis of the previous section of this chapter. The kiwis have convergently filled the mammalian mole niche, and three other bird lineages have convergently filled mammalian ungulate-herbivore niches: the recently extinct moas filled the antelope niche of a browsing ungulate, and the kakapo parrots and takahe rails have filled the sheep niche of a grazing ungulate (Daugherty et al. 1993). The suite of ecological roles filled by mammals elsewhere in the world are thus filled by birds in New Zealand.

New Zealand split off early in the fragmentation of the Gondwanan supercontinent in the Mesozoic, and was geographically isolated from subsequent mammalian evolution in the Cenozoic following the destruction of the dinosaur-dominated ecosystems of the Earth in the end-Cretaceous mass extinction. There are no native ground-dwelling mammals on New Zealand, and there are only two species of bats, whose ancestors flew there across the waters (Roots 2006). In the absence of mammalian moles, antelopes, and sheep, the birds convergently evolved the ecological roles of these animals. It is as if these ecological roles exist in the absence of either mammals or birds, and that they would be convergently found by evolution within some other phylogenetic lineage if they had not been discovered by birds in New Zealand and mammals in other regions of the world.

The peculiar fauna of New Zealand therefore suggest that ecological convergence exists at the level of ecosystems and communities as well as at the level of individual ecological roles. In the absence of mammalian predators, other phylogenetic lineages have filled predatory niches in New Zealand—not only the Weka rail, but also ancient reptiles, giant molluscs, and giant myriapod arthropods. These nonmammal predators have convergently evolved many mammalian traits—such as long life spans, low reproductive rates, and large body sizes—that are not characteristic of normal molluscs and arthropods (Daugherty et al. 1993). The predatory tuatara, *Sphenodon punctatus*, is a reptilian "living fossil" (Benton 2005, 237–239) that is not a lizard (although it resembles a lizard) but rather is the sole remaining survivor of a lepidosaurian sister group to the Squamata, the Sphenodontia (see appendix). The tuatara has a life span of over 70 years, and reproduces only every four years after taking 13 years to reach sexual maturity (Daugherty et al. 1993), a life history more similar to that of a mammal than of a reptile. Its ancestors survived the end-Cretaceous mass extinction, while the New Zealand non-avian dinosaurs did not. The giant predatory snail *Paryphanta busbyi*, which has a shell 115 millimeters in diameter, can live up to 40 years and takes 15 years to reach sexual maturity (Daugherty et al. 1993). The giant predatory centipede *Cormocephalus rubriceps* can grow to the length of 250 millimeters, and its females brood only about 20 eggs (Daugherty et al. 1993). In the omnivore ecological role, the cricket-like giant weta insect, *Deinacrida heteracantha*, has converged on the rat niche, reaching the comparable body length of 150 millimeters and a weight of 70 grams. Like rats, they forage at night and congregate in diurnal shelters (Daugherty et al. 1993).

The New Zealand islands are very unusual, but not unique. The island of Madagascar has not been separated from the African mainland as long as New Zealand has been isolated by water, but ecosystem convergences have occurred there as well, on a smaller scale. There are no laurasiatherian hedgehogs and shrews on Madagascar. Instead, afrosoricid mammals have convergently evolved these same ecological roles in the greater hedgehog tenrec, *Setifer setosus*, and in the lesser long-tailed shrew tenrec, *Microgale longicaudata*. On the African mainland itself, the laurasiatherian otter niche and mole niche have been convergently filled by the afrosoricid giant otter "shrew," *Potamogale velox*, and the cape golden "mole," *Chrysochloris asiatica* (Lecointre and Le Guyader 2006). Thus, the same suite of ecological niches—hedgehog, shrew, otter, mole—exist in African ecosystems, but they are not filled by hedgehogs, shrews, otters, or moles.

The classic, most frequently cited example of ecosystem convergence is provided by the marsupial fauna of Australia and Tasmania. The isolated marsupial mammals of these regions have independently evolved

Table 4.8
Convergent evolution of ecological-analog compositions of marsupial-dominated ecosystems in isolated Australian and Tasmanian regions and placental-dominated ecosystems in the rest of the world

---

1 Convergent ecological analog: LARGE AMBUSH PREDATOR ("great cat niche")
Placental-mammal ecosystems: Lion (Eutheria: Felidae; *Panthera leo*)
Australian ecosystem: Marsupial lion (Marsupialia: Thylaconidae; *Thylacoleo carnifex* †Pleistocene)

2 Convergent ecological analog: SMALL AMBUSH PREDATOR ("small cat niche")
Placental-mammal ecosystems: Wild cat (Eutheria: Felidae; *Felis sylvestris*)
Australian ecosystem: Marsupial quoll "cat" (Marsupialia: Dasyuridae; *Dasyurus viverrinus*)

3 Convergent ecological analog: PURSUIT PREDATOR ("wolf niche")
Placental-mammal ecosystems: Wolf (Eutheria: Canidae; *Canis lupus*)
Australian ecosystem: Tasmanian wolf (Marsupialia: Thylacinidae; *Thylacinus cynocephalus* †Holocene)

4 Convergent ecological analog: NOCTURNAL FORAGING PREDATOR ("wolverine niche")
Placental-mammal ecosystems: Wolverine (Eutheria: Mustelidae; *Gulo gulo*)
Australian ecosystem: Tasmanian devil (Marsupialia: Dasyuridae; *Sarcophilus harrisi*)

5 Convergent ecological analog: ANT EATER ("anteater niche")
Placental-mammal ecosystems: Anteater (Eutheria: Myrmecophagidae; *Myrmecophaga tridactyla*)
Australian ecosystem: Numbat (Marsupialia: Myrmecobiidae; *Myrmecobius fasciatus*)

6 Convergent ecological analog: FOSSORIAL INSECTIVORE ("mole niche")
Placental-mammal ecosystems: Mole (Eutheria: Talpidae; *Talpa europea*)
Australian ecosystem: Marsupial mole (Marsupialia: Notoryctidae; *Notoryctes typhlops*)

7 Convergent ecological analog: SMALL COMMUNAL-NESTING OMNIVORE ("rat niche")
Placental-mammal ecosystems: Black rat (Eutheria: Muridae; *Rattus rattus*)
Australian ecosystem: Brown marsupial rat (Marsupialia: Dasyuridae; *Antechinus stuartii*)

8 Convergent ecological analog: TINY COMMUNAL-NESTING OMNIVORE ("mouse niche")
Placental-mammal ecosystems: Mouse (Eutheria: Muridae; *Mus musculus*)
Australian ecosystem: Fat-tailed marsupial mouse (Marsupialia: Dasyuridae; *Sminthopsis crassicandata*)

9 Convergent ecological analog: GLIDING ARBOREAL OMNIVORE ("flying squirrel niche")
Placental-mammal ecosystems: Flying squirrel (Eutheria: Sciuridae; *Glaucomys volans*)
Australian ecosystem: Marsupial flying opossum (Marsupialia: Petauridae; *Petaurus australis*)

10 Convergent ecological analog: SEMI-FOSSORIAL HERBIVORE ("woodchuck niche")
Placental-mammal ecosystems: Woodchuck (Eutheria: Sciuridae; *Marmota monax*)
Australian ecosystem: Wombat (Marsupialia: Phascolomidae; *Phascolomys ursinus*)

11 Convergent ecological analog: MID-SIZED FAST-RUNNING BROWSING HERBIVORE ("deer niche")
Placental-mammal ecosystems: Red deer (Eutheria: Cervidae; *Cervus elaphus*)
Australian ecosystem: Red kangaroo (Marsupialia: Macropodidae; *Macropus rufus*)

Table 4.8
(continued)

12 Convergent ecological analog: LARGE HERDING MIXED-FEEDING HERBIVORE ("bison niche")
Placental-mammal ecosystems: Bison (Eutheria: Bovidae; *Bison bison*)
Australian ecosystem: Diprotodon (Marsupialia: Diprotodontidae; *Diprotodon optatum* †Pleistocene)

13 Convergent ecological analog: SMALL MIXED-FEEDING HERBIVORE ("rabbit niche")
Placental-mammal ecosystems: Rabbit (Euarchontoglires: Lagomorpha: Leporidae; *Oryctolagus cuniculus*)
Australian ecosystem: Wallaby (Marsupialia: Macropodidiae; *Macropus agilis*)

Note: The geological age of extinct species is marked with a †. For data sources, see text.

an ecological analog structure, a series of 13 ecological roles, that mirror those found in placental mammalian ecosystems in other regions of the world (table 4.8). The isolated marsupials convergently evolved both of the two principal placental-mammal predatory strategies: the stalk-and-ambush strategy of the placental felids, in which the stalked prey animal is not aware that it is in danger until the split second in which the felid pounces upon it, and the pursuit-and-charge strategy of the placental canids, in which the prey animal is well aware of its pack pursuers, who run it to exhaustion and then charge upon it en masse. Both the large marsupial lion and the Tasmanian wolf are extinct (the latter only recently), but the small marsupial quoll "cat" still survives (Turner 1997; Lecointre and Le Guyader 2006). In addition, the marsupials have convergently evolved the nocturnal foraging predator niche, with the voracious Tasmanian devil filling the role of the placental wolverine.

Two insectivore niches have been convergently evolved by the Australian marsupials, with the numbat filling the anteater niche and the marsupial mole filling the mole niche. These two highly specialized ecological roles have independently appeared a surprising number of times in ecosystems in widely separated regions of the Earth (see tables 2.4, 4.4, and 4.7).

Three omnivore ecological analogs have been convergently evolved by the isolated Australian fauna, with the brown marsupial rat and the fat-tailed marsupial mouse respectively filling the placental rat niche and mouse niche (Lecointre and Le Guyader 2006), and the marsupial flying opossum has independently evolved the ecological role of the placental flying squirrel (table 4.8).

Last, Australian marsupials have independently evolved four herbivore ecological roles that are analogs of those found in placental-mammal ecosystems: the wombat not only fills the ecological niche of the woodchuck, it even resembles a woodchuck. The red kangaroo does not look like a deer, yet "it lives roughly the same way" (Benton 2005, 313). In addition, both the deer and the kangaroo have independently evolved the ruminant-stomach system of digestion (see table 2.10). The wallaby is the native Australian player of the rabbit ecological role, a role that is now increasingly being filled by the invasive European rabbit (and to the detriment of the wallaby). The Australian diprotodons are now gone, dying out in the extinction of many large land animals in the Pleistocene, but these bison-sized marsupials once formed great herds (Benton 2005) of mixed-feeding herbivores, eating the twigs and leaves of shrubs as well as grass, like the once-great herds of bison in placental ecosystems (table 4.8).

As in the previous discussion of the peculiar fauna of New Zealand, it is as if these 13 ecological roles exist in the absence of either placental or marsupial mammals and, since the placentals were not present in Australia, the marsupials independently discovered them in their own separate evolution. This impression can be tested by considering the pattern of animal evolution on yet another isolated, continental-sized island: South America. South America is not an island today, but for most of the 65 million years of the Cenozoic it was surrounded by water, cut off from both Africa to the east (from which it split apart in the fragmentation of the supercontinent Gondwana in the Mesozoic) and from North America to the north (Benton 2005).

During their period of isolation, the mammals of South America independently evolved a series of at least 12 ecological roles that are analogs of those found in terrestrial ecosystems in other regions of the world (table 4.9). Some of these convergences are spectacular: not only did the ambush-predator niche independently evolve in North and South America, as it did in Australia, but this ecological role was played by highly specialized saber-toothed placental cats in North America and almost identical-appearing saber-toothed marsupial "cats" in South America (as discussed in chapter 2, table 2.7). And not only did the fast-running, browsing herbivore niche independently evolve in North and South America, but in both regions the animals also convergently reduced the number of digits on their feet down to one large hoofed toe—*Pliohippus pernix* in the north and *Thoatherium minusculum* in the south—and at the same time, in the Miocene. The two animals never met one another, yet they existed at the same time and were remarkably similar in appearance.

Many of the South American ecological convergences were made by meridiungulate mammals, a group of mammals that are now extinct (see appendix). Completely in isolation, the meridiungulates convergently filled the ecological niches of wild pigs, woodchucks, mastodons, horses, rabbits, camels, hippopotamuses, beavers, and tapirs with their own analogs (Prothero and Schoch 2002).

It is interesting that 13 ecological niches in Australia and 12 ecological niches in South America, each evolving in isolation, converged with niches in other regions of the world. However, by comparing the lists of ecological analogs given in tables 4.8 and 4.9, we can see that the two sets of niches listed are not all the same. In fact, only 5 of the 12 (or 13) in the two sets are the same: the great cat, anteater, woodchuck, deer, and rabbit niches. The South American mammals evolved a series of ecological roles that are analogs of those independently evolved in North America, but that appear not to have been discovered by the Australian marsupials. These include the elephant, camel, hippopotamus, tapir, beaver, and wild pig niches. Amphibious herbivores and large frugivores, in particular, appear to be missing in Australian ecosystem evolution. These absent ecological roles may well be due to the aridity of Australian ecosystems and its effect on the Australian flora. Yet if that is the reason, why did the Australian marsupials not evolve the camel niche, an ecological role that independently appeared in both South and North American ecosystems?

In the reverse comparison, several predator niches appear to be missing in South American ecosystem evolution. Why did the South American marsupials not evolve the wild cat, wolf, and wolverine niches? Why were there no South American marsupial mole, rat, or flying squirrel analogs? Some of these absences may be artificial, and simply due to the vagaries of fossil preservation. All of the Australian animals listed in table 4.8 either are alive today or have only relatively recently become extinct (that is, going back in time at most about two million years, to the beginning of the Pleistocene). In contrast, the patterns of South American ecosystem evolution examined in table 4.9 extend back in time to the Eocene, some 55 million years into the past (Prothero and Schoch 2002). Thus, some of these missing niches may in fact have been filled in the past, but their fossil species either have not been found yet or were never preserved in the fossil record in the first place.

Table 4.9
Convergent evolution of ecological-analog compositions of terrestrial ecosystems in isolated South American regions and terrestrial ecosystems in the rest of the world in the Cenozoic

1 Convergent ecological analog: LARGE SABER-TOOTHED AMBUSH PREDATOR ("sabertooth tiger niche")
North American ecosystem: Sabertooth tiger (Eutheria: Felidae; *Smilodon fatalis* †Pleistocene)
South American ecosystem: Marsupial sabertooth tiger (Marsupialia: Thylacosmilidae; *Thylacosmilus atrox* †Pliocene)

2 Convergent ecological analog: ANT EATER ("anteater niche")
Asian ecosystem: Giant pangolin (Laurasiatheria: Manidae; *Manis gigantea*)
South American ecosystem: Giant anteater (Xenarthra: Myrmecophagidae; *Myrmecophaga tridactyla*)

3 Convergent ecological analog: FORAGING OMNIVORE ("wild pig niche")
North American ecosystem: Musk-hog peccary (Cetartiodactyla: Suina: Tayassuidae; *Pecari tajacu*)
South American ecosystem: Notoungulate "warthog" (Meridiungulata: Notoungulata: Toxodonta: Isotemnidae; *Thomashuxleya rostrata* †Eocene)

4 Convergent ecological analog: SUBTERRANEAN HERBIVORE ("gopher niche")
North American ecosystem: Plains pocket gopher (Rodentia: Sciuromorpha: Geomyidae; *Geomys bursarius*)
South American ecosystem: Chilean coruro (Rodentia: Hystricomorpha: Octodontidae; *Spalacopus cyanus*)

5 Convergent ecological analog: SEMIFOSSORIAL HERBIVORE ("woodchuck niche")
North American ecosystem: Woodchuck (Rodentia: Sciuridae; *Marmota monax*)
South American ecosystem: Archaeohyrax (Meridiungulata: Notoungulata: Typotheria: Archaeohyracidae; *Archaeohyrax concentricus* †Eocene)

6 Convergent ecological analog: LARGE HERDING AND BROWSING HERBIVORE ("elephant niche")
North American ecosystem: Mastodon (Afrotheria: Proboscidea: Mammutidae; *Mammut americanum* †Pleistocene)
South American ecosystem: Pyrotherian "mastodon" (Meridiungulata: Pyrotheria: Pyrotheriidae; *Pyrotherium romeri* †Oligocene)

7 Convergent ecological analog: ONE-TOED, FAST-RUNNING BROWSING HERBIVORE ("horse niche")
North American ecosystem: Western horse (Perissodactyla: Equidae; *Pliohippus pernix* †Miocene)
South American ecosystem: Litoptern "horse" (Meridiungulata: Litopterna: Proterotheriidae; *Thoatherium minusculum* †Miocene)

8 Convergent ecological analog: SMALL MIXED-FEEDING HERBIVORE ("rabbit niche")
North American ecosystem: Eastern cottontail rabbit (Euarchontoglires: Lagomorpha: Leporidae; *Sylvilagus floridanus*)
South American ecosystem: Notoungulate "rabbit" (Meridiungulata: Notoungulata: Typotheria: Hegetotheriidae; *Propachyrucos ameghinorum* †Oligocene)

9 Convergent ecological analog: SEMI-ARID-ADAPTED FORAGING HERBIVORE ("camel niche")
North American ecosystem: Western camel (Cetartiodactyla: Camelidae; *Camelops hesternus* †Pleistocene)
South American ecosystem: Litoptern "camel" (Meridiungulata: Litopterna: Macraucheniidae; *Macrauchenia patagonica* †Pleistocene)

Table 4.9
(continued)

10 Convergent ecological analog: LARGE HERDING AMPHIBIOUS HERBIVORE ("hippopotamus niche")
North American ecosystem: Anthracotherium "hippopotamus" (Cetartiodactyla: Anthracotheriidae; *Anthracotherium magnus* †Oligocene)
South American ecosystem: Notoungulate "hippopotamus" (Meridiungulata: Notoungulata: Toxodonta: Toxodontidae; *Toxodon platensis* †Pleistocene)

11 Convergent ecological analog: SMALL AMPHIBIOUS HERBIVORE ("beaver niche")
North American ecosystem: Canadian beaver (Euarchontoglires: Rodentia: Castoridae; *Castor canadensis*)
South American ecosystem: Notoungulate "beaver" (Meridiungulata: Notoungulata: Typotheria: Mesotheriidae; *Mesotherium cristatum* †Pleistocene)

12 Convergent ecological analog: LARGE FRUGIVORE ("tapir niche")
North American ecosystem: Tapir (Perisodactyla: Tapiridae; *Tapirus californicus* †Pleistocene)
South American ecosystem: Astropotherian "tapir" (Meridiungulata: Astrapotheria: Astrapotheriidae; *Astrapotherium magnum* †Miocene)

Note: The geological age of extinct species is marked with a †. For data sources, see text.

Isochronous and Heterochronous Ecosystem Convergence

The pattern of ecosystem convergence seen in the Australian example, table 4.8, would support the hypothetical analogy of a convergent terrestrial ecological play being in existence on the Earth at the present time, with its ecological roles being filled by a cast of marsupial mammals in Australia and by a cast of placental mammals in other regions of the Earth. Such a pattern of ecosystem convergence can be technically termed *isochronous*, that is, at the same point in time. In contrast, the pattern of ecosystem convergence seen in the South American example, table 4.9, suggests that the scripted roles in the ecological play have remained the same in time, that is, these roles have not significantly changed in the past 55 million years. The wild pig niche existed in the Eocene, and was filled by a notoungulate "warthog" in South America, and the one-toed litoptern "horse" ran in the grasslands of the Miocene, much as modern horses do today (Prothero and Schoch 2002). Such a pattern of ecosystem convergence is *heterochronous*, that is, occurring in different periods of time.

Has the script, the cast of ecological roles, in the terrestrial ecosystem play remained unchanged for the past 55 million years? Or is this apparent heterochronous ecosystem convergence simply due to the fact that the roles are all being played by the same types of animals during this span of the Cenozoic? The extinct South American meridiungulates were laurasiatherian mammals, after all, and thus related to the modern perisodactyl and cetartiodactyl mammals that are still playing these ecological roles today (see appendix).

One way to examine the question of the reality of heterochronous ecosystem convergence is to go even further back in time, to a time before terrestrial ecosystems were dominated by mammals of any type at all, to a time when the dinosaurs reigned instead. The Mesozoic and Cenozoic eras of time are separated by the second-greatest ecological disruption of terrestrial ecosystems in Earth's history (McGhee et al. 2004). Surely the dinosaur-dominated ecosystems of the Mesozoic must have been fundamentally different from the mammal-dominated ecosystems of the Cenozoic—the script of ecological roles in the Mesozoic must have described a completely different ecological play.

Not so. The dinosaurs of the Mesozoic independently evolved a series of at least 12 ecological roles that are analogs of those found in Cenozoic terrestrial ecosystems (table 4.10). Since the Mesozoic world was so different from the Cenozoic, I have expanded the ecological comparisons between these two eras of time to include fluvial, marine, oceanic, and aerial ecological roles, and not just the roles of ground-dwelling terrestrial animals. All in all, a series of at least 19 ecological roles existed in the Mesozoic world that are analogs of those played by Cenozoic animals (table 4.10).

Dinosaurs independently evolved the ecological roles of lions, wild cats, wolves, wolverines, anteaters, elephants, deer, bison, rhinoceroses, glyptodonts, goats, and ground sloths. However, since the dinosaurs filled these ecological niches long before the mammals, it is more correct to say that the Cenozoic mammals have convergently refilled the ecological niches of allosaurs, coelophysises, velociraptors, troodonts, alvarezsaurids, sauropods, hypsilophodonts, hadrosaurs, triceratopses, ankylosaurs, pachycephalosaurs, and therizinosaurs. Although some of these two groups of animals resemble each other morphologically (triceratopses and rhinoceroses, ankylosaurs and glyptodonts, therizinosaurs and ground sloths), the others appear radically different—the mammalian predators are all quadrupeds, whereas the dinosaurian predators walked on their hind legs only. Yet they all were ecological analogs of each other, making their living in roughly the same way. The goats, for example, have a ruminant stomach system to process plant material, while the pachycephalosaurs had a gizzard-like gastric mill to accomplish the same purpose. The similarity to ungulate mammals in the herbivorous ecological roles of many of the ornithopod dinosaurs, such as the hypsilophodonts and hadrosaurs, has prompted Carrano et al. (1999, 256) to conclude: "Although late Mesozoic and late Cenozoic terrestrial ecosystems were profoundly different in terms of both animal and plant taxa, there may be universal constraints on the ecological roles played by large herbivores, resulting in convergence in morphology (and, by implication, behavioral ecology) between groups as taxonomically distinct as dinosaurs and mammals." Thus, Carrano et al. (1999) suggest that Mesozoic hadrosaurs and Cenozoic ungulates perhaps provide an example of both ecological and behavioral convergence (we shall consider behavioral convergence in animals in detail in chapter 6).

Not all dinosaurian ecological roles are filled by Cenozoic mammals, and not all mammalian roles were filled by dinosaurs. Although the otter is a fluvial piscivore, the gavialid crocodile is a much closer modern ecological analog to the toothy spinosaur (table 4.10). The dinosaurs were exclusively terrestrial animals, but if we examine marine and oceanic habitats, we find that other Mesozoic animals independently evolved ecological roles that are now played by Cenozoic mammals: the plesiosaur, ichthyosaur, and pliosaur ecological niches have been convergently refilled by Cenozoic sea lions, porpoises, and killer whales. In the air, the insectivorous pterosaur and frugivorous pterosaur niches have been independently reevolved by the microbats and the frugivorous bats. And the modern pelican, itself an avian dinosaur, convergently refilled the niche of the Mesozoic pteranodon and even resembles a small pteranodon in appearance. Mammals also have convergently evolved flying piscivores, such as the greater bulldog bat, *Noctilio leporinus*, although it looks nothing like a pteranodon (excepting the wings).

In these 19 ecological roles (table 4.10), the Cenozoic ecosystem play is scripted in the same way as the Mesozoic ecosystem play. The roles of the ecological play have not changed, although the casts of actors have changed dramatically following the end-Cretaceous mass extinction. Yet, just as in the comparison of the Australian and South American ecosystems, there are missing roles in the Mesozoic. In the terrestrial realm, where are the dinosaurian moles, rats, woodchucks, rabbits, camels, and beavers? Were these missing niches, these ecological roles, in fact filled by dinosaurs, whose fossil species have yet to be discovered? Or are these ecological roles newly evolved by the Cenozoic mammals? If the latter, then the Cenozoic ecosystem is not entirely convergent upon the Mesozoic—it would, instead, be a "revised" play. Given the ancientness of the dinosaur-dominated ecosystem, some 65 to 220 million years ago, we may

Table 4.10
Convergent evolution of ecological-analog compositions of Mesozoic dinosaur-dominated ecosystems and Cenozoic mammal-dominated ecosystems

---

1 Convergent ecological analog: LARGE AMBUSH PREDATOR ("great cat niche")
Cenozoic ecosystem: Lion (Mammalia: Felidae; *Panthera leo*)
Mesozoic ecosystem: Allosaur (Dinosauria: Allosauridae; *Allosaurus fragilis* †Jurassic)

2 Convergent ecological analog: SMALL AMBUSH PREDATOR ("small cat niche")
Cenozoic ecosystem: Wild cat (Mammalia: Felidae; *Felis sylvestris*)
Mesozoic ecosystem: Coelophysis (Dinosauria: Coelophysidae; *Coelophysis rhodesiensis* †Triassic)

3 Convergent ecological analog: PURSUIT PREDATOR ("wolf niche")
Cenozoic ecosystem: Wolf (Mammalia: Canidae; *Canis lupus*)
Mesozoic ecosystem: Velociraptor (Dinosauria: Dromaeosauridae; *Velociraptor mongoliensis* †Cretaceous)

4 Convergent ecological analog: NOCTURNAL FORAGING PREDATOR ("wolverine niche")
Cenozoic ecosystem: Wolverine (Mammalia: Mustelidae; *Gulo gulo*)
Mesozoic ecosystem: Troodont (Dinosauria: Troodontidae; *Saurornithoides mongoliensis* †Cretaceous)

5 Convergent ecological analog: ANT EATER ("anteater niche")
Cenozoic ecosystem: Anteater (Mammalia: Myrmecophagidae; *Myrmecophaga tridactyla*)
Mesozoic ecosystem: "Desert bird" alvarezsaurid (Dinosauria: Alvarezsauridae; *Shuvuuia deserti* †Cretaceous)

6 Convergent ecological analog: LARGE HERDING BROWSING HERBIVORE ("elephant niche")
Cenozoic ecosystem: Elephant (Mammalia: Elephantidae; *Loxodonta africana*)
Mesozoic ecosystem: Sauropod (Dinosauria: Titanosauridae; *Titanosaurus madagascariensis* †Cretaceous)

7 Convergent ecological analog: MID-SIZED FAST-RUNNING BROWSING HERBIVORE ("deer niche")
Cenozoic ecosystem: Deer (Mammalia: Cervidae; *Cervus elaphus*)
Mesozoic ecosystem: Hypsilophodont (Dinosauria: Hypsilophodontidae; *Hypsilophodon foxii* †Cretaceous)

8 Convergent ecological analog: LARGE HERDING MIXED-FEEDING HERBIVORE ("bison niche")
Cenozoic ecosystem: Bison (Eutheria: Bovidae; *Bison bison*)
Mesozoic ecosystem: Hadrosaur (Dinosauria: Hadrosauridae; *Parasaurolophus walkeri* †Cretaceous)

9 Convergent ecological analog: LARGE HORNED GRAZING HERBIVORE ("rhinoceros niche")
Cenozoic ecosystem: Rhinoceros (Mammalia: Rhinocerotidae; *Rhinoceros unicornus*)
Mesozoic ecosystem: Triceratops (Dinosauria: Ceratopsidae; *Triceratops albertensis* †Cretaceous)

10 Convergent ecological analog: LARGE ARMORED GRAZING HERBIVORE ("glyptodont niche")
Cenozoic ecosystem: Glyptodont (Mammalia: Glyptodontidae; *Doedicurus clavicaudatus* †Pleistocene)
Mesozoic ecosystem: Ankylosaur (Dinosauria: Ankylosauridae; *Ankylosaurus magniventris* †Cretaceous)

Table 4.10
(continued)

11 Convergent ecological analog: MID-SIZED RUMINANT GRAZER ("goat niche")
Cenozoic ecosystem: Goat (Mammalia: Bovidae; *Capra hircus*)
Mesozoic ecosystem: Pachycephalosaur (Dinosauria: Pachycephalosauridae; *Pachycephalosaurus wyomingensis* †Cretaceous)

12 Convergent ecological analog: GIANT FRUGIVORE ("ground sloth niche")
Cenozoic ecosystem: Giant ground sloth (Mammalia: Megatheriidae; *Nothrotheriops shastensis* †Pleistocene)
Mesozoic ecosystem: "Sloth-claw" therizinosaur (Dinosauria: Therizinosauridae; *Nothronychus mckinleyi* †Cretaceous)

13 Convergent ecological analog: FLUVIAL PISCIVORE ("gavialid niche")
Cenozoic ecosystem: Gavialid crocodile (Archosauria: Crocodilidae; *Gavialis gangeticus*)
Mesozoic ecosystem: Spinosaur (Dinosauria: Spinosauridae; *Spinosaurus maroccanus* †Cretaceous)

14 Convergent ecological analog: SHALLOW-MARINE PISCIVORE ("sea-lion niche")
Cenozoic ecosystem: Sea lion (Mammalia: Otariidae; *Zalophus californianus*)
Mesozoic ecosystem: Plesiosaur (Diapsida: Sauropterygia: Cryptocleididae; *Cryptocleidus oxoniensis* †Jurassic)

15 Convergent ecological analog: SMALL OPEN-OCEAN CARNIVORE ("porpoise niche")
Cenozoic ecosystem: Porpoise (Mammalia: Phocaenidae; *Phocaena phocaena*)
Mesozoic ecosystem: Ichthyosaur (Diapsida: Ichthyosauria: Ichthyosauridae; *Ichthyosaurus platyodon* †Jurassic)

16 Convergent ecological analog: LARGE OPEN-OCEAN CARNIVORE ("killer whale niche")
Cenozoic ecosystem: Killer whale (Mammalia: Delphinidae; *Orcinus orca*)
Mesozoic ecosystem: Pliosaur (Diapsida: Sauropterygia: Rhomaleosauridae; *Rhomaleosaurus megacephalus* †Jurassic)

17 Convergent ecological analog: FLYING INSECTIVORE ("bat niche")
Cenozoic ecosystem: Bat (Mammalia: Vespertilionidae; *Myotis myotis*)
Mesozoic ecosystem: "Frog-jaw" pterosaur (Ornithodira: Pterosauria: Aneurognathidae; *Batrachognathus volans* †Jurassic)

18 Convergent ecological analog: FLYING FRUGIVORE ("fruit bat niche")
Cenozoic ecosystem: Fruit bat (Mammalia: Pteropodidae; *Rousettus aegyptiacus*)
Mesozoic ecosystem: "Bakony dragon" pterosaur (Ornithodira: Pterosauria: Azhdarchidae; *Bakonydraco galaczi* †Cretaceous)

19 Convergent ecological analog: FLYING PISCIVORE ("pelican niche")
Cenozoic ecosystem: Pelican (Aves: Pelecanidae; *Pelecanus occidentalis*)
Mesozoic ecosystem: Pteranodon (Ornithodira: Pterosauria: Pteranodontidae; *Pteranodon longiceps* †Cretaceous)

Note: The geological age of extinct species is marked with a †. For data sources, see text.

never be able to reconstruct all the complexities of its structure, particularly for small animals whose fragile bones have a very low probability of preservation in the fossil record.

Thus far we have examined and compared lists of ecological roles that have independently evolved in ecosystems that are separated from one another in space or time. An alternative approach to examining the phenomenon of convergent ecosystem evolution is to examine the ecological-role composition of ecosystems that have evolved in equivalent physical environmental regions, or biomes, such as tundra regions, desert regions, and so on. The hypothesis to be tested here is whether ecosystems that develop in equivalent biomes will also develop equivalent compositions of ecological roles, or niches. Several isochronous studies have examined Mediterranean-style climatic regions of the Mediterranean, of course, and equivalent biomes in Chile and California (Went 1971; Mooney 1977; Peet 1978). The studies included in Mooney (1977) documented extensive ecosystem convergence in the types of ecological roles played by both plants and vertebrate animals, but much less convergence in the roles played by invertebrate animals, particularly insects. On the other hand, Milewski and Bond (1982) document the convergent evolution of the same seed-dispersal ecological roles in insects in Mediterranean-style climatic regions in Australia and South Africa (particularly myrmecochory, which we considered in the last chapter).

Extensive ecosystem convergence is not a phenomenon confined to terrestrial biomes. McKinney (2007) and McKinney et al. (2007) document the heterochronous convergent evolution of Paleozoic-structured marine ecosystems in the unusual oligotrophic (low-nutrient level) biomes in the Northern Adriatic Sea of Europe. Only 2 percent of the species present are survivors of the ancient Paleozoic faunas; the other species are all modern, yet they have converged on lifestyles that are 250 million years out of date! These include 30-centimeter-high polychaete worms, *Sabella spallanzanii*, that have convergently refilled the ecological niche of Paleozoic crinoids, and numerous bivalve molluscs, such as the scallop *Aequipecten opercularis*, that have convergently refilled the ecological niche of Paleozoic brachiopod shellfish. While the ecological roles are the same as those in the Paleozoic, the players of those roles are modern marine species.

A Periodic Table of Niches?

In the previous section of the chapter we examined data suggesting that the script of ecological roles in nature seems to exist in the absence of the identity of the players, whether they are dinosaurs or mammals (table 4.10), placentals or marsupials (table 4.8), and that these players have convergently filled those roles independently in different positions in both space and time. While stressing the importance of the phenomenon of ecological convergence, Pianka (1978: 300) also cautions that "evolutionary convergence can easily be read into a situation by placing undue emphasis upon superficial similarities but failing to appreciate fully the inevitable dissimilarities between pairs of supposed ecological equivalents."

Thus, an alternative approach to the comparison of existent ecosystems would be to attempt to compare existent ecosystems with nonexistent ecosystems; then we could seek to understand what it is about existent ecosystems that has led organisms to construct them repeatedly rather than constructing theoretically possible but nonexistent ecosystems. That is, we can apply the analytical techniques of theoretical morphology (McGhee 2007) to ecology. As Maclaurin and Sterelny (2008, 128–129) argue, "one way of thinking about ecological diversity is in terms of a phenomenological ecospace. The dimensions of that space . . . define a space of possibilities," analogous to a theoretical morphospace; but they also caution that "[a]s with morphospace, though, a total ecospace is of high and somewhat arbitrary dimensionality" (Maclaurin and Sterelny 2008, 111).

Pianka (1978) in fact used this approach in his attempt not only to create an empty niche—to specify an ecological role a priori without reference to any living organism—but also to arrange those empty niches into a "periodic table of niches," a space of potential ecological roles, again analogous to a theoretical morphospace. He used two dimensions to construct that space: a trophic-niche dimension and a life-history-niche dimension (table 4.11). The trophic-niche dimension spans the possible trophic spectrum of autotrophy (primary producers) to heterotrophy (herbivores and carnivores). The life-history-niche dimension spans the possible spectrum from *r*-selected organisms (those with high reproductive rates, low parental care for offspring, and short life spans) to *K*-selected organisms (those with low reproductive rates, high parental care for offspring, and long life spans).

Within this space, Pianka (1978) thus created nine permutation positions, empty niches, or potential ecological roles, in the absence of any actual living organisms. The next step in the analysis is to ask: are these potential niches actually empty in nature? The answer is no—Pianka (1978) discovered that actual living organisms have evolved each of the nine potential ecological roles (table 4.11). Pianka (1978) then added an

Table 4.11
The ecological periodic table of niches

Ecological life-history niche	Trophic niche		
	Primary producers	Herbivores	Carnivores
r-selected organisms	Annual plants	Aphids, Caterpillars	Mantids
Midpoint of *r*-*K* continuum	Shrubs	Lemmings	Weasels
K-selected organisms	Perennial plants (especially trees)	Deer, Bison	Cougars, Wolves

Note: Modified from Pianka (1978).

additional dimension to the analysis: the dimension of space utilization. This dimension spans the spectrum of organisms that exploit space in only two dimensions to those that exploit space in three (table 4.12). As in the previous analysis (table 4.11), he discovered that the additional six potential empty niches of three-dimensional space utilization by heterotrophs were in fact filled by actual animals in nature (table 4.12).

However, the three ecological niches of three-dimensional space exploitation by primary producers are not filled—Pianka (1978) had created possible but nevertheless nonexistent ecological roles (table 4.12), although he did not comment further upon these possibilities. In the spirit of theoretical morphology, it is possible to predict what such an organism might look like: a plant that has evolved a gas-filled bladder (or multiple bladders) that allows it to float in the air. Such an organism could easily photosynthesize in midair (in fact, it could avoid competition for light from other plants by doing just that, and thus would have a selective advantage over ground-dwelling plants), but it would also require water and nutrients. Water is obtainable from water vapor in clouds or from the atmosphere itself in humid regions of the Earth. Nutrients are a problem, though. Such a hypothetical plant would probably have to be carnivorous, obtaining its necessary nutrients by preying on flying insects or other animals, in a manner convergent on that of ground-dwelling carnivorous plants (table 4.1).

A balloon-like, floating carnivorous plant has never evolved on Earth; this potential ecological role remains empty (table 4.12). Why? The reason is probably the extreme instability of such a habitat in the Earth's existent weather patterns; sessile ground-dwelling plants have had to evolve complex mechanisms to solve their own serious weather-related

Table 4.12
Pianka's inclusion of organisms that exploit space in three dimensions (3D), as opposed to just two dimensions (2D), in his periodic table of niches

Ecological life-history niche	Trophic niche		
	Primary producers	Herbivores	Carnivores
r-selected organisms	2D: Annual plants	2D: Aphids, Caterpillars	2D: Mantids
	3D: —	3D: Bees, Butterflies	3D: Dragonflies
Midpoint of *r*-*K* continuum	2D: Shrubs	2D: Lemmings	2D: Weasels
	3D: —	3D: Squirrels, Fruit bats	3D: Flycatchers, Insectivorous bats
K-selected organisms	2D: Perennial plants (especially trees)	2D: Deer, Bison	2D: Cougars, Wolves
	3D: —	3D: Parrots	3D: Eagles, Falcons

Note: Modified from Pianka (1978). Note that the three possible ecological niches of three-dimensional space exploitation by primary producers are not filled.

problems, as discussed in chapter 3. These problems might be exacerbated, perhaps fatally, for a hypothetical floating plant with no control over where it might be carried by the winds; if blown into desert regions, it could die of dehydration; if blown out over the ocean, it could perish by being unable to feed on flying insects. But might such a plant exist on another world, a world with weather patterns radically different from those found on Earth? The concept of entirely possible ecological roles, roles that are nevertheless nonexistent on Earth, has implications for the concept of convergent evolution itself, implications that will be examined in detail in chapter 7.

Pianka (1978) noted in passing that a potential periodic table of niches for marine organisms would differ from that for terrestrial organisms in that there are relatively few *K*-selected marine organisms. Bambach (1983) used a similar approach to Pianka (1978) to create an ecospace for marine organisms by using a similar trophic-niche dimension, but he abandoned the *r*-*K* life-history-niche dimension of Pianka (1978) in favor of a habitat-mobility-niche dimension. In table 4.13, I have simplified Bambach's original trophic-niche dimension by creating a detritivore niche, one that combines the organic-detritus collection mechanisms of filter feeding and deposit feeding. I have further simplified the habitat-mobility-niche dimension to cover the spectrum from pelagic (habitats

Table 4.13
Occupation of marine-niche ecospace in the Cambrian

Habitat-mobility niche	Trophic niche		
	Detritivores	Herbivores	Carnivores
Pelagic	Trilobites	—	—
Epibenthic			
1. Sessile	Brachiopods, Eocrinoids	—	—
2. Mobile	Trilobites, Ostracodes	Monoplacophorans, Ostracodes	—
Endobenthic			
1. Mobile, shallow	Trilobites, Bristle worms, Brachiopods	—	Bristle worms
2. Mobile, deep	—	—	—

Note: Modified from Bambach (1983).

in the water column) to epibenthic (habitats on the sea bottom) to endobenthic (habitats within the sediments below the sea bottom). On the other hand, I have made his original analysis a bit more complicated by exhausting all possible permutations of these ecological states along these two ecological dimensions and considering all of these permutations as possible ecological roles (table 4.13). In his original analysis, Bambach (1983) omitted some of these permutations, a point I will return to later.

Further, Bambach (1983) added an important new dimension not present in Pianka's analyses: geological time. He then proceeded to analyze the pattern of marine ecosystem evolution for the past 545 million years since the evolution of multicellular heterotrophs, that is, the animals (autotrophs were not included in the analysis). Of the 15 possible ecological roles in Bambach's ecospace, only six were evolved by animals in the Cambrian—the other nine niches were empty (table 4.13). By the middle of the Paleozoic, animals had evolved 11 of the possible ecological roles, and only four niches remained empty (table 4.14). And by the Mesozoic, only three possible ecological roles remained unfilled (table 4.15).

Note that although the permutations of ecological roles are exactly the same in tables 4.13, 4.14, and 4.15, the animals playing those roles change through time. Trilobites played the pelagic-detritivore role in the Cambrian but were replaced by conodonts and hemichordates in the mid-Paleozoic, who themselves were replaced by crustaceans and

Table 4.14
Occupation of marine-niche ecospace in the Middle Paleozoic

Habitat-mobility niche	Trophic niche		
	Detritivores	Herbivores	Carnivores
Pelagic	Conodonts, Hemichordates	—	Cephalopods, Placoderm fishes, Sea scorpions
Epibenthic			
1. Sessile	Brachiopods, Crinoids, Bryozoans, Sponges	—	Cnidarians
2. Mobile	Agnathan fishes, Monoplacophorans, Gastropods	Echinoids, Gastropods, Ostracodes	Cephalopods, Crustaceans, Starfish
Endobenthic			
1. Mobile, shallow	Trilobites, Bivalves, Bristle worms	—	Merostomes, Bristle worms
2. Mobile, deep	Bivalves	—	Bristle worms

Note: Modified from Bambach (1983).

Table 4.15
Occupation of marine-niche ecospace in the Mesozoic

Habitat-mobility niche	Trophic niche		
	Detritivores	Herbivores	Carnivores
Pelagic	Crustaceans, Gastropods	Osteichthyans	Osteichthyans, Chondrichthyans, Marine reptiles, Cephalopods
Epibenthic			
1. Sessile	Bivalves, Bryozoans, Barnacles, Sponges	—	Cnidarians
2. Mobile	Bivalves, Gastropods, Crustaceans	Gastropods, Polyplacophorans, Crustaceans	Gastropods, Crustaceans, Starfish
Endobenthic			
1. Mobile, shallow	Bivalves, Echinoids, Sea cucumbers	—	Gastropods, Crustaceans, Bristle worms
2. Mobile, deep	Bivalves, Bristle worms	—	Bristle worms

Note: Modified from Bambach (1983).

gastropods in the Mesozoic. Brachiopods were important sessile epibenthic detritivores throughout the Paleozoic (tables 4.13 and 4.14), and indeed are known as the shellfish of the Paleozoic. Their ecological role was taken over by the bivalve molluscs (table 4.15) following the end-Permian mass extinction, the greatest ecological disruption in Earth history, both in the sea and on the land (McGhee et al. 2004).

Marine reptiles, especially ichthyosaurs and pliosaurs, were important players of the pelagic-carnivore role in the Mesozoic (table 4.15). Today that same ecological role still exists, but it is now being played by marine mammals, such as porpoises and killer whales. That last conclusion we had already reached by comparing the actual existent Mesozoic and Cenozoic ecosystems (table 4.10). Bambach's analysis is importantly different, however, in that he created the pelagic-carnivore niche not by examining lists of actual animals and their ecologies, but by considering the possible permutations of ecological states along a spectrum of two different ecological dimensions, trophic-niche and habitat-mobility-niche. This is the same analytic procedure that is used in theoretical morphology (McGhee 2007), and Bambach and his colleagues are now engaged in the explicit construction of a theoretical ecospace that is analogous to a theoretical morphospace (Bush et al. 2007).

Originally, however, Bambach (1983, 726) considered some of the permutations of his two ecological dimensions to be "not biologically practical adaptive strategies." These include the three possible but currently empty niches that are listed in table 4.15. Two of these roles belong to mobile endobenthic herbivores, an ecological role unfilled in marine ecosystems. Yet this same ecological role has been evolved in terrestrial ecosystems (the subterranean herbivore, or gopher, niche; see table 4.9). The absence of animals filling this niche in marine ecosystems probably reflects the different morphologies and ecologies of marine plants, which generally do not have extensive root systems, and terrestrial plants, which do. Thus, while this same ecological role may be nonfunctional in marine ecosystems, it is clearly functional in terrestrial ecosystems.

The last empty niche listed in table 4.15 is the ecological role of a sessile epibenthic herbivore. This ecological-dimension permutation may indeed represent an impossible ecological role. Sessile epibenthic carnivores exist—the cnidarians—who simply wait for prey animals to come within reach of their tentacles. A sessile herbivore has never evolved on Earth, mainly because on Earth the plants themselves are sessile and thus would never move to be within reach of the sessile herbivore. But might such an animal exist on another world, a world that has evolved mobile plants? Might this odd ecological role be possible after all, if only theoretically? In chapter 7 we will explore in more detail the implications of theoretically possible, but nevertheless nonexistent, ecological roles on Earth for the concept of convergent evolution.

The phenomenon of ecosystem convergence supports the idea that there may be "universal constraints on the ecological roles" played in those ecosystems (Carrano et al. 1999, 256), just as the phenomenon of morphological convergence argues for evolutionary constraint on the spectrum of possible forms available to animals and plants, as discussed in chapters 2 and 3. In the case of ecosystems, the nature of these potential universal constraints is far from clear, and is part of the larger debate concerning the degree of integration and structure of ecosystems themselves (for an overview of this larger question, see Maclaurin and Sterelny 2008, 106–131). The creation of theoretical ecospaces could add much more rigor to the debate concerning ecosystem convergence by revealing the total spectrum of what is ecologically possible, and what is not, in potential ecosystems. The works of Pianka (1978), Bambach (1983), and Bush et al. (2007) are first steps in this direction, an analytic approach to theoretical morphospace that needs much more attention from ecologists.

Thus far in this book, we have considered the phenomena of convergence in individual animal and plant forms and, on a larger scale, in the ecological roles of entire assemblages of animal and plant species. In the next chapter we shall consider convergence on much smaller scales; that is, in the very molecules that make up living animals and plants.

5 Convergent Molecules

Recent advances in protein chemistry suggest that at least one set of biological forms—the basic protein folds—is determined by physical laws similar to those giving rise to crystals and atoms. . . . If it does turn out that a substantial amount of higher biological form is natural, then the implications will be radical and far-reaching. It will mean that physical laws must have had a far greater role in the evolution of biological form than is generally assumed . . . underlying all the diversity of life is a finite set of natural forms that will recur over and over again anywhere in the cosmos where there is carbon-based life.
—Denton and Marshall (2001, 417)

Convergent Molecules?

The source code for all life on Earth is contained in the deoxyribonucleic-acid molecule, DNA. DNA codes for RNA molecules, ribonucleic acids, and RNA codes for the assembly of amino acids into proteins, the building blocks of life. Each of the essential molecules of life has a finite number of possible states. Since DNA and RNA molecules contain only four different nucleotides, the code for Earth life is a base-four system. As such, the probability, p, of convergent molecular evolution of the same nucleotide at the same site in two different DNA or RNA molecules, via random mutation, is one in four ($p = 0.25$ per site). Protein molecules contain only 20 different amino acids; thus, the probability of the convergent molecular evolution of the same amino acid at the same site in two different protein molecules is one in twenty ($p = 0.05$ per site). Proteins are complex molecules, and the function of the protein is often determined not only by its amino acid composition but also by the complex manner in which the molecule is folded. Basic protein folds also exist in a finite number of geometries, and the number of fold geometries that are potentially functional are only a tiny subset of all possible folds (Axe 2004). Denton and Marshall (2001) estimate that the number of different protein-fold geometries that exist in life on Earth are between 500 and 1,000; therefore, the probability of independently evolving protein molecules with the same fold geometry is one in five hundred ($p = 0.002$) to one in a thousand ($p = 0.001$).

Even though some of these probabilities are quite high, many biologists consider the occurrence of convergent molecular evolution in nature to be quite unlikely. Biological molecules are typically very large, and the possibility of many independent site convergences is improbable. For example, even though the probability of the convergent evolution of the same nucleotide at the same site in two different DNA molecules is one in four ($p = 0.25$), the probability of five such independent site convergences occurring in two molecules is only two in 10,000 ($p = p_1 \times p_2 \times p_3 \times p_4 \times p_5 = 0.0002$). Numerous independent site convergences are thus unlikely.

But what if the probabilities are not independent? Castoe et al. (2009, 8990) point out that mitochondrial DNA codes for metabolic genes that are functionally related, and that "directional selection may thus tend to affect many mitochondrial genes at once, leading to large-scale convergence if similar selective events occur in different lineages." Weinreich et al. (2006) point out that pervasive biophysical pleiotropic effects, or intermolecular interactions, in proteins constrain much of protein evolution. In two separate protein studies, they point out that only 15 to 29 percent of all possible mutational trajectories in these proteins are accessible to natural selection, and thus the probability that evolution will convergently occur along "largely identical mutational trajectories" is quite high (Weinreich et al. 2006, 113). Further theoretical analyses of mutational pathways in molecular adaptive landscapes reveal that few pathways are evolvable, leading to the implication that "evolution might be more reproducible than is commonly perceived, or [might] even be predictable" (Poelwijk et al. 2007, 386). Rokas and Carroll (2008, 1943) stress that the twin roles of positive selection for particular amino acid substitutions in proteins and purifying selection against others "constrains substitutions to a small number of functionally equivalent amino acids," and that the action of natural selection has therefore produced "frequent and widespread parallel evolution of protein sequences."

DNA

The analysis of convergent evolution at the level of the DNA molecule is difficult at present due to the lack of data. Although we have entered the "age of complete genomes," Castoe et al. (2010) point out that only in the past five to seven years have substantial numbers of mitochondrial genome data sets become available for the vertebrates, and that complete nuclear genome data sets number only in the tens for these highly derived animals (bacterial genomes are densely sampled, but these organisms have ancient phylogenies and thus the number of changes separating their genomes are large, making the detection of convergent evolution difficult; Castoe et al. 2010).

The few data that do exist are startling. Vision in vertebrate animals is due to the activation of the visual pigments in the eye by impacting photons. The visual pigments are composed of opsin proteins and a chromophore. Most vertebrates have a single type of chromophore, but five groups of opsin proteins exist: rhodopsin (RH1) and RH1-like (RH2), which are maximally activated by light in the blue-green part of the spectrum (about 480 nm to 530 nm); short-wavelength-sensitive opsin-1 (SWS1), which is maximally activated by violet-blue wavelengths (about 410 nm to 490 nm); short-wavelength-sensitive opsin-2 (SWS2), which is maximally activated by ultraviolet-violet wavelengths (about 355 nm to 440 nm); and a long-wavelength- to medium-wavelength-sensitive opsin (LWS/MWS), which is maximally activated by green-red wavelengths (about 490 nm to 570 nm; Zhang 2003; Bowmaker and Hunt 2006).

The DNA coding for the opsin proteins has been intensively investigated, and early studies suggested that the evolution of red-sensitive opsins in humans and some fish independently evolved from green-sensitive opsins by the same nucleotide substitutions in the encoding DNA (Yokoyama and Yokoyama 1990). Subsequent studies of the evolution of trichromatic color vision in primates revealed that the precise same three substitutions in DNA nucleotides that coded for LWS/MWS opsins independently evolved in Old World primates (Catarrhini) and New World primates (Platyrrhini; table 5.1). Parallel evolution in the LWS/MWS opsin amino acid sequences in both primate groups (Hunt et al. 1998) resulted in the replacement of serine with alanine at site 180 (substituting G for T in the first position in the DNA codon for serine), the replacement of tyrosine with phenylalanine at site 277 (substituting T for A in the second position in the DNA codon for tyrosine), and the replacement of threonine with alanine at site 285 (substituting G for A in the first position in the DNA codon for threonine).

These DNA coding changes resulted in a 30-nanometer spectral shift in the maximum activation of the LWS/MWS opsin, from 530 to 560

Table 5.1
Convergent evolution of identical nuclear DNA molecules coding for photosensitive opsin proteins

---

1 Convergent molecule and function: LWS/MWS OPSIN–ENCODING DNA (30 nm spectral shift in vision produced by the identical three nucleotide substitutions in the DNA coding for the long-wavelength/medium-wavelength-sensitive cone opsin protein)

Convergent lineages:

1.1 Capuchin monkey (Primates: Simiiformes: Platyrrhini: Cebidae; *Cebus apella*)

1.2 Human (Primates: Simiiformes: Catarrhini: Hominidae; *Homo sapiens*)

2 Convergent molecule and function: SWS1 OPSIN–ENCODING DNA (66 nm spectral shift in vision produced by the identical nucleotide substitution in the DNA coding for the short-wavelength-sensitive cone opsin protein)

Convergent lineages:

2.1 Tammar wallaby (Mammalia: Marsupialia: Diprotodontia: Macropodidae; *Macropus eugeni*)

2.2 Cow (Mammalia: Eutheria: Laurasiatheria: Cetartiodactyla: Ruminantia: Bovidae; *Bos taurus*)

3 Convergent molecule and function: SWS1-ultraviolet OPSIN–ENCODING DNA (35 nm spectral shift in vision produced by the identical nucleotide substitution in the DNA coding for the ultraviolet-wavelength-sensitive cone opsin protein)

Convergent lineages:

3.1 Rhea (Aves: Paleognathae: Struthioniformes 2: Rheidae; *Rhea americana*)

3.2 Herring gull (Aves: Neognathae: Neoaves: Ciconiiformes: Laridae; *Larus argentatus*)

3.3 Zebra finch (Aves: Neognathae: Neoaves: Passeriformes: Passeridae; *Taeniopygia guttata*)

3.4 Gray parrot (Aves: Neognathae: Neoaves: Psittaciformes: Psittacidae; *Psittacus erithacus*)

4 Convergent molecule and function: SWS1-violet OPSIN–ENCODING DNA (spectral tuning shifts in vision produced by the identical nucleotide substitutions in the DNA coding for the violet-wavelength-sensitive cone opsin protein)

Convergent lineages (three identical site changes):

4.1.1 African clawed frog (Tetrapoda: Batrachomorpha: Lissamphibia: Batrachia: Anura: Pipidae; *Xenopus laevis*)

4.1.2 Human (Tetrapoda: Reptiliomorpha: Amniota: Synapsida: Mammalia: Eutheria: Euarchontoglires: Primates: Hominidae; *Homo sapiens*)

Convergent lineages (one identical site change):

4.2.1 Zebra finch (Amniota: Sauropsida: Archosauromorpha: Ornithodira: Dinosauria: Saurischia: Theropoda: Aves: Passeriformes: Passeridae; *Taeniopygia guttata*)

4.2.2 Budgerigar (Aves: Psittaciformes: Psittacidae; *Melopsittacus undulatus*)

4.2.3 Human (Amniota: Synapsida: Mammalia: Eutheria: Euarchontoglires: Primates: Hominidae; *Homo sapiens*)

Convergent lineages (one identical site change):

4.3.1 Malawi fish (Osteichthyes: Actinopterygii: Teleostei: Perciformes: Cichlidae; *Metriaclima zebra*)

4.3.2 African clawed frog (Osteichthyes: Sarcopterygii: Tetrapoda: Batrachomorpha: Lissamphibia: Batrachia: Anura: Pipidae; *Xenopus laevis*)

4.3.3 Chicken (Tetrapoda: Reptiliomorpha: Amniota: Sauropsida: Archosauromorpha: Ornithodira: Dinosauria: Saurischia: Theropoda: Maniraptora: Aves: Galliformes: Phasianidae; *Gallus gallus*)

5 Convergent molecule and function: RH1 RHODOPSIN–ENCODING DNA (10 nm spectral shift in vision produced by the identical nucleotide substitution in DNA coding for the RH1 rhodopsin protein)

Convergent lineages:

5.1 Longfin Baikal sculpin (Osteichthyes: Actinopterygii: Teleostei: Scorpaeniformes: Cottocomephoridae; *Cottocomephorus inermis*)

5.2 Baikalian deepwater sculpin (Osteichthyes: Actinopterygii: Teleostei: Scorpaeniformes: Abyssocottidae; *Abyssocottus korotneffi*)

Note: For data sources, see text.

nanometers. Hunt et al. (1998, 3299) suggest that these changes were adaptive, the functional result of visual foraging for yellow and orange fruits against a green foliage background, and that "the separate origin of trichromacy in New and Old World primates would indicate that the selection of these three [amino acid] sites is the result of convergent evolution."

The catarrhine and platyrrhine primates are closely related—both are members of the simiiform clade (table 5.1)—and the two lineages diverged only some 35 million years ago in the Eocene (Benton 2005). One could thus argue that these two lineages demonstrate parallel molecular evolution; that is, independent parallel nucleotide substitutions in similar DNA molecules. In contrast, the marsupial mammals and placental mammals diverged some 125 million years ago, in the Early Cretaceous (Benton 2005). Yet even in these two ancient, independent lineages of mammals, convergent evolution in opsin-encoding DNA has been demonstrated (table 5.1).

Deeb et al. (2003) report that the identical nucleotide substitution in DNA coding for the SWS1 opsin amino acid sequence occurred independently in the marsupial tammar wallaby and in the placental cow: replacement of phenylalanine by tyrosine at site 86 (substituting A for T in the second position in the DNA codon for phenylalanine). This DNA coding change resulted in a 66-nanometer spectral shift in the maximum activation of the SWS1 opsin, from 358 to 424 nanometers. The tammar wallaby grazes on grass and low tree branches in the late afternoon, and Deeb et al. (2003) suggest that the convergent evolution in SWS1-encoding DNA observed in it and in the placental grazing ungulates enables the animals to visually distinguish features in grasses with differing shades of green.

Primates are unique among mammals in having trichromatic color vision, in that most mammals have only dichromatic vision. In contrast, some fish, reptiles, and birds have tetrachromatic vision, and may be able to see twice the number of colors that even trichromatic primates can see (Ödeen and Håstad 2003). Many of these same animals can see ultraviolet light, which has much shorter wavelengths than the 400-nanometer violet light that we are able to see. Ödeen and Håstad (2003) have demonstrated that ultraviolet vision evolved four times independently in the birds (table 5.1) by the identical nucleotide substitution in DNA coding for the SWS1 opsin amino acid sequence: replacement of serine by cysteine at site 90 (substituting G for C in the second position in the DNA codon for serine). This DNA coding change results in a 35-nanometer spectral shift, from 406 down to 371 nanometers, in the maximum activation of the SWS1 opsin. Ödeen and Håstad (2003, 859) suggest that the observed convergent evolution is adaptive, although its precise function is not clear: the difference in ultraviolet and violet vision "is quite dramatic and changes not only the perception of objects that reflect light solely in the UV or violet ranges but also the perception of objects that reflect both UV/violet and longer wavelengths. This should have important consequences for foraging, habitat use, social signaling, and mate choice."

An exhaustive analysis of the evolution of ultraviolet vision in the vertebrates (Shi and Yokoyama 2003; Zhang 2003) has revealed even more convergent molecular evolution in the SWS1 opsin–encoding DNA (table 5.1). Such phylogenetically divergent lineages as frogs and primates have convergently evolved the same three nucleotide substitutions in DNA coding for the SWS1 opsin: the replacement of phenylalanine by leucine at site 49 (substituting A or G for T or C in the third position in the DNA codon for phenylalanine), replacement of threonine by proline at site 93 (substituting C for A in the first position in the DNA codon for threonine), and replacement of serine by threonine at site 118 (substituting A for T in the first position in the DNA codon for serine). Primates and two different groups of avian dinosaurs have independently evolved an SWS1 opsin that replaced alanine with glycine at site 114 (substituting G for C in the second position in the DNA codon for alanine). And the very distantly related fish, frog, and chicken lineages have independently evolved an SWS1 opsin that replaced leucine with valine at site 116 (substituting G for C in the first position in the DNA codon for leucine).

The SWS1 opsin protein contains over 300 amino acids, yet Shi and Yokoyama (2003) argue that only nine amino acid positions are involved in spectral shifting. For all the opsin proteins, Bowmaker and Hunt (2006, R484) argue that "there are only a limited number of sites within opsin that can be altered without producing a non-functional pigment." They demonstrate that natural selection targets these sites in deepwater fish, producing "classic examples of spectral tuning of visual pigments within specific opsin classes which correlate with photic environments" (Bowmaker and Hunt 2006, R486). Adaptation of vision to different light intensities in different water depths extends to the rods of the eye as well as the color-sensitive cones; for example, Bowmaker and Hunt (2006) show that two separate species of deepwater Lake Baikal fish independently evolved the same RH1 rhodopsin protein by replacing aspartic acid with asparagine at site 83 (substituting A for G in the first position in the DNA codon for aspartic acid) in the RH1-encoding DNA, producing a spectral shift from 505 to 495 nanometers (table 5.1).

Convergent evolution has also been demonstrated to occur in the genes encoding for photoreceptor proteins that are not primarily used in vision, but in the maintenance of circadian rhythms in animals and heliotropic movement in plants (Cashmore et al. 1999). Charles Darwin himself discovered that his plants ceased their daily movement if a filter was placed between them and the sun, a filter that removed blue wavelength light. It is now known that cryptochrome genes, *cry*, encode for cryptochrome proteins, photoreceptors that respond to light in the 400-to-500-nanometer wavelength. These genes are present throughout the green eukaryotes, from algae to land plants. Cashmore et al. (1999, 763) demonstrate that they have independently evolved in the metazoa: "cryptochromes represent an example of repeated evolution, a special case of convergent evolution in which a new genetic function arises independently in two different lineages from orthologous (or paralogous) genes. . . . This phenomenon contrasts with classic convergent evolution, where the ancestral genes are unrelated." As discussed in chapter 1, the phenomenon in a strict sense would be termed "parallel evolution."

Other examples of convergent molecular evolution come from the analysis of digestive proteins. Five independent lineages of animals have convergently evolved stomachal-fermentation systems, as discussed in chapter 2 (see table 2.10). Three of these lineages have also experienced convergent molecular evolution of the enzymes found in those specialized stomachs (table 5.2), in which a conventional lysozyme was independently recruited for digestive purposes in the stomach (Stewart et al. 1987; Kornegay et al. 1994; Zhang and Kumar 1997). The production of the stomach lysozyme was independently produced in these three lineages by the identical two nucleotide substitutions in the DNA coding for amino acid sequences in the stomach lysozomes: replacement of asparagine with aspartic acid at site 75 (substituting G for A in the first position of the DNA codon for asparagine), and replacement of aspartic acid with asparagine at site 87 (substituting A for G in the first position of the DNA codon for aspartic acid). Interestingly, a third parallel nucleotide substitution occurred in the stomach lysozymes of the hoatzin, a bird, and the ungulate mammals—the replacement of alanine with glycine at site 76 (substituting G for C in the second position of the DNA codon for alanine)—which does not occur in the langur monkey. This

Table 5.2
Convergent evolution of identical nuclear DNA and mitochondrial DNA molecules coding for digestive proteins

1 Convergent molecule and function: STOMACH LYSOZYME–ENCODING DNA (conversion of conventional lysozymes to digestive stomach lysozymes produced by the identical two nucleotide substitutions in the DNA coding for the stomach lysozymes)

Convergent lineages:

1.1 Hoatzin (Amniota: Sauropsida: Archosauromorpha: Ornithodira: Dinosauria: Saurischia: Theropoda: Aves: Opisthocomiformes: Opisthocomidae; *Opisthocomus hoatzin*)

1.2 Cow (Amniota: Synapsida: Therapsida: Mammalia: Eutheria: Laurasiatheria: Cetartiodactyla: Ruminantia: Bovidae; *Bos taurus*)

1.3 Langur monkey (Mammalia: Eutheria: Euarchontoglires: Primates: Cercopithecidae; *Presbytis entellus*)

2 Convergent molecule and function: PANCREATIC RIBONUCLEASE–ENCODING DNA (decrease in pH for the maximum ribonucleolytic activity of the enzyme produced by the identical three nucleotide substitutions in the DNA coding for the pancreatic RNase)

Convergent lineages:

2.1 Asian douc langur monkey (Primates: Cercopithecidae: Colobinae; *Pygathrix nemaeus*)

2.2 African guereza monkey (Primates: Cercopithecidae: Colobinae; *Colobus guereza*)

3 Convergent molecules and function: METABOLIC PROTEIN–ENCODING MITOCHONDRIAL DNA (44 nonrandom parallel amino acid substitutions across all 13 mtDNA genes coding for oxidative phosphorylation metabolic proteins)

Convergent lineages:

3.1 Australian bearded dragon (Lepidosauria: Squamata: Iguania: Agamidae; *Pogona vitticepes*)

3.2 Blind snake (Lepidosauria: Squamata: Scleroglossa: Autarchoglossa: Anguimorpha: Serpentes: Typhlopidae; *Typhlops diardi*)

Note: For data sources, see text.

suggests a closer convergence in stomachal-fermentation function in the hoatzins and ungulates, as the statistical analyses of Zhang and Kumar (1997) found that the patterns of amino acid substitutions at these sites in the enzyme were significantly unusual and could be taken as evidence for the action of nonneutral natural selection. Similar to the convergent molecular evolution in opsin proteins, Kornegay et al. (1994, 927) suggest that there may be a limited number of ways to convert a conventional vertebrate lysozyme-*c* gene into a stomach lysozyme gene, "a set of specific structural changes that are necessary and sufficient to convert a lysozyme for stomach function even when different genes have been separated by more than 300 million years of independent evolution," as in the case of the avian dinosaurs and the placental mammals (table 5.2).

Further study of the stomachal-fermenting colobine monkeys reveals parallel evolutionary changes in pancreatic ribonuclease–encoding DNA that occurred independently in Asian colobine monkeys and African colobine monkeys (table 5.2), two lineages that diverged some 13 million years ago (Prud'homme and Carroll 2006). Zhang (2006) reports that three identical nucleotide substitutions in the DNA coding for the pancreatic RNase amino acid sequence independently occurred in these two lineages: replacement of arginine with glutamine at site four (substituting A for G in the second position in the DNA codon for arginine), replacement of lysine by glutamic acid at site six (substituting G for A in the first position in the DNA codon for lysine), and replacement of arginine by tryptophan at site 39 (substituting T for C or A in the first position in the DNA codon for arginine). This DNA coding change resulted in the lowering of the pH of the maximum ribonucleolytic activity of the enzyme from 7.4 to 6.3 (Zhang 2006). The end result is a pancreatic RNase with maximum activities at more acid conditions than usual, conditions that match those of the acidic environment of the colobine monkey's small intestine. Zhang (2006, 819) concludes that "the new genes acquired enhanced digestive efficiencies through parallel amino acid replacements driven by darwinian selection" and that "molecular evolution has a certain degree of repeatability and predictability under the pressures of natural selection."

As in the nuclear genome data sets, initial analyses of some of the mitochondrial genome data sets that do exist have produced startling results. Castoe et al. (2009, 2010) have documented that 44 amino acid substitutions have occurred in parallel in all 13 mitochondrially encoded metabolic proteins in agamid lizards and snakes, an unexpected result that they summarize by stating: "convergent molecular evolution in mitochondria can occur at a scale and intensity far beyond what has been documented previously" (Castoe et al. 2009, 8986). Although they are both squamate reptiles, the agamid lizards and the snakes are only distantly related: the agamid lizards are iguanians, a basal group of squamates, whereas the snakes are highly derived anguimorphs, located many nodes up in the squamate clade (table 5.2; see also the appendix).

Further, Castoe et al. (2009, 2010) argue that the observed convergent evolution of mitochondrial DNA, mtDNA, is driven by natural selection. The convergent sites occur primarily at the first and second positions in the mtDNA codons, as is the case in the previously discussed cases of nuclear DNA convergence. The affected mtDNA codes for core oxidative phosphorylation proteins, which are some of the most highly conserved proteins in the genome. As in the previously discussed cases of convergent molecular evolution, Castoe et al. (2009, 2010) argue that there appear to be a limited number of ways in which amino acid sequences can be modified and yet remain functional in these proteins, and that sustained selective pressure may preferentially target the mtDNA encoding these proteins if there are only a few possible ways to achieve the same functional innovation. In addition, Naylor et al. (1995) and Yee (1999) have argued that the mtDNA molecule itself may be limited in its number of nucleotide permutations. They point out that the functional requirement for hydrophobicity in many mitochondrial proteins biases the second codon position in the encoding mtDNA to either T or C, and that molecular evolution may be constrained to cycle between these two nucleotide states at the second codon position. In summary, Castoe et al. (2010, 67) conclude that "molecular convergence can happen en masse in nature, affecting multiple genes . . . [and] the result implies that the protein adaptive landscape is sometimes highly constrained."

The extent of the newly discovered convergence in mtDNA has profound implications for molecular phylogenetic analyses, implications that Castoe et al. (2009, 8990) find worrisome: "contrary to widespread belief, . . . non-neutral convergence can be a major force in molecular evolution, and . . . it should be considered more seriously as a cause of phylogenetic incongruence among datasets." They point out that mtDNA molecular phylogenies would place agamid lizards and snakes in the same clade, whereas both morphological data and nuclear genome data place the agamid lizards with the basal Iguania and the snakes many nodes away in the highly derived Anguimorpha (table 5.2). But what about the numerous cases in nature where the morphological data and the nuclear genome data clash? In those cases, it is generally assumed that similarities in morphological traits have arisen by convergent evolution and are not synapomorphies, because the nuclear genome data can be trusted to be almost free from molecular convergences. What if this is not true? Other studies have identified parallel amino acid replacements across families of closely linked genes, such as the histocompatibility complex in humans (Yeager and Hughes 1999) and toxin-resistant sodium channels in puffer fish (Jost et al. 2008), that are also argued to be due to nonrandom natural selection.

In an intriguing study, Rodríguez-Trelles et al. (2003) document the convergent evolution of the aldehyde oxidase gene in two independent gene-duplication events combined with identical amino acid replacements in the duplicated DNA (table 5.3). The aldehyde oxidase gene, *Ao*, which codes for the aldehyde oxidoreductase enzyme, AOX, is known to have evolved by gene duplication of the xanthine dehydrogenase gene,

Table 5.3
Convergent evolution of identical nuclear DNA molecules coding for aldehyde oxidoreductase enzymes

Convergent molecule and function: ALDEHYDE OXIDOREDUCTASE–ENCODING DNA (identical 10 nucleotide substitutions in the DNA coding for the aldehyde oxidoreductase enzyme that catalyzes the oxidation of aldehydes into acids) *Convergent lineages:* 1 Asian rice (Eukarya: Bikonta: Chlorobiota: Embryophyta: Euangiosperms: Monocotyledons: Poaceae; *Oryza sativa*) 2 Puffer fish (Eukarya: Unikonta: Metazoa: Vertebrata: Osteichthyes: Actinoptergyii: Tetraodontidae; *Fugu rubripes*)

Note: For data sources, see text.

Xdh, in eukaryotes before the evolution of multicellular life. The AOX enzyme is present in metazoans up to the evolution of the primitive chordates, as it is still present in the urochordate sea peach *Ciona intestinalis*, after which point it was evidently lost. The AOX enzyme was then independently reevolved in the osteichthyan fishes by a second duplication event in the *Xdh* gene, producing a new *Ao* gene, and is present in subsequent vertebrates, including humans.

What is particularly interesting in these gene duplication events is the fact that "after each duplication, the *Ao* duplicate underwent a period of rapid evolution during which identical sites across the two molecules . . . were subjected to intense positive Darwinian selection" (Rodríguez-Trelles et al. 2003, 13413), shifting the substrate acted upon by the enzyme from hypoxanthine in the ancestral gene coding to aldehydes in the new genes. In total, 10 sites in the DNA of the two independent *Ao* genes have identical nucleotide substitutions, as seen in the match of the two independent AOX enzyme amino acid sequences. The two convergent *Ao* genes "originated from *Xdh* by duplications separated by approximately one billion years (i.e., the time span from the origin of multicellular eukaryotes to the last common ancestor of vertebrates)," yet the very same nucleotide substitutions still occurred in the DNA of the independent gene duplicates due to natural selection for the same enzymatic function (Rodríguez-Trelles et al. 2003, 13417).

At present, relatively few examples of adaptive convergent molecular evolution at the level of DNA nucleotide sequences are known. However, Castoe et al. (2010) point out that (1) we are only now beginning to acquire large, complete genome data sets that can be analyzed for convergence, (2) systematic surveys for molecular convergence are usually not conducted, and (3) it is difficult to distinguish random, neutral molecular convergence from nonneutral, adaptive molecular convergence. They suggest that the phenomenon may be much more widespread than believed, and that "greater detection of adaptive convergence events in nature may also provide important insight into the diversity of protein adaptive landscapes in nature, and help to explain why in some cases these landscapes appear highly constrained" (Castoe et al. 2010, 69).

Last, Stern and Orgogozo (2009, 747) also argue that the phenomenon of parallel genetic evolution is much more common than believed, as in their study they examined 350 different mutations that produce phenotypic variation and found that "more than half of these represent cases of parallel genetic evolution." Noting that some of these clearly involve DNA coding for gene products (such as photosensitive optical protcins and digestive proteins that we have considered previously; see tables 5.1 and 5.2, respectively), they point out that "gene function explains part but not all of the observed pattern of parallel genetic evolution. In several cases, parallelism has been observed even though mutations in a large number of genes can produce similar phenotypic changes."

Stern and Orgogozo (2008, 2009) and Stern (2010) point out that there are two fundamentally different types of DNA coding for genes: one DNA region that codes for the gene product (protein or RNA), and another adjacent *cis*-regulatory DNA region that encodes the instructions determining when and where the gene product will be produced. They point out that, given the molecular data presently at hand, the overwhelming number of mutations that produce morphological variation between species occur in these *cis*-regulatory regions, and not in the gene product regions. Stern and Orgogozo (2009, 750) further argue that "[a]lthough mutations are thought to occur randomly in the genome, the distribution of mutations that cause biological diversity appears to be highly nonrandom" (that is, preferentially in *cis*-regulatory DNA), and that the "genetic basis of phenotypic evolution thus appears to be somewhat predictable." Convergent molecular evolution is here seen to be a product of preferential mutational changes in a limited subset of *cis*-regulatory DNA regions, which they term "hot spots": "evolutionary relevant mutations are expected to accumulate in a few hot-spot genes and even in particular regions within a single gene" that control development (Stern and Orgogozo 2009, 750).

To summarize this section of the chapter, at the DNA level of molecular evolution we have examined examples of (1) the convergent evolution of identical nuclear DNA molecules and (2) the convergent evolution of identical mitochondrial DNA molecules, both produced by identical nucleotide substitutions occurring in the molecules of very distantly related organisms. The DNA molecule possesses only four nucleotides, and has a relatively simple geometry: the double helix. In the next section we shall examine the convergent evolution of the complex geometries and chemical compositions of the huge protein molecules.

Proteins and Protein Functions

The independent evolution of two convergent protein molecules with the same amino acid sequences does not necessarily mean that the DNA coding for those two convergent proteins is itself exactly the same. This is due to the fact that most amino acids are coded for by more than one DNA codon; for example, the six codons TCT, TCC, TCA, TCG, AGT, and AGC all code for the same amino acid, serine. Identical amino acid sequences may be encoded by two quite different DNA nucleotide sequences. Thus, the phenomenon of convergent evolution in protein molecules is at least semi-independent of the phenomenon of convergent evolution at the level of the DNA molecules.

Indeed, Chen et al. (1997) have demonstrated the convergent evolution of an antifreeze protein molecule in two independent groups of cold-water fishes that are encoded by totally different DNA (Logsdon and Doolittle 1997). The convergent molecule is an antifreeze glycoprotein, AFGP, found in paracanthopterygian gadiform fish in the Arctic waters of the Northern Hemisphere of the Earth; it is also found in acanthopterygian perciform fish in Antarctic waters on the opposite side of the planet (table 5.4). The two groups of fishes are adapted for existence in subfreezing temperatures, and have evolved independently in isolation at the two poles of the planet, geographically separated by the wide expanse of temperate- and tropical-temperature waters in between.

The AFGP of the Antarctic notothenioid fish is encoded by a gene that is largely constructed of previously noncoding DNA, combined with bits of DNA coding for a pancreatic trypsinogen gene that is used in digestion. This gene may have been modified for preventing freezing in the intestinal fluid, a function that was later expanded throughout the circulatory system in the evolution of the new AFGP gene, whereas the AFGP gene of the Arctic cod fish shows no DNA nucleotide sequence identity with the trypsinogen gene, and arose from a completely unrelated genomic locus (Chen et al. 1997; Logsdon and Doolittle 1997). Nonetheless, the Antarctic notothenioid and Arctic cod AFGP molecules

Table 5.4
Convergent evolution of identical antifreeze protein molecule in independent lineages, and of different molecules that serve the same antifreeze molecular function

A Convergent molecule (same molecule, different convergent lineages)

1 Convergent protein and function: AFGP (blood serum glycoprotein used to lower the freezing temperature of tissue to prevent cell rupture)

Convergent lineages:

1.1 Artic cod (Osteichthyes: Actinopterygii: Teleostei: Euteleosti: Paracanthopterygii: Gadiformes: Gadidae; *Boreogadus saida*)

1.2 Antarctic notothenioid (Osteichthyes: Actinopterygii: Teleostei: Euteleosti: Acanthopterygii: Perciformes: Notothenidae; *Dissostichus mawsoni*)

2 Convergent protein and function: AFP Type II (blood serum protein used to lower the freezing temperature of tissue to prevent cell rupture)

Convergent lineages:

2.1 Atlantic herring (Osteichthyes: Actinopterygii: Teleostei: Clupeomorpha: Clupeiformes: Clupeidae; *Clupea harengus*)

2.2 American smelt (Osteichthyes: Actinopterygii: Teleostei: Euteleosti: Protocanthopterygii: Osmeriformes: Osmeridae; *Osmerus mordax*)

2.3 Sea raven (Osteichthyes: Actinopterygii: Teleostei: Euteleosti: Acanthopterygii: Scorpaeniformes: Cottidae; *Hemitripterus americanus*)

B Convergent molecular function (same function, different convergent molecules): Molecular function: Antifreeze protection

Convergent molecules:

1 AFP Type I

Lineage: Winter flounder (Osteichthyes: Actinopterygii: Teleostei: Euteleosti: Acanthopterygii: Pleuronectiformes: Pleuronectidae; *Pleuronectes americanus*)

2 AFP Type III

Lineage: Ocean pout (Osteichthyes: Actinopterygii: Teleostei: Euteleosti: Acanthopterygii: Perciformes: Zoarcidae; *Macrozoarces americanus*)

3 AFP Type IV

Lineage: Longhorn sculpin (Osteichthyes: Actinopterygii: Teleostei: Euteleosti: Acanthopterygii: Scorpaeniformes: Cottidae; *Myoxocephalus octodecimspinosus*)

4 AFP Cf

Lineage: Spruce budworm (Arthropoda: Mandibulata: Hexapoda: Lepidoptera: Tortricidae; *Choristoneura fumiferana*)

5 AFP Tm

Lineage: Yellow mealworm beetle (Arthropoda: Mandibulata: Hexapoda: Coleoptera: Polyphaga: Tenebrionidae; *Tenebrio molitor*)

Note: For data sources, see text.

are nearly identical in amino acid composition and are comprised mainly of threonine-alanine-alanine repeats, usually encoded by ACA-GCT/G-GCA codons in the Antarctic notothenioids, whereas the Arctic cod use DNA codons rarely observed in the notothenioid fish (Logsdon and Doolittle 1997). The convergent amino acid triplet-repeating structure of AFGP appears to be required for antifreeze-ice interaction to occur, and probably evolved in the notothenioids about 10 to 14 million years ago, as the Antarctic Ocean began to freeze, and about two to three million years ago in the cod, during the Arctic glaciation (Chen et al. 1997; Logsden and Doolittle 1997).

Subsequent work has revealed the convergent evolution of yet another antifreeze molecule: AFP Type II. This molecule evolved independently in cold-water clupeomorph fish (Atlantic herring) and in two clades of euteleostian fishes, the protocanthopterygian American smelts and the acanthopterygian sea ravens (table 5.4). Unlike in the case of AFGP, the convergent evolution of AFP Type II in these three fish groups appears to have been a case of parallel independent changes in the same C-type lectin genes ancestrally found in all three groups (Fletcher et al. 2001).

Convergence in antifreeze protein function does not always produce convergent evolution, or parallel evolution, of the same antifreeze molecules, as in the case AFGP or AFP Type II. Consider the convergent evolution of wings in insects, pterosaurs, birds, and bats (see table 2.2), all of which serve the common function of generating lift for powered flight. The structure of the wing is different in each animal group, but the function is the same. Fletcher et al. (2001) point out that three additional antifreeze proteins have evolved in the cold-water fishes—AFP Types I, III, and IIV—and Davies et al. (2002) discuss the evolution of two more antifreeze proteins that have evolved in the arthropods, AFP Cf and Tm (table 5.4).

The amino acid sequences in these additional antifreeze proteins are different, but aspects of their geometries are very similar (Davies et al. 2002), and their functions are the same. In particular, Davies et al. (2002, 932) point out that the helical geometries of AFP Cf and AFP Tm are such that "their ice-binding sites are virtually superimposable" even though AFP Cf is a left-coiled β-helix and AFP Tm is a right-coiled β-helix. Davies et al. (2002, 932) describe this geometric result as a "remarkable example of convergent evolution" in that the "two ranks of threonine residues line up perfectly because the threonines are in the same rotameric configuration" in the helices.

Preventing tissue freezing is a specialized problem in a restricted ecological setting, namely the polar regions of the Earth. Transporting oxygen from one region of tissues to another is a functional challenge that is ubiquitous to life itself. Three different types of proteins have convergently evolved to fulfill the same oxygen-binding function: hemoglobin, hemocyanin, and hemerythrin (table 5.5). Of these three, hemoglobin is by far the most widespread in life on Earth (Hardison 1996). All three molecules are ancient: Bailly et al. (2008) argue that α-helical bundles were the earliest protein folds to emerge in the evolution of life, as they are well adapted to binding metal ions. Thus, the globins and the hemerythrins represent two ancient families of molecules that evolved the functions of sequestering reduced iron and sensing and controlling oxygen concentrations, as corrosive oxygen was lethal to the earliest

Table 5.5
Convergent evolution of similar oxygen-binding protein molecules in independent lineages, and of different molecules that serve the same oxygen-binding molecular function

A Convergent molecule (same molecule, different convergent lineages):

1 Convergent protein and function: HEMOCYANIN (protein that uses copper to bind oxygen for the function of oxygen transport)

Convergent lineages:

1.1 Common octopus (Protostomia: Lophotrochozoa: Eutrochozoa: Mollusca: Cephalopoda: Octopodidae; *Octopus vulgaris*)

1.2 Horseshoe crab (Protostomia: Ecdysozoa: Arthropoda: Cheliceriformes: Limulidae; *Limulus polyphemus*)

2 Convergent protein and function: HEMERYTHRIN (co-option of the ancient nonheme oxygen-sensor molecule hemerythrin for the function of oxygen transport)

Convergent lineages:

2.1 Terebratulid brachiopod (Protostomia: Lophotrochozoa: Lophophorata: Brachiopoda: Cancellothyrididae; *Terebratulina retusa*)

2.2 Serpulid bristle worm (Protostomia: Lophotrochozoa: Eutrochozoa: Spiralia: Sipuncula: Sipunculidae; *Sipunculus nudus*)

2.3 Priapulid worm (Protostomia: Ecdysozoa: Introverta: Priapulida: Priapulidae; *Priapulus caudatus*)

3 Convergent protein and function: two-CCP-domain HAPTOGLOBIN (modified hemoglobin-binding plasma protein with the ability to form high-molecular-weight oligomers)

Convergent lineages:

3.1 Cow (Eutheria: Laurasiatheria: Cetartiodactyla: Ruminantia: Bovidae; *Bos taurus*)

3.2 Human (Eutheria: Euarchontoglires: Primates: Catarrhini: Hominidae; *Homo sapiens*)

B Convergent molecular function (same function, different convergent molecule): Molecular function: oxygen transport; convergent molecule: HEMOGLOBIN

Lineage: Earthworm (Protostomia: Lophotrochozoa: Eutrochozoa: Spiralia: Annelida: Lumbricidae; *Lumbricus terrestris*)

Note: For data sources, see text.

forms of anaerobic life (Bailly et al. 2008). With the evolution of aerobic life, transport of oxygen within the organism for metabolic purposes is a critical function, and van Holde et al. (2001, 15566) hypothesize that oxygen-transport proteins developed "in several independent ways, hemoglobins from myoglobins, hemerythrins from myohemerythrins, and the two kinds of hemocyanins from two different classes of phenol oxidases."

Hemocyanins evolved twice, independently, in the clade of the molluscs (lophotrochozoans) and the clade of the arthropods (ecdysozoans; table 5.5). In the molluscs, Class 3a hemocyanins are used for oxygen transport in the blood of cephalopods, such as the octopus, and of gastropods, including both marine and terrestrial snails. In the arthropods, Class 3b hemocyanins are present in the blood of cheliceriformes, such as the horseshoe crab, and mandibulate crustaceans, such as the true crabs (van Holde et al. 2001).

The hemocyanins bind oxygen with copper (thus, the blood of an octopus is bluish-green when it is oxygenated, not red like ours), but copper used as an oxygen-binding metal is not as efficient as iron. Both hemoglobin and hemerythrin use iron, but the latter is a nonheme protein (despite its deceptive name) unlike hemocyanin and hemoglobin, and is less efficient in oxygen transport than hemoglobin. The hemerythrins are patchily present throughout life, from bacteria to archaea to the eukarya, and are used in a variety of functions, the most ancient of which is oxygen sensing (Bailly et al. 2008). Only late in the evolution of life have three groups of protostome invertebrates—the brachiopods, serpulids, and priapulids (table 5.5)—independently co-opted the hemerythrin protein for the primary function of oxygen transport. Blood sufficiently enriched in hemerythrin is pale violet when oxygenated, not blue-green as for hemocyanin or red as for hemoglobin.

Last, mammals possess a specialized plasma protein, haptoglobin, that can be used to inhibit hemoglobin's oxidative activity, if needed. A variant of this protein, two-CCP-domain haptoglobin, independently evolved in the ruminant mammals and, curiously, in humans (table 5.5). Because the two-CCP-domain haptoglobin is found throughout the clade of the ruminants, the mutation producing this protein occurred early in the Cenozoic for these mammals (Wicher and Fries 2007). The same protein evolved much more recently in humans, for it is not present in our close primate cousins. In both cows and humans, the convergent duplication of the gene segment coding for the CCP domain of the α-chain in the haptoglobin molecule resulted in a much longer α-chain in the new protein, which contains two CCP domains in tandem. Wicher and Fries (2007) point out that the new protein can form much higher-molecular-weight oligomers than the conventional protein, and suggest that the different binding properties of this molecule is useful in preventing infection in the ruminant mammals. Its function in humans is less clear, but data do exist that indicate humans with the mutated haptoglobin molecule are more resistant to life-threatening streptococcus infections, and possibly to malarial infections (Wicher and Fries 2007).

Preventing tissue death by freezing or oxygen starvation is a matter of life or death, yet convergent molecular evolution also occurs in situations that are not so dire. Mundy (2005) reviews the recent discovery that melanic polymorphism in three groups of distantly related wild birds has evolved in parallel, in that the plumage color variations seen in these birds are produced by point mutations in amino acid substitution in the same gene coding for the MC1R protein (table 5.6). Many birds are well known for the dazzling colors of their plumage, and so it was unexpected to find parallel changes in the same gene to be producing the color variations in such distantly related species. Mundy (2005, 1638) concludes by noting that the "examples discussed here clearly only represent the tip of the iceberg as far as avian plumage colouration genetics in the wild is concerned. Over 300 avian species exhibit plumage polymorphisms"; thus, he implies that more examples of parallel evolution of the MC1R protein may soon be discovered.

Likewise, Protas et al. (2005) have shown that mutations in the protein-encoding sequences of the same gene, the one coding for

Table 5.6
Convergent evolution of similar melanin-controlling protein molecules

1 Convergent protein and function: MC1R (protein producing plumage color variations in birds)
Convergent lineages:
1.1 Snow goose (Aves: Neognathae: Galloanserae: Anseriformes: Anatidae; *Anser caerulescens*)
1.2 Bananaquit (Aves: Neognathae: Neoaves: Passeriformes: Coerebidae; *Coereba flaveola*)
1.3 Arctic skua (Aves: Neognathae: Neoaves: Charadriiformes: Stercorariidae; *Stercorarius parasiticus*)

2 Convergent protein and function: OCA2 (protein producing albinism in fish)
Convergent lineages:
2.1 Molino cave populations of the Mexican tetra, *Astyanax mexicanus*
2.2 Pachón cave populations of the Mexican tetra, *Astyanax mexicanus*
2.3 Japonés cave populations of the Mexican tetra, *Astyanax mexicanus*

Note: For data sources, see text.

OCA2, produces albinism in three independent populations of cave fish (table 5.6). Animals living in the perpetually dark environment of caves routinely lose the function of their eyes and experience pigment loss. This same evolutionary trend repeatedly occurs across phylogenetically diverse lineages, in very distantly related organisms such as spiders, isopods, fish, and salamanders. As a result, it was a surprise when, in studying pigment loss in the Mexican tetra cavefish in two widely separated caves in Mexico, Molino and Pachón, Protas et al. (2005, 109) discovered that "albinism evolved independently in these two caves, by convergent evolution in the same gene." Preliminary analyses indicate that yet another independent cave population of these fish, Japonés, also has experienced melanin loss by convergent evolution of the gene producing OCA2. The independent evolution of albinism was an expected given, but the fact that three independent populations of cavefish experienced mutations in the very same gene to produce that albinism was not.

An even more unexpected demonstration of convergent evolution in the same gene has been discovered by Colosimo et al. (2005) in 14 independent freshwater populations of the three-spine stickleback fish species *Gasterosteus aculeatus* (table 5.7). The *Gasterosteus aculeatus* species complex contains two strikingly different end-member phenotypes: a heavily armored ecomorph, typically possessing a row of over 30 bony armor plates extending along the length of the body of the fish, and a lightly armored ecomorph, possessing at most nine plates only at the anterior end of the fish. The heavily armored ecomorph is found exclusively in marine waters, and the lightly armored ecomorph in freshwater streams and lakes. These end-member phenotypes are so different that they were classified as separate species by the great French morphologist Georges Cuvier in 1829, but subsequent observation has demonstrated that the two ecomorphs freely interbreed and hybridize (Colosimo et al. 2005).

Of particular interest is the fact that the lightly armored ecomorph repeatedly and independently evolved from members of the heavily armored marine ecomorph that invaded the new freshwater stream and lake habitats which had been created around the world following the retreat and melting of the glaciers from 20,000 to 10,000 years ago. The greater maneuverability and flexibility of the lightly armored ecomorph is believed to be of selective advantage in freshwater predatory regimes, as opposed to open-ocean marine ones (Colosimo et al. 2005).

Table 5.7
Convergent evolution of the EDA protein in independent freshwater populations of the three-spine stickleback fish species *Gasterosteus aculeatus*

Convergent protein and function: EDA (protein used in constructing rows of bony armor plates on the bodies of stickleback fish)
Convergent populations:
A European populations
1 Schwale (Germany) freshwater *G. aculeatus*
2 Loch Fada (Scotland) freshwater *G. aculeatus*
3 Norway stream (Norway) freshwater *G. aculeatus*
4 Blautaver (Iceland) freshwater *G. aculeatus*
B Eastern North American population
5 Olmstead Park (Massachusetts) freshwater *G. aculeatus*
C Western North American populations
6 Santa Clara River (California) freshwater *G. aculeatus*
7 Friant (California) freshwater *G. aculeatus*
8 San Francisco Creek (California) freshwater *G. aculeatus*
9 Conner Creek (Washington) freshwater *G. aculeatus*
10 Paxton Lake (Canada) freshwater *G. aculeatus*
11 Salmon River (Canada) freshwater *G. aculeatus*
12 Wallace Lake (Alaska) freshwater *G. aculeatus*
13 Alaska stream (Alaska) freshwater *G. aculeatus*
D Japanese population
14 Nakagawa Creek (Japan) freshwater *G. aculeatus*

Note: For data sources, see text.

The protein EDA is used in constructing the body armor of *Gasterosteus aculeatus*, and Colosimo et al. (2005) have demonstrated that four mutations in the *Eda* protein-encoding gene repeatedly and independently occur in 13 freshwater populations of lightly armored ecomorphs around the world. One exception to the convergent evolution of the freshwater stickleback *Eda* gene is the Nakagawa Creek population in Japan (table 5.7), a population with a lightly armored phenotype that "is due to an independently derived allele of *Eda*" different from those of the other 13 freshwater populations examined (Colosimo et al. 2005, 1929).

Other examples of the convergent evolution of structural proteins are due not to convergent changes in the same gene in independent lineages, but to identical structural repeats in the amino acid sequences comprising those proteins. Four different types of fibroelastic structural proteins have independently evolved in the arthropods and vertebrates, all of which have multiple glycine-glycine-leusine-glycine-tyrosine amino acid repeats in the protein (table 5.8). The amino acid sequence convergences

Table 5.8
Convergent evolution of similar structural protein molecules

1 Convergent proteins and functions: OOTHECIN, CHORION CLASS B, LAMPRIN, and ELASTIN (fibroelastic structural proteins used in the construction of chordate cartilages, arthropod silks, and insect egg cases)

Convergent lineages:

1.1 American cockroach (Bilateria: Protostomia: Ecdysozoa: Arthropoda: Hexapoda: Blattodea: Blatellidae; *Periplaneta americana*)

1.2 Domestic silk moth (Arthropoda: Hexapoda: Lepidoptera: Bombycidae; *Bombyx mori*)

1.3 Sea lamprey (Bilateria: Deuterostomia: Chordata: Vertebrata: Petromyzontiformes: Petromyzontidae; *Petromyzon marinus*)

1.4 Norwegian rat (Chordata: Vertebrata: Osteichthyes: Sarcopterygii: Reptiliomorpha: Amniota: Synapsida: Therapsida: Mammalia: Eutheria: Muridae; *Rattus norvegicus*)

2 Convergent proteins and functions: QID74 and BR3 (proteins used to construct cell walls by filamentous fungus and to construct external tube walls by midges)

Convergent lineages:

2.1 Mycoparasitic filamentous fungus (Unikonta: Fungi: Eumycetes: Ascomycota: Hypocreaceae; *Trichoderma harzianum*)

2.2 Midge (Unikonta: Metazoa: Protostomia: Arthropoda: Hexapoda: Chironomidae; *Chironomus tentans*)

3 Convergent protein and function: APOLIPOPROTEIN(a) (adaptive function uncertain; possibly protects against certain infectious diseases by inhibiting plasminogen activation. Otherwise similar to lipoproteins that bind fibrin and deliver cholesterol in reconstructing damaged tissue)

Convergent lineages:

3.1 European hedgehog (Eutheria: Laurasiatheria: Eulipotyphles: Erinaceidae; *Erinaceus europeus*)

3.2 Human (Eutheria: Euarchontoglires: Primates: Catarrhini: Hominidae; *Homo sapiens*)

Note: For data sources, see text.

seen in these proteins are “driven by structural or functional properties imparted to the protein by the shared sequence,” producing the ability of these “predominantly hydrophobic proteins to self-organize into a polymeric, fibrillar matrix” (Robson et al. 2000, 1751). Twenty-eight of the amino acids in the cockroach oothecin protein are in precisely the same sequence in the sea lampry lamprin protein, an amino acid sequence convergence that “may represent one of the best examples of primary sequence convergence so far identified” (Robson et al. 2000, 1751). These two animals are very distantly related: the cockroach is a protostomous bilaterian, and the sea lampry is a deuterostome.

Rey et al. (1998, 6212) report an “unexpected homology” found in comparing a fungal and an insect protein: fully 25.3 percent of the amino acid sequence is identical in the fungal Qid74 protein and the insect BR3 protein (table 5.8). The fungal protein is used in cell wall construction, whereas the insect protein is secreted in the midge's saliva, and is used in constructing external tube walls within which the insect lives. Rey et al. (1998, 6215) note that the two genes coding for the two proteins, one fungal and one insect, had to have originated independently of each other, and that the "ancestor of each new gene created *de novo* has to be a noncoding DNA sequence that abounds in every eukaryotic genome"; furthermore, the observed protein convergence "was not between the genes but between the similar noncoding repeats whence the two [new] genes sprung quite independently of each other."

More enigmatic is the convergent evolution of the protein apolipoprotein(a), or apo(a), in the hedgehog and in humans (table 5.8). Apo(a) is part of the low-density lipoprotein complex (LDL) that is a known risk factor in the development of atherosclerosis in humans, hence its research interest in medicine. It is known that apo(a) is found in Old World primates (Catarrhini), but it was a surprise to find the same protein in the very distantly related European hedgehogs: Lawn et al. (1997, 11992) conclude that it is "a remarkable example of 'parallel gene evolution.' . . . By apparent remodeling of a plasminogen-like gene, hedgehog and human ancestors independently evolved an apo(a) protein" that functions identically in both organisms. Unfortunately, some of those functions increase the risk of cerebral stroke in humans, which is one of the reasons the protein was under study in the first place.

Last, we have the convergent evolution of protein function in which multiple independent proteins with their own independent functions are co-opted to serve the convergent protein function, a process known as gene sharing. Convergent evolution of protein function by gene sharing was first discovered in the crystallin proteins of the eye lens in animals: "The term 'gene sharing' was used initially to generalize the finding that crystallins with a structural, optical function in the lens may also be expressed in other tissues [in the body], where they have a metabolic, nonrefractive function. Thus, the protein encoded in the *identical* gene may perform entirely different functions depending upon its expression pattern . . . gene regulation is a critical factor for gene sharing, leading to the use of a metabolic [enzyme] or stress protein as a structural crystallin. It is ironic that the abundantly expressed crystallins, long considered inert bricks as it were, squirreled away in a tiny transparent tissue . . . have become landmarks for multifunctional uses of widely expressed proteins that illustrate the dynamism of evolutionary change!" (Piatigorsky 2007, 55–56; author's emphasis).

Subsequent analysis has revealed the convergent evolution of J-crystallins, Ω-crystallins, S-crystallins, and drosocrystallins in invertebrate animals (table 5.9). The α-, β-, and γ-crystallins are found in the eye lenses of all known vertebrates, but the vertebrate βγ-crystallins are believed to have evolved from the Ci-βγ-crystallin of the urochordates (table 5.9), and thus are not independent originations (Piatigorsky 2007). Within the vertebrates, δ-crystallins independently evolved in the sauropsid clade, and are present in modern reptiles and avian dinosaurs. Other crystallins are more restricted in their phylogenetic distribution, and are called "taxon-specific." These include the ρ-, π-, ι-, and ρB-crystallins in some frogs and geckos, τ-crystallins in some fish and turtles, and ε-crystallins in the archosaurs (table 5.9). Within the mammals, υ-crystallins independently evolved in the monotremes, μ-crystallins in the marsupials, and η-, ζ-, and λ-crystallins in separate branches of the placental mammal clade (table 5.9). In conclusion, Piatigorsky (2007, 65) states: "the diversity and taxon-specificity of lens crystallins throughout the animal kingdom are consistent with overlapping functions by different proteins, indicative of convergent evolution for crystallin roles."

To summarize this section of the chapter, at the protein level of molecular evolution we have examined examples of (1) the convergent evolution of similar amino acid sequences in unrelated proteins, (2) the convergent evolution of similar structural geometries in proteins with different amino acid sequences, and (3) the convergent evolution of similar protein function by gene sharing, in which the same protein-encoding gene is used for two entirely different functions depending upon its regulatory expression. Proteins converge in every possible way.

Enzymes and Enzyme Functions

Enzymes are specialized protein molecules that catalyze the chemical reactions of other molecules. Theoretically, it is the catalytic function itself that is subject to natural selection, not the specific molecule producing that function, hence opening the possibility that different types of molecular structures might be selected if they could be modified to produce the same catalytic function. Thus, the phenomenon of convergent evolution in enzyme function is at least semi-independent of the phenomenon of convergent evolution of identical protein molecules.

We have previously examined the convergent evolution of some enzymes, such as stomach lysozyme (table 5.1) and aldehyde oxidore-

Table 5.9
Convergent evolution of crystallin protein molecules by gene sharing

Convergent molecular function: Construction of a transparent eye lens that optimizes the refractive index for focusing and image formation

Convergent molecules:

1 J1-, J2-, J3-CRYSTALLINS
Lineage: Box jellyfish (Metazoa: Cnidaria: Cubozoa: Carybdeidae; *Tripedalia cystophora*)

2 Ω/L-CRYSTALLIN
Lineage: Sea scallop (Metazoa: Bilateria: Protostomia: Lophotrochozoa: Mollusca: Bivalvia: Pectinidae; *Placopecten magellanicus*)

3 SL11/Lops4-, S-CRYSTALLINS
Lineage: Octopus (Protostomia: Lophotrochozoa: Mollusca: Cephalopoda: Octopodidae; *Octopus vulgaris*)

4 DROSOCRYSTALLIN
Lineage: Fruit fly (Protostomia: Ecdysozoa: Arthropoda: Hexapoda: Drosophilidae; *Drosophila melanogaster*)

5 Ci-βγ-CRYSTALLIN
Lineage: Sea squirt (Bilateria: Deuterostomia: Urochordata: Ascidiacea: Cionidae; *Ciona intestinalis*)

6 τ-CRYSTALLIN
Lineage: Sea lamprey (Deuterostomia: Chordata: Vertebrata: Petromyzontiformes: Petromyzontidae; *Petromyzon marinus*)

7 ρ-CRYSTALLIN
Lineage: Leopard frog (Chordata: Osteichthyes: Sarcopterygii: Tetrapoda: Batrachomorpha: Lissamphibia: Anura: Ranidae; *Rana pipiens*)

8 π-, ι-, ρB-CRYSTALLINS
Lineage: Striped day gecko (Tetrapoda: Reptiliomorpha: Amniota: Sauropsida: Lepidosauromorpha: Scleroglossa: Gekkonidae; *Phelsuma lineata*)

9 ε-CRYSTALLIN
Lineage: Crocodile (Sauropsida: Archosauromorpha: Crurotarsi: Crocodylidae; *Crocodylus niloticus*)

10 δ1-, δ2-CRYSTALLINS
Lineage: Chicken (Sauropsida: Archosauromorpha: Ornithodira: Dinosauria: Saurischia: Theropoda: Maniraptora: Aves: Galliformes: Phasianidae; *Gallus gallus*)

11 υ-CRYSTALLIN
Lineage: Platypus (Amniota: Synapsida: Therapsida: Mammalia: Monotremata: Ornithorhynchidae; *Ornithorhynchus anatinus*)

12 μ-CRYSTALLIN
Lineage: Kangaroo (Mammalia: Marsupialia: Diprotodontia: Macropodidae; *Macropus rufus*)

13 η-CRYSTALLIN
Lineage: Elephant shrew (Mammalia: Eutheria: Afrotheria: Macroscelidea: Macroscelididae; *Elephantulus rozeti*)

14 ζ-CRYSTALLIN
Lineage: Camel (Mammalia: Eutheria: Laurasiatheria: Cetartiodactyla: Camelidae; *Camelus dromedarius*)

15 λ-CRYSTALLIN
Lineage: Rabbit (Mammalia: Eutheria: Euarchontoglires: Lagomorpha: Leporidae; *Oryctolagus cuniculus*)

Note: For data sources, see text.

ductase (table 5.3), that is caused by the convergent evolution of the DNA directly coding for those proteins. Therefore, in these instances the convergent enzyme function is a result of convergent evolution of the same protein producing that function. In a case of "convergent evolution of similar enzymatic function on different protein folds," Bork et al. (1993, 31) argue that the "enzymatic function of sugar phosphorylation appears to have evolved independently . . . by convergent evolution" on three different protein geometries: the hexokinase, ribokinase, and galactokinase families of protein folds (table 5.10). Although proteins in each of these three families have different three-dimensional folds and strikingly different sequence patterns, "each catalyzes chemically equivalent reactions on similar or identical substrates" (Bork et al. 1993, 31). Moreover, enzymatic functions for specific sugars have independently evolved within the three kinase families: fructoskinase specificity has independently evolved in both the hexokinase and ribokinase families, and glucose specificity has evolved several times independently within the hexokinase family. Bork et al. (1993, 38) argue that two different types of molecular convergence are exhibited by the kinase enzymes: (1) from different protein structures to the same enzymatic function, and (2) from

Table 5.10
Convergent evolution of different enzyme molecules that serve the same metabolic catalytic function

1 Convergent enzyme function: SUGAR PHOSPHORYLATION (enzymes that catalyze the first step in sugar metabolism)

Convergent molecules:

1.1 HEXOKINASE family

Lineage: Human (Eukarya: Unikonta: Metazoa: Deuterostomia: Vertebrata: Hominidae; *Homo sapiens*)

1.2 RIBOKINASE family

Lineage: E. coli bacterium (Eubacteria: Proteobactia: γ-Proteobacteria; *Escherichia coli*)

1.3 GALACTOKINASE family

Lineage: Lactobacillus (Eubacteria: Firmicutes: *Bacillus* group; *Lactobacillus helveticus*)

2 Convergent enzyme function: β-ELIMINATION MECHANISM (enzymes that catalyze the β-elimination of sugar uronic acids)

Convergent molecules:

2.1 POLYGALACTURONIC ACID LYASE Pel10A

Lineage: Cellvibrio soil bacterium (Eubacteria: Proteobactia: γ-Proteobacteria; *Cellvibrio japonicus*)

2.2 POLYGALACTURONIC ACID LYASE Pel1C

Lineage: Erwinia plant-pathogen bacterium (Eubacteria: Proteobactia: γ-Proteobacteria; *Erwinia chrysanthemi*)

Note: For data sources, see text.

different functional specificities to similar functional specificities within the same protein structure.

In examining enzymes that catalyze the β-elimination of sugar uronic acids, Charnock et al. (2002, 12067) found that the "active center provides a stunning example of convergent evolution" between two bacterial enzymes (table 5.10). The catalytic center of the *Cellvibrio* soil bacterium's Pel10A enzyme is "isostructural with the catalytic center of the totally unrelated family PL-1 enzyme Pel1C" of the *Erwinia* plant-pathogen bacterium (Charnock et al. 2002, 12067), an "essentially identical disposition of six active-center groups despite no topological similarity between these enzymes" in a case of "convergent evolution of catalytic mechanism" (Charnock et al. 2002, 12070).

Serine protease enzymes serve a variety of important functions, from digestive to blood clotting. Reardon and Farber (1995) demonstrate that three very different serine proteases have converged on the same enzymatic function, peptide bond hydrolysis, in bacteria, plants, and animals (table 5.11). In their analysis of the α/β barrel proteins, they cite these three serine proteases as "the best example of convergent evolution to a similar active site," noting that the "three proteins have entirely different 3-dimensional structures" and yet the "catalytic triad is essentially identical in the three proteins" (Reardon and Farber 1995, 501). They further suggest that this convergence is due more to physics than selection: "convergent molecular evolution occurs when there is only one energetically reasonable pathway for a chemical reaction," such as peptide bond hydrolysis (Reardon and Farber 1995, 497).

Tripp et al. (2001) document the convergent evolution of carbonic anhydrases in the archaea, plants, and animals (table 5.11). The α-, β-, and γ-classes of carbonic anhydrase found in these independent lineages have no significant amino acid sequence identity, and their very different fold geometries "underscore their independent origins. Despite gross structural differences, the active sites of all three classes function with [a] single zinc atom . . . [and] all three classes employ a two-step iso-mechanism" (Tripp et al. 2001, 48615).

We have previously examined the convergent evolution of the stomach lysozymes (table 5.1), used in the digestion of cellulose. Another difficult molecule to digest is chitin. Shen and Jacobs-Lorena (1999) document the convergent evolution of chitinases in animals and plants (table 5.11). Arthropods use chitinase to digest the chitin of their exoskeletons during molting, whereas plants use chitinase to defend themselves against fungal and insect infections by breaking down the structural chitin

Table 5.11
Convergent evolution of different enzyme molecules that serve the same catalytic function

1 Convergent enzyme function: PEPTIDE BOND HYDROLYSIS (enzymes that catalyze the hydrolysis of a peptide bond using a Ser nucleophile)

Convergent molecules:

1.1 SUBTILISIN

Lineage: Hay bacillus (Eubacteria: Firmicutes: *Bacillus* group; *Bacillus subtilis*)

1.2 WHEAT SERINE CARBOXYPEPTIDASE

Lineage: Bread wheat (Eukarya: Bikonta: Chlorobionta: Tracheophyta: Angiospermae: Poaceae; *Triticum aestivum*)

1.3 CHYMOTRYPSIN

Lineage: Cow (Eukarya: Unikonta: Metazoa: Deuterostomia: Vertebrata: Bovidae; *Bos taurus*)

2 Convergent enzyme function: CO_2 HYDRATION (enzymes that catalyze the reversible hydration of carbon dioxide)

Convergent molecules:

2.1 γ-Class CARBONIC ANHYDRASE

Lineage: Methanosarcinalean archaeon (Archaea: Euryarchaeota: Methanosarcinales; *Methanosarcina thermophila*)

2.2 β-Class CARBONIC ANHYDRASE

Lineage: Pea (Eukarya: Bikonta: Chlorobiota: Tracheophyta: Angiospermae: Fabaceae; *Pisum sativum*)

2.3 α-Class CARBONIC ANHYDRASE

Lineage: Human (Eukarya: Unikonta: Metazoa: Deuterostomia: Vertebrata: Hominidae; *Homo sapiens*)

3 Convergent enzyme function: CHITIN DIGESTION (digestive enzymes that break down glycosidic bonds in chitin)

Convergent molecules:

3.1 Plant CHITINASE CBD

Lineage: Tobacco (Bikonta: Chlorobiota: Tracheophyta: Angiospermae: Solanaceae; *Nicotiana tabacum*)

3.2 Animal CHITINASE CBD

Lineage: Domestic silk moth (Unikonta: Metazoa: Protostomia: Arthropoda: Hexapoda: Bombycidae; *Bombyx mori*)

Note: For data sources, see text.

present in these organisms. The plant and animal chitinase proteins are very different in structure, and the chitin binding domains (CBDs) of the two enzymes do not share significant amino acid sequence similarity; nevertheless, the fact that "plant and invertebrate CBDs share a common core structure and chitin-binding mechanism while being unrelated in evolution, argue[s] in favor of convergent evolution" of these molecules (Shen and Jacobs-Lorena 1999, 343).

Last, convergent evolution also exists in enzymatic functions involving RNA molecules, as well as the proteins. Within the cell, the aminoacyl-tRNA synthetases are responsible for the synthesis of the substrates for ribosomal translation of messenger RNA, with 20 aminoacyl-tRNA synthetases corresponding to the 20 amino acids used in protein synthesis. Terada et al. (2002, 257) document the convergent evolution of two lysyl-tRNA synthetases in independent archaean and bacterial lineages (table 5.12), and point out that the "identical, sophisticated reaction occurs within two unrelated architectures" of the lysyl-tRNA synthetases-I (LysRS-I) and lysyl-tRNA synthetases-II (LysRS-II) molecules; furthermore, the "functional convergence of the two enzymes . . . is illustrated by the mirror images of LysRS-I and LysRS-II structures. . . . LysRS-I and LysRS-II can be docked simultaneously on the $tRNA^{Lys}$ molecule without any steric clashes" (Terada et al. 2002, 260).

The "hammerhead" macromolecular ribozyme structure occurs sporadically in both viruses and animals (table 5.13), and Salehi-Ashtiani and Szostak (2001) argue that the hammerhead ribozyme has evolved independently many times. Their experiments with this molecule "firmly suggest that the hammerhead ribozyme is the simplest RNA motif capable of catalysing RNA strand cleavage at rates between 1.0 and

Table 5.12
Convergent evolution of different enzyme molecules that serve the same RNA catalytic function

Convergent enzyme function: LYSYL-tRNA SYNTHESIS (enzymes that catalyze the synthesis of the substrate for $tRNA^{Lys}$)

Convergent molecules:

1 LYSYL-tRNA SYNTHETASE-I

Lineage: Thermococcalean archaeon (Archaea: Euryarchaeota: Thermococcales; *Pyrococcus horikoshii*)

2 LYSYL-tRNA SYNTHETASE-II

Lineage: E. coli bacterium (Eubacteria: Proteobactia: γ-Proteobacteria; *Escherichia coli*)

Note: For data sources, see text.

Table 5.13
Convergent evolution of the same ribozyme molecular structure

Convergent enzyme and function: HAMMERHEAD RIBOZYME (enzyme that catalyzes RNA strand cleavage at rates between 0.1 and 1.0 per minute)

Convergent lineages:

1 Cave cricket (Bilateria: Protostomia: Ecdysozoa: Arthropoda: Hexapoda: Orthoptera: Rhaphidophoridae; *Dolichopoda linderi*)

2 Newt (Bilateria: Deuterostomia: Chordata: Osteichthyes: Sarcopterygii: Batrachomorpha: Urodela: Salamandridae; *Notophthalamus viridescens*)

Note: For data sources, see text.

1.0" per minute (Salehi-Ashtiani and Szostak 2001, 82). They argue that their "results suggest that the evolutionary process may have been channelled . . . towards repeated selection of the simplest solution to a biochemical problem" (Salehi-Ashtiani and Szostak 2001, 82), and that "purely chemical constraints (that is, the ability of only certain sequences to carry out particular functions) can lead to the repeated evolution of the same macromolecular structures" (Salehi-Ashtiani and Szostak 2001, 84).

In summary, we have examined examples of (1) the convergent evolution of the same enzyme function produced by the convergent evolution of the same protein producing that function, (2) the convergent evolution of the same catalytic core structure in enzyme molecules that have no similarity in amino acid sequence or in the three-dimensional structure of the molecules, and (3) the convergent evolution of the same macromolecular structure in unrelated enzymes. Enzymes converge in every possible way.

Convergent Evolution of Cellular and Tissue Structures?

The self-assembly of molecules into crystals is a natural process driven by the laws of physics and, as we believe the laws of physics to be invariant in space, the same molecules will self-assemble into the same crystal structure throughout the universe. That is, a salt crystal (NaCl) that forms on Mars will have the identical crystal structure as a salt crystal that forms on the Earth. Denton and Marshall (2001) have argued that the same might be true for protein folds: the same amino acids will self-assemble into the same protein-fold geometries throughout the universe, as they are driven by the laws of physics to do so. They imply that convergent molecular evolution may be driven more by the laws of physics than by natural selection; in essence, natural selection has no choice. Selection for a given function will necessarily result in convergent molecular evolution.

Denton et al. (2002) further argue that, above the molecular level of protein-fold geometries, two other microbiological levels exhibit evidence of self-assembly that is determined by the laws of physics, similar to the nonbiological self-assembly of molecules into crystals. First are the cellular microtubules, whose forms "are not specified directly in the DNA, nor does any genetic program direct their assembly;" rather, they are generated by the dynamic interactions of a few basic components of the cell (Denton et al. 2002, 339). Second is cell form itself which, when experimentally altered by microsurgery, can "recover its proper form by searching a conformational space for its correct conformation, just like a folding protein" (Denton et al. 2002, 340). Therefore, like the molecules composing them, the macromolecular structures of the cell itself converge.

West et al. (1999) assert that biological convergence in form due to the laws of physics extends upward from the cell to entire tissue systems, such as capillary, intestinal, and lung geometries and surface areas in animals and root, branch, and leaf geometries and surface areas in plants. They note that biological form follows an allometric quarter-power scaling, rather than the expected third-power from Euclidean geometric scaling, and argue that this follows from the fractal-like architecture of all vascular branching systems in living organisms: "The vast majority of organisms exhibit scaling exponents very close to 3/4 for metabolic rate and 1/4 for internal times and distances. These are the maximal and minimal values, respectively, for the effective surface area and linear dimensions for a volume-filling fractal-like network . . . it is testimony to the severe geometric and physical constraints on metabolic processes, which have dictated that all of these organisms obey a common set of quarter-power scaling laws" (West et al. 1999, 1679). They continue: "[u]nlike the genetic code, which has evolved only once in the history of life, fractal-like distribution networks that confer an additional effective fourth dimension have originated many times" (West et al. 1999, 1679).

In summary, convergent evolution occurs across the entire spectrum of molecules that make up life. We have studied examples of the convergent evolution of identical nucleotide substitutions in nuclear and mtDNA molecules of distantly related organisms, similar amino acid sequences in unrelated protein molecules, similar structural geometries in proteins with different amino acid sequences, and similar protein functions by gene sharing; the convergent evolution of the same enzyme function produced by the convergent evolution of the same protein producing that function; the same catalytic core structure in enzyme molecules that have no similarity in amino acid sequence or in the three-dimensional structure of the molecules; and the convergent evolution of the same macromolecular structure in unrelated enzymes. The effects of functional constraint can be seen throughout molecular evolution, in that selection for a given function repeatedly produces convergent molecular compositions and/or molecular geometries. The number of functional molecular evolutionary pathways available to life is not endless but is quite restricted, and convergent evolution is the direct result.

In the next chapter we shall examine what, at first glance, might seem to be even more improbable than the convergent evolution of molecules. Animals have minds; we all think (some animals more than others). Can it be that convergence extends to minds, and even to the ways in which we think?

6 Convergent Minds

Discussions of the evolution of intelligence have focused on monkeys and apes because of their close evolutionary relationships to humans. Other large-brained social animals, such as corvids, also understand their physical and social worlds. . . . Because corvids and apes share these cognitive tools, we argue that complex cognitive abilities evolved multiple times in distantly related species with vastly different brain structures in order to solve similar socioecological problems. . . . [T]his convergent evolution of cognition has not been built on a convergent evolution of brains.
—Emery and Clayton (2004, 1903)

Convergent Minds?

How can minds be said to converge? We consider our mental states to be a function of our brain structures and sensory inputs. At first glance, then, organisms with radically different brain structures would be expected to have radically different minds. Take, for example, the brains of a magpie and a human. One is a bird brain, possessed by an avian dinosaur in the sauropsid clade of amniotes, and the other is a primate brain, possessed by a placental mammal in the synapsid clade of amniotes. The sauropsid and synapsid lineages diverged back in the Carboniferous, and have evolved along independent pathways ever since. Magpie brains and human brains are structurally very different, separated by a vast chasm of 340 million years of independent evolution. Yet it can be demonstrated that, in many ways, magpies think like humans (Emery and Clayton 2004; Prior et al. 2008). Our minds have converged.

Problem-Solving Behavior

It was long thought that the tool was the unique invention of the human mind. Only the human mind had the ability to understand the limits to accomplishing a task using the body alone, and thus to conceive of extending those limits by using an external object to accomplish that task. We now know that this long-held belief is not true; other minds have made that conceptual leap as well. Quite a few other minds—no less than 12 independent lineages of animals use tools (table 6.1).

Within our own lineage, we can trace the manufacture of stone tools back some 2.6 million years ago to our ancestor *Homo habilis*, aptly named the "handyman" (evidence of tool usage in our lineage goes as far back as *Australopithecus afarensis*, some 3.39 million years ago, but it is not clear if *A. afarensis* manufactured stone tools or simply used naturally occurring sharp stones as tools; see McPherron et al. 2010).

Table 6.1
Convergent evolution of tool-using and tool-making behavior

1 Convergent behavior and function: TOOL USAGE (using an existing external object as a tool to attain an immediate goal)

Convergent lineages:

1.1 Black-and-yellow mud wasp (Bilateria: Protostomia: Ecdysozoa: Arthropoda: Mandibulata: Hexapoda: Hymenoptera: Vespoidea: Sphecidae; *Sceliphron caementarium*)

1.2 Forest ant (Arthropoda: Mandibulata: Hexapoda: Hymenoptera: Formicidae: Myrmicinae; *Aphaenogaster rudis*)

1.3 Woodpecker finch (Bilateria: Deuterostomia: Chordata: Osteichthyes: Sarcopterygii: Reptiliomorpha: Amniota: Sauropsida: Archosauromorpha: Dinosauria: Saurischia: Theropoda: Maniraptora: Aves: Neognathae: Neoaves: Passeriformes: Emberizidae; *Cactospiza pallida*)

1.4 Egyptian vulture (Aves: Neognathae: Neoaves: Falconiformes: Accipitridae; *Neophron percnopterus*)

1.5 Asian elephant (Amniota: Synapsida: Therapsida: Mammalia: Eutheria: Afrotheria: Proboscidea: Elephantidae; *Elephas maximus*)

1.6 Sea otter (Mammalia: Eutheria: Laurasiatheria: Carnivora: Mustellidae; *Enhydra lutris*)

1.7 Polar bear (Mammalia: Eutheria: Laurasiatheria: Carnivora: Ursidae; *Ursus maritimus*)

1.8 Bottlenose dolphin (Mammalia: Eutheria: Laurasiatheria: Cetartiodactyla: Cetacea: Odontoceti: Delphinidae; *Tursiops truncatus*)

1.9 Black-striped capuchin monkey (Mammalia: Eutheria: Euarchontoglires: Primates: Platyrrhini: Cebidae; *Cebus libidinosus*)

1.10 Yellow baboon (Mammalia: Eutheria: Euarchontoglires: Primates: Catarrhini: Cercopithecoidea: Cercopithecidae; *Papio cynocephalus*)

2 Convergent behavior and function: TOOL CONSTRUCTION (modifying an existing external object or objects to create a tool to attain an immediate goal)

Convergent lineages:

2.1 New Caledonian crow (Amniota: Sauropsida: Archosauromorpha: Dinosauria: Saurischia: Theropoda: Maniraptora: Aves: Neognathae: Neoaves: Passeriformes: Corvidae; *Corvus moneduloides*)

2.2 Orangutan (Amniota: Synapsida: Therapsida: Mammalia: Eutheria: Euarchontoglires: Primates: Catarrhini: Hominoidea: Hominoidae: Pongidae; *Pongo pygmaeus*)

Note: For data sources, see text.

Before the invention of these Paleolithic tools, species in the Hominini certainly must have created tools out of wood and other organic matter, tools that decomposed and thus are not preserved in the fossil record. The classic studies of field anthropologists in Africa, beginning with Jane Goodall (Wade 2006), have demonstrated to us that our close Panini cousins, the chimpanzees (*Pan troglodytes*) and the bonobos (*Pan paniscus*), also create and use tools made of wood and other organic matter; thus, tool construction is found throughout the Homininae. Since then, gorillas (*Gorilla gorilla*) and now orangutans (*Pongo pygmaeus*) have been observed making tools (van Schaik et al. 1996), and so tool-construction behavior is a synapomorphy for the entire clade of the Hominoidae, all of the great apes, not just ourselves.

Rather than being a unique product of the human mind, is tool construction a unique product of the great ape mind? Not so. Both laboratory experimentation and field observation have demonstrated that corvid birds, such as the New Caledonian crow, make tools (table 6.1). When in the wild, these birds are particularly adept at making hook tools by carefully choosing suitable twigs and then modifying them to form hooks, and will do the same thing with metal wires in laboratory conditions (Emery and Clayton 2004). We are not unique; the vastly different brains of these avian dinosaurs can also conceive of taking an external object and transforming it into another object for the purpose of solving a problem or achieving a goal. Our minds have converged.

Ten other lineages of animals are known to use tools, but at present are not known to make them (table 6.1). Tool usage is a somewhat simpler mental concept, in which the animal looks for an existing object to use as a tool but does not conceive of modifying that object into another object to use as a tool. Old World monkeys (Catarrhini) and New World monkeys (Platyrrhini) have independently evolved tool-using behavior; capuchin monkeys pick up and carry stones with the express purpose of using them later to hammer nuts, and baboons do the same in hammering hard fruit, as well as using sticks for various purposes (Baber 2003). Nonprimate minds have also converged on hammer-using behavior: sea otters use stones to hammer mollusc shells, Egyptian vultures use stones to hammer ostrich eggs, and polar bears have been observed to drop stones on seals to stun them (Baber 2003). Even the vastly different, and tiny, mind of the wasp has convergently evolved hammer-using behavior: the black-and-yellow mud wasp will collect and carry "stones" (to them; small grains of sand to us) in their mandibles, and use these to hammer mud into burrows in which it has laid its eggs to protect the eggs from predators (Baber 2003). Other insects, such as the forest ant, collect and use leaves as containers in which to carry food. In chapter 4 we considered the case of the finch that has converged on the ecological role of the woodpecker (see table 4.3). The woodpecker finch searches for and carries away a tool, a cactus spine, which it uses to extract wood-boring insects. And last, the mighty elephant has been routinely observed to carry and use branches to scratch itself (Baber 2003), and the ocean-dwelling bottlenose dolphin collects and uses sponges to protect its skin from abrasion while foraging on the sea bottom (Hansell and Ruxton 2008). Animals with extremely different brains, scattered across very distantly related phylogenetic lineages, have independently evolved tool-using behavior (table 6.1).

Taking a different perspective, Hansell and Ruxton (2008, 77) argue that entirely too much emphasis has been placed on tool behavior in animals because it is always interpreted in terms of human evolution: "Those who feel that tools are 'special' might be correct—but they might be more special to researchers than to the animals that use them." As they point out, a great deal of attention was given to the discovery that gorillas make tools, while at the same time researchers ignored the fact that these animals build complicated nest structures out of modified branches and twigs on a daily basis. Likewise, relatively few birds make or use tools (table 6.1), but the majority of birds build nests, some of which are structurally very complex.

Before considering the convergent evolution of architectural behavior in animals, let us briefly review the evolution of architectural behavior in the most diversified animal architect on the planet at present, namely humans. Our ancestors, like our living cousins the chimpanzees and bonobos, surely built nests in trees, habitats constructed to provide a safe and comfortable place to sleep, high above the ground and thus out of the reach of ground-dwelling predators. In contrast, many of our ground-dwelling ancestors looked for preexisting natural structures, such as caves, that could be used as habitats. Caves and other naturally occurring habitats are rare, and with further growth in human population numbers, humans took the next step: the construction of an artificial habitat on the ground. Two potential construction materials are readily available: vegetative matter, such as wood, and dirt. Humans began to build habitats made of modified wood, an architectural behavior still widely in use today, or with adobe, which many people also still use today. Only later did we begin to build habitats with cut stone, fired bricks (which are much more durable than adobe), and concrete (artificial stone). At present, we have added metal, glass, and plastic to the construction materials used for our habitats.

Other animals have evolved the exact same sequence of architectural behavior, and they did so long before we existed (table 6.2). Hansell (2005, 33) argues that "[a]rchitecture results from the application of behaviour to materials"; for this reason, I have omitted the behavioral step where animal species simply hunt for preexisting structures that can be used for habitats, like hermit crabs that use snail shells, and will move directly to animals that actively gather and modify material to construct a habitat. Like humans, other animal species construct habitats with material that is either vegetative or earthen. Unlike humans, however, animals have not evolved the capability of working with materials like metal, glass, or plastic in their construction, although some do steal and inventively use our worked materials in their habitats!

Hansell (2005) identifies five architectural behaviors found in nest-building birds, listed here in terms of increasing behavioral complexity: stacking, entangling, Velcro-fastening, stitching, and weaving. At present there exists no comprehensive phylogenetic analysis of how many independent times each of these behaviors have evolved in separate groups of birds. Thus, I will discuss as many convergent examples that I am aware of, knowing that future phylogenetic analyses will almost certainly expand that list.

The eagle (table 6.2) is an example of a bird that constructs its nest by first searching for and collecting branches of the desired length and diameter, and then transporting those branches back to the tree it has chosen for the nest. The eagle simply stacks the branches atop one another until a bowl-shaped nest has been created. This same nest-building architectural behavior is used by some ground-nesting birds as well, such as the swan (table 6.2). In general, many of the large tree-dwelling or ground-dwelling birds use stacking to build their nests: storks (Cicioniidae), pelicans (Pelicanidae), and herons (Ardeidae), in addition to eagles and swans (Hansell 2005).

Nests built by entangling architectural behavior are sturdier than those built by simple stacking, as the ends of the branches or twigs in these nests are actively bent and pulled back into the core of the nest by the bird. The familiar wood pigeon (table 6.2) is an example of a bird that constructs its nest using entangling (Hansell 2005). The same method is used by the gorilla (table 6.2) and chimpanzee (and presumably many of our ancestors), who search out a spot in the trees with an upright fork and then bend branches back into this foundation while holding them in

Table 6.2
Convergent evolution of architectural behaviors

1 Convergent behavior and function: CONSTRUCTION USING VEGETATIVE MATERIAL (collecting and modifying vegetative material to build a wicker-like bowl, globe, baglike, or tentlike construction for use as a protective shelter)

Convergent lineages:

1.1 Veined octopus (Bilateria: Protostomia: Lophotrochozoa: Mollusca: Cephalopoda: Coleoidea: Octopodidae; *Amphioctopus marginatus*)

1.2 Arctic yellow jacket (Ecdysozoa: Arthropoda: Mandibulata: Hexapoda: Hymenoptera: Vespoidea: Vespidae: Polistinae; *Dolichovespula norwegica*)

1.3 Neotropical paper wasp (Arthropoda: Mandibulata: Hexapoda: Hymenoptera: Vespoidea: Sphecidae; *Microstigmus comes*)

1.4 Mute swan (Bilateria: Deuterostomia: Chordata: Osteichthyes: Sarcopterygii: Reptiliomorpha: Amniota: Sauropsida: Archosauromorpha: Dinosauria: Saurischia: Theropoda: Maniraptora: Aves: Neognathae: Galloanserae: Anseriformes: Anatidae; *Cygnus olor*)

1.5 Bald eagle (Aves: Neognathae: Neoaves: Falconiformes: Accipitridae; *Haliaeetus leucocephalus*)

1.6 Wood pigeon (Aves: Neognathae: Neoaves: Columbiformes: Columbidae; *Columba palumbus*)

1.7 European penduline tit (Neognathae: Neoaves: Passeriformes: Remizidae; *Remiz pendulinus*)

1.8 Long-tailed tit (Aves: Neognathae: Neoaves: Passeriformes: Aegithalidae; *Aegithalos caudatus*)

1.9 Gray-backed camaroptera (Aves: Neognathae: Neoaves: Passeriformes: Cisticolidae; *Camaroptera brevicondata*)

1.10 Village weaverbird (Aves: Neognathae: Neoaves: Passeriformes: Ploceidae: Ploceinae; *Ploceus cucullatus*)

1.11 Montezuma oropendola (Aves: Neognathae: Neoaves: Passeriformes: Fringillidae: Icterinae; *Gymnostinops montezuma*)

1.12 Great bowerbird (Aves: Neognathae: Neoaves: Passeriformes: Ptilonorhynchidae; *Chlamydera nuchalis*)

1.13 Dusky-footed wood rat (Amniota: Synapsida: Therapsida:Mammalia: Eutheria: Euarchontoglires: Rodentia: Myomorpha: Cricetidae; *Neotoma fuscipes*)

1.14 European harvest mouse (Mammalia: Eutheria: Euarchontoglires: Rodentia: Myomorpha: Muridae; *Micromys minutus*)

1.15 Gray squirrel (Mammalia: Eutheria: Euarchontoglires: Rodentia: Sciuromorpha: Sciuridae; *Sciurus carolinensis*)

1.16 Peter's tent-making bat (Mammalia: Eutheria: Laurasiatheria: Chiroptera: Microchiroptera: Phyllostomidae; *Uroderma bilobatum*)

1.17 Short-nosed fruit bat (Mammalia: Eutheria: Laurasiatheria: Chiroptera: Megachiroptera: Pteropodidae; *Cynopterus sphinx*)

1.18 Gorilla (Mammalia: Eutheria: Euarchontoglires: Primates: Catarrhini: Hominoidea: Hominidae: Gorillinae; *Gorilla gorilla*)

2 Convergent behavior and function: CONSTRUCTION USING ANIMAL-GENERATED MATERIAL (collection and modification of animal-generated material to build a woven bowl, globe, baglike, or tentlike construction for use as a protective shelter)

Convergent lineages:

2.1 Common tailorbird (Amniota: Sauropsida: Archosauromorpha: Dinosauria: Saurischia: Theropoda: Maniraptora: Aves: Neognathae: Neoaves: Passeriformes: Cisticolidae; *Orthotomus sutorius*)

Table 6.2
(continued)

2.2 Chaffinch (Aves: Neognathae: Neoaves: Passeriformes: Fringillidae; *Fringilla coelebs*)
2.3 Modern human (Amniota: Synapsida: Therapsida: Mammalia: Eutheria: Euarchontoglires: Primates: Catarrhini: Hominoidea: Hominidae: Homininae; *Homo sapiens*)

3 Convergent behavior and function: CONSTRUCTION USING EARTHEN MATERIAL (collecting and modifying earthen material to build an adobe-like dome, cylindrical, pyramidal, or globelike construction for use as a protective shelter)
Convergent lineages:
3.1 Neotropical mud wasp (Bilateria: Protostomia: Ecdysozoa: Arthropoda: Mandibulata: Hexapoda: Hymenoptera: Vespoidea: Sphecidae; *Trigonopsis cameronii*)
3.2 Malasian hover wasp (Arthropoda: Mandibulata: Hexapoda: Hymenoptera: Vespoidea: Vespidae: Stenogastrinae; *Liostenogaster flavolineata*)
3.3 Ceylonese potter wasp (Arthropoda: Mandibulata: Hexapoda: Hymenoptera: Vespoidea: Vespidae: Eumeninae; *Paraleptomenes mephitis*)
3.4 Warlike termite (Arthropoda: Mandibulata: Hexapoda: Isoptera: Termitidae: Macrotermitinae; *Macrotermes bellicosus*)
3.5 Cliff swallow (Bilateria: Deuterostomia: Chordata: Osteichthyes: Sarcopterygii: Reptiliomorpha: Amniota: Sauropsida: Archosauromorpha: Dinosauria: Saurischia: Theropoda: Maniraptora: Aves: Neognathae: Neoaves: Passeriformes: Hirundinidae; *Hirundo pyrrhonota*)
3.6 White-necked rockfowl (Aves: Neognathae: Neoaves: Passeriformes: Picathartidae; *Picathartes gymnocephalus*)
3.7 Rufous hornero (Aves: Neognathae: Neoaves: Passeriformes: Furnariidae; *Furnarius rufus*)
3.8 North American beaver (Amniota: Synapsida: Therapsida: Mammalia: Eutheria: Euarchontoglires: Rodentia: Castorimorpha: Castoridae; *Castor canadensis*)
3.9 Modern human (Mammalia: Eutheria: Euarchontoglires: Primates: Catarrhini: Hominoidea: Hominidae: Homininae; *Homo sapiens*)

Note: For data sources, see text.

place with their feet; finally, they tuck leafy twigs growing around the rim of the nest back into the core of the nest (von Frisch 1974). Unlike birds, however, apes construct their sleeping nests on a daily basis.

Many plants have convergently evolved seeds with the tiny barbs and hooks that provided the inspiration for the modern fastening product Velcro (see table 3.15). Some birds use this same fastening principle in building their nests: they actively seek out materials with tiny projecting processes and position these in the nest so that they ensnare each other, thereby creating even sturdier nests than those built using entangling. The long-tailed tit (table 6.2) uses small-leaved mosses for Velcro, and the related bushtit (*Psaltiparus minimus*) uses lichen (Hansell 2005). These nests are so sturdy that they can be totally enclosed into a baglike or socklike nest, with an opening at one end, as seen in the nests built by European penduline tits (table 6.2; von Frisch 1974). Rather than using Velcro-fastening, other birds actually stitch their nests together, pushing thin plant fiber through leaf margins, grasping it on the other side, and pushing it through again until separate leaves are stitched together. The gray-backed camaroptera (table 6.2) is an example of a bird that uses stitching behavior to build its nests (Hansell 2005).

Last, the most complex architectural behavior seen in birds is actual weaving, a suite of multiple behaviors including creating "loops, half-hitches, hitches, bindings, slip knots, and overhand knots, as well a more regular over-and-under pattern of weft through warp that defines human weaving" (Hansell 2005, 73–74), as can be seen in the nests built by the village weaverbird (table 6.2). Unlike most bird architectural behavior, which appears to be totally instinctual, weaving involves a learning component: young weaverbirds must practice and observe older weaverbirds in order to perfect the weaving technique (Hansell 2005). Yet this most complex architectural behavior in birds is not unique: "Weaving has arisen independently in two taxa of birds, the Old World weaver birds (Ploceinae) and the New World oropendolas, orioles, and caciques (Icterini, Fringillidae)" (table 6.2; Hansell 2005, 74).

In addition to using vegetative material, many birds also use animal-generated materials such as silk and mammal hair to construct nests. Hansell (2005) notes that 56 percent of passeriform families of birds actively search for arthropod silk, particularly spider silk, to use as strong threads in their nest building; this is seen in the stitching-constructed nests of the common tailorbird (table 6.2). Other birds, such as the chaffinch (table 6.2), use mammal hair. Gould and Gould (2007, 194) document the case of a Carolina wren (*Thryothorus ludovicianus*) that constructed its nest using human-made fishing line, rather than spider silk, to build its nest—incorporating the fishing hooks and lead sinkers into the nest! Humans have converged on these avian habitat architectures in the building of tents constructed of woven wool and other animal hairs.

The birds are avian dinosaurs that evolved in the Jurassic, some 100 million years before the destruction of the dinosaurian ecosystem and its replacement by the mammalian, as discussed in chapter 4. In the Cenozoic Era, mammalian architectural behavior has converged on that of the more ancient avian, as seen in the entangling-constructed nests built by modern apes. The European harvest mouse (table 6.2) builds a globular-shaped enclosed nest, with an entrance hole on one side, suspended in the branches of shrubs or the stems of oats (von Frisch 1974). Gray squirrels (table 6.2) build similar, but larger enclosed globular nests in trees; these myomorph and sciuromoph rodents have converged on the architecture of the weaverbird.

The tent-making bats have evolved an architectural behavior not seen in birds. These bats seek out suitably placed large leaves in trees and then bite into the veins of the leaves, creating a line of perforation along which the leaf folds down on either side under the pull of gravity (Hansell 2005). The bats use different bite techniques for different leaf types in order to create the three-dimensional tent structure. Odd as this architectural behavior is, it is not unique. Tent-making behavior has independently evolved in both Neotropical microbats, like Peter's tent-making bat in Costa Rica, and in Paleotropical megabats, like the short-nosed fruit bat in India (table 6.2; Kunz et al. 1994).

The wasps have also evolved an architectural behavior not seen in birds. They use vegetative material in their nest constructions, but it is a material that they manufacture themselves—paper. These wasps chew plant material to produce pulped and blended fragments that adhere together, and that are further held together by the wasp's dried spit, to form sheets of paper that are both light in weight but also strong in tension (Hansell 2005). Complex paper-constructed nests are built by both social wasps, such as the vespid Arctic yellow jacket, and solitary wasps, such as the sphecid Neotropical paper wasp (table 6.2). Humans have converged on these wasp architectures in the constructing of rice-paper shoji that are used for lightweight walls in many Japanese buildings.

Like humans, animals also use vegetative material to build habitats on the ground level, but these architectures are not as common as those that are constructed with earthen material. The dusky-footed wood rat (table 6.2) builds lodges of branches and twigs that can be up to one meter high, and that contain as many as five separate chambers or rooms that are used for different purposes, such as a nursery, a living room, and a lavatory (von Frisch 1974). The bowerbirds of Australia and New Guinea (table 6.2) are famous for their elaborate ground constructions that, unlike all the other architectural examples discussed here, are not used as functional protective habitats but as artistic constructions designed to attract mates. These display stages—and stage it is—may be furnished with "curtains" made of leaves or vines, decorated with "floodlight" borders of flowers, colored stones, berries, or snail shells. They have enormous (for the size of the bird) arched stage roofs of branches that are sometimes painted with berry juice, and are fronted by "theatre aisle" walkways leading to the stage. If a female appears, the male gives an elaborate performance of dancing and singing on the stage. If the female is unimpressed, she walks away; if she is attracted, she joins the male in the dance.

Have the bowerbirds convergently invented art? Gould and Gould (2007, 221–222) argue that bowerbird behavior "may be very close to crossing some important line between the mental experiences of humans and other animals. Bowerbirds can indulge in this frivolity, if that is what it is, because there are few predators and little competition in their habitat much of the year, and many species have a very long or continuous breeding season—factors that have allowed artistic creativity to flourish among humans as well. . . . At the cognitive level, these creations are the most complex seen in birds. Only beavers and humans undertake work with more steps and greater flexibility in design, materials, and execution."

Perhaps even more striking, however, is the recent discovery that the veined octopus (table 6.2), a marine mollusc species, has innovatively seized human-made vegetative objects to build habitats: coconut half-shells. In the last several decades, an abundance of this new building material appeared in the octopus's shallow marine realm, as humans eat the coconuts and then discard the clean and lightweight shell halves into the waters surrounding their coastal communities in Indonesia. Over 20 individual veined octopi have been observed collecting, transporting, and assembling the coconut shell halves into protective habitats to hide in. When flushed from the shells by human scuba divers, the octopi quickly return and reassemble the shell-constructed habitats, leaving Finn et al. (2009, R1070) to conclude that "even marine invertebrates engage in behaviours that we once thought the preserve of humans."

Returning to the terrestrial realm, and on the ground level, many humans lived in caves first (in essence, holes in the ground), and then began to build adobe habitats (in essence, artificial caves). Many birds have followed this same behavioral pathway in their evolution, with the exception that their adobe habitats are built on vertical cliff faces rather than on horizontal ground surfaces. Molecular phylogenies of swallows and martins (Hirundinidae; table 6.2) reveal the multiple independent evolution of the behavioral pathway of first dwelling in preexisting cliff cavities, then constructing open mud cups attached to cliff faces as nests, and finally constructing enclosed adobe globes attached to cliff faces as nests (Gould and Gould 2007). In contrast to the small adobe nests of the cliff swallow, the picathartid white-necked rockfowl (table 6.2) builds a two-kilogram adobe nest, even though the bird itself weighs only 200 grams (Hansell 2005). Gould and Gould (2007, 184) comment: "The invention of adobe cups that then develop into enclosed cavities, so evident in the progression from barn to cliff swallows, is also seen in some birds that build in trees." The most striking example of a tree-dwelling adobe builder is the rufous hornero of South America (table 6.2), nicknamed the "ovenbird" because its vertical adobe nest looks like a baker's oven (von Frisch 1974). The globe-shaped nest contains a large interior room that is used as a brooding chamber and a narrow, hall-like entrance passage that opens into a ten-centimeter-diameter hole to the outside. Both the male and female birds cooperate in its construction, which takes them over two weeks to complete (von Frisch 1974).

Wasps have evolved a similar behavioral pathway in their habitat constructions, as noted by Hansell (2005, 232): "I have identified two alterations in nest building behaviour that are associated with the transition from solitary to social behaviour of wasps. . . . The first is the transition from an excavated to a constructed nest. . . . This occurred in both the Sphecidae and Vespidae (Eumeninae). The second change was in the choice of the building material from mud to paper." Both paper-building behavior and adobe-building behavior is convergent in wasps (table 6.2). An interesting transition between excavating tunnels for below-ground habitats to constructing above-ground adobe habitats is seen in the eumenin funnel wasp (*Paralastor emarginatus*) of Australia. This wasp first digs a tunnel, lines it with mud, and then extends these mud walls above ground to build an external cylinder, which it increases in diameter to build a downward-pointing funnel opening (Gould and Gould 2007). The funnel serves the purpose of concealing the entrance to the tunnel from parasites while the female is hunting for caterpillars to paralyze and place in the tunnel as food for her later-developing larvae. Solitary sphecid wasps, such as the Neotropical mud wasp (table 6.2), do no excavating but build adobe brood chambers entirely above ground. While adobe building is the norm for many sphecid wasps, the eusocial Malasian hover wasp (Stenogastrinae) also builds with adobe, as does the Ceylonese potter wasp (Eumeninae) within a subfamily of wasps that normally builds with paper (table 6.2; Hansell 2005).

Termites also have converged on the extension of their below-ground excavations to above-ground adobe-building behavior (table 6.2). Compared to the size of their inhabitants, termite adobe dwellings are huge: the Australian compass termite, *Amitermes meridionalis*, builds above-ground nests that are up to 5 meters high and 3 meters long, but only around 30 centimeters thick. The species is given the name "compass" because its slablike buildings are always oriented with the long direction north-south, such that the habitat always has a broad side warmed by the sun (either to the east or the west, depending upon the time of day), and an opposite side that is cool and shaded (von Frisch 1974). The temperature difference between the two broad sides of the habitat produces air flow from the cool side of the habitat through to the warm side—the termites have invented buildings with air conditioning! And not just one type of air-conditioning system—the much-studied African warlike termite (table 6.2) builds five-meter-high pyramidal-shaped dwellings with hard, cement-like adobe outer walls and an interior adobe nest that is suspended within the outer walls by a series of pillars, arches, and buttresses. The inner dwelling is surrounded by an air space, through which warm air rises from the bottom of the building, creating a lower-pressure region that then draws cooler, denser air into the interior of the building from the outside. The source of heat at the bottom of the dwelling is the fermentation in fungus gardens arranged in special adobe chambers—the termites have thus also invented agriculture (a point that we shall consider in more detail below). A large habitat of the warlike termite can contain as many as two million termites. The various air-conditioning systems that termites build also serve the purpose of flushing the carbon-dioxide-rich air created by their breathing out the top of the building; oxygen-enriched air is drawn in through the ventilation holes in the side of the building (von Frisch 1974).

Australian compass termites and African warlike termites inhabit generally arid regions; in wet tropical regions, termites change the architecture of their dwellings to deal with the problem of frequent rainfall. The termite species *Cubitermes fungifaber* builds tall, cylindrical-shaped adobe dwellings with multiple roofs that are slanted downward and extend out beyond the diameter of the main habitat (Gould and Gould 2007). These multiroofed towers are reminiscent of the pagodas build by humans in Asia.

Last, beavers have evolved diversified architectural behaviors that converge on those of humans: they build with logs, adobe, and stone. The beavers are the largest rodents alive today, and may weigh up to 30 kilograms (von Frisch 1974). The North American beaver habitat, or lodge, is constructed of both vegetative and earthen material; it is architecturally very similar to the log cabins constructed by early European settlers in North America (table 6.2). The beavers first build a domelike habitat with cut tree branches and logs, and then carefully plaster and caulk the openings between the branches in the walls of the lodge with mud and clay. Unlike a log cabin, however, the door to the lodge is under water, and the lodge itself is located on an artificially created island in a pond or lake. Both of these features serve to protect the beavers' dwelling from invasion by predators. These round dome-shaped dwellings, isolated from the shore out in a lake, are reminiscent of the crannogs built by prehistoric humans in Scotland.

The pond or lake that surrounds the beavers' lodge is often the creation of the beavers themselves: "beavers are experts not only in the building of dwellings but also in hydro-engineering, and have performed tremendous feats in this line long before man attempted anything of the kind" (von Frisch 1974, 268). Blocking a flowing stream or river with a dam is not an easy task, yet beavers do so by first ramming cut branches and logs into the river bed, supporting these against the flow of the river with forked branches pointing upstream, and adding further anchoring by bringing heavy stones to hold the dam superstructure in place. They further excavate the river bed on the upstream side of the dam, which not only reduces the speed of water flow but also provides the beavers with the earthen material they need to finish sealing the dam. The finished dam is higher in the center than at the termini, thus forcing the dammed river to flow over the dam near the two river banks. The beavers then actively control the level of the lake by either lowering or raising these two overflow sluices—they frequently lower the level of the lake during winter to created a breathing space under the ice of the frozen surface of the lake, an air space that extends the entire area of the lake, providing them with a swimming region protected from the winter cold. Von Frisch (1974) reports that the largest beaver dam on record was 700 meters long, and was strong enough to support the weight of human riders on horseback.

Do convergent architectures reveal convergent minds? Gould and Gould (2007, 271–272) argue that "[m]ental activity is, by its nature, private; what goes on in the brain has to be inferred. In tracing the evolution of cognitive strategies, the most tangible evidence is found among animals that build. . . . These abilities seem to have evolved independently in several different groups, but always apparently in about the same order, and to serve analogous ends." That is, given the same problem—such as building a protective habitat—animals with radically different brains have evolved the same architectural behaviors. Those analogous behaviors thus reflect analogous mental activities and cognitive strategies taking place in independent lineages, which is convergent mental evolution.

On the other hand, do humans really think like termites when they build an adobe habitat? Gould and Gould (2007, 5) also note that "[w]e know that animal building, like most tasks animals accomplish, depends on many different neural mechanisms. For most creatures, though, instinct rather than learning seems to guide behavior." This conclusion was revealed in the pioneering work with animal instinctive behavior by Konrad Lorenz, Niko Tinbergen, and Karl von Frisch, work for which they were awarded the 1973 Nobel prize in physiology or medicine. Behavioral evolution by the trial-and-error process of natural selection requires very long periods of time and generations of reproductive cycles. Behavioral evolution by learning, in contrast, is very rapid: one generation simply teaches the next generation the new behavioral innovations discovered by the parent generation. Cognitive strategies are transmitted by teachers, not by genes. Yet stereotypic instinctive behaviors and flexible learned behaviors have both converged on the same result in architecture (table 6.2).

The discipline of theoretical morphology (McGhee 2007) takes a different perspective on the question of convergent architectural behavior by examining the geometries of the architecture itself: Are there only a limited number of ways to build structures? If so, convergent behavior must necessarily result because independent lineages of organisms have a limited number of building options to discover. The convergent usage of the cantilever and the arch by such disparate organisms as mud wasps, termites, cliff swallows, and humans results from the fact that cantilevers and arches are necessary in the construction of adobe architectures. Convergent minds are here the product of the functional constraints of geometry. The implications of this theoretical morphologic perspective on convergent evolution will be explored in more detail in chapters 7 and 8, but I would point out here that we should not be surprised to find cantilevers and arches on Earth-like worlds elsewhere in the universe, produced by totally alien minds.

To conclude this section of the chapter, let us consider one last extremely complex problem-solving behavior: agriculture. Modern humans, *Homo sapiens*, are the only vertebrate species that has evolved agriculture. Agricultural behavior is convergent, however, in that several populations of humans made the transition from hunter-gatherers to agriculturalists independently of each other around the world about 10,000 years ago: in the Tigris and Euphrates valleys in the Near East, in South Asia, in Central Asia, and in Central America (Gupta 2004).

Was agriculture clearly an idea whose time had come, beginning about 10,000 years ago, as a unique behavioral trait of the human species? Not so. In contrast to the single human species, the hexapod arthropods have independently evolved agriculture nine separate times in different lineages (table 6.3). The transition from hunter-gatherer to agriculturalist was made once by the ants, once by the termites, and no less than seven separate times by the ambrosia beetles (Farrell et al. 2001; Mueller et al. 2005). And they did so long before humans existed: the platypodine ambrosia beetles evolved agricultural behavior 60 million years ago

Table 6.3
Convergent evolution of agricultural behaviors

1 Convergent behavior and function: CROP AGRICULTURE (cultivation of crops for nourishment)

Convergent lineages:

1.1 Attine ant (Bilateria: Protostomia: Ecdysozoa: Arthropoda: Mandibulata: Hexapoda: Hymenoptera: Formicidae: Myrmicinae: Attini: Paleoattini; *Apterostigma auriculatum*)

1.2 Xyleborine ambrosia beetle (Arthropoda: Mandibulata: Hexapoda: Coleoptera: Curculionidae: Xyleborini; *Dryocoetoides cristatus*)

1.3 Hyorrhynchine ambrosia beetle (Arthropoda: Mandibulata: Hexapoda: Coleoptera: Curculionidae: Hyorrhynchini; *Sueus niisimai*)

1.4 Platypodine ambrosia beetle (Arthropoda: Mandibulata: Hexapoda: Coleoptera: Curculionidae: Platypodini; *Australoplatypus incompertus*)

1.5 Bothrosternine ambrosia beetle (Arthropoda: Mandibulata: Hexapoda: Coleoptera: Curculionidae: Bothrosternini; *Cseninus lecontei*)

1.6 Xyloterine ambrosia beetle (Arthropoda: Mandibulata: Hexapoda: Coleoptera: Curculionidae: Xyloterini; *Indocryphalus pubipennis*)

1.7 Scolytoplatypodine ambrosia beetle (Arthropoda: Mandibulata: Hexapoda: Coleoptera: Curculionidae: Scolytoplatypodini; *Scolytoplatypus macgregori*)

1.8 Corthyline ambrosia beetle (Arthropoda: Mandibulata: Hexapoda: Coleoptera: Curculionidae: Corthylini; *Gnathotrupes quadrituberculatus*)

1.9 Black-winged subterranean termite (Arthropoda: Mandibulata: Hexapoda: Isoptera: Termitidae: Macrotermitinae; *Odontotermes formosanus*)

1.10 Modern human (Bilateria: Deuterostomia: Chordata: Osteichthyes: Sarcopterygii: Reptiliomorpha: Amniota: Synapsida: Therapsida: Mammalia: Eutheria: Euarchontoglires: Primates: Catarrhini: Hominoidea: Hominidae: Homininae; *Homo sapiens*)

2 Convergent behavior and function: ANIMAL HUSBANDRY (cultivation of animals for nourishment)

Convergent lineages:

2.1 Japanese ant (Bilateria: Protostomia: Ecdysozoa: Arthropoda: Mandibulata: Hexapoda: Hymenoptera: Formicidae: Formicinae; *Formica yessensis*)

2.2 Modern human (Bilateria: Deuterostomia: Chordata: Osteichthyes: Sarcopterygii: Reptiliomorpha: Amniota: Synapsida: Therapsida: Mammalia: Eutheria: Euarchontoglires: Primates: Catarrhini: Hominoidea: Hominidae: Homininae; *Homo sapiens*)

Note: For data sources, see text.

(Farrell et al. 2001), the ants did so 50 million years ago (Schultz and Brady 2008), and the termites evolved agricultural behavior between 34 to 24 million years ago (Mueller et al. 2005).

Agricultural behavior is incredibly complex. It includes (1) preparing the substrate to be used for growing the crops, (2) planting the crops, (3) monitoring the growth and potential disease status of the crops, (4) protecting the crops from disease, (5) protecting the crops from crop-eating species other than the farmers, (6) weeding invasive species out of the crops, (7) using chemical herbicides for weed control, (8) using microbes for biological pest control, (9) using microbial symbionts to procure nutrients for the crops, and (10) the sustainable harvesting of the crops for food (Mueller et al. 2005). This list is not a list of agricultural behaviors found in humans—it is a list of behaviors found in ants! In fact, Mueller et al. (2005, 564) argue that "[b]ecause of the universality of crop diseases in both human and insect agriculture, it may be fruitful to examine the . . . solutions that have evolved convergently in insect agriculture for possible application to human agriculture." Here we should note that ant agriculturalists use microbes for biological pest control, whereas most human agriculturalists (thus far) use chemical pesticides.

There is one major difference in human and hexapod agriculture: hexapods cultivate fungus crops, whereas humans usually cultivate plant crops (although we do also farm mushrooms). The most advanced hexapod agriculturalists have in fact evolved a new way to digest cellulose: rather than digesting it internally with the usage of anaerobic bacteria and specialized stomachs like many vertebrate animals (itself a convergent trait; see table 2.10), advanced hexapods use fungus gardens. Indeed, Schultz and Brady (2008, 5435) argue that a colony of leaf-cutter ants is "the ecological equivalent of a large mammalian herbivore in terms of collective biomass, lifespan, and quantity of plant material consumed."

Fungi are the only major group of organisms that are able to digest cellulose in the presence of free oxygen (Gould and Gould 2007). They also do not need light, as they are not photoautotrophs like plants, but they do need heat and humidity for efficient digestion. Both ants and termites have independently evolved the behaviors needed to create warm, moist agricultural chambers for fungus gardens, the ants underground and the termites in their above-ground adobe buildings. The ants are particularly important in analyzing the evolution of agricultural behavior, in that living ants still use five distinctly different agricultural systems that can be demonstrated to have evolved in a sequence of increasing agricultural sophistication (Schultz and Brady 2008). "Lower agriculture," in which a wide range of fungal species are cultivated, first appeared in paleoattine ants some 50 million years ago (table 6.3). About 30 million years later, there was a bifurcation in agricultural behavior. "Yeast agriculture," in which neoattine ants specialize in cultivating leucocoprineaceous fungi in the single-celled yeast phase, evolved 20 million years ago, and "coral fungus agriculture," in which paleoattine ants specialize in cultivating nonlecocoprineaceous fungi in the clade of Pterulaceae, evolved 15 million years ago. Also about 20 million years ago a separate clade of neoattine ants evolved "domesticated agriculture," in which the fungal species have coevolved so much with their ant cultivators that they are no longer capable of existing on their own outside of the ant gardens. These domesticated fungi also produce nutritious swollen hyphal tips, or gongylidia, that the ants harvest for food (Schultz and Brady 2008). Last, some 12 million years ago the leaf-cutting neoattine ant farmers evolved. The leaf-cutter ants specialize in cutting and processing fresh vegetation for substrates on which to grow their fungus crops, unlike all the other ants, who use dead organic detritus. These ants, who in effect "eat" living plants, are the dominant "herbivores" in the New World tropics in terms of the quantity of the plant material that they remove from the ecosystem (Schultz and Brady 2008).

Unlike the versatile ant farmers, termite farmers (table 6.3) cultivate species of a single genus of fungus, *Termitomyces*, which they have domesticated (Mueller et al. 2005). Termite domestic fungi produce vegetative nodules that are harvested by the termites for food, similar to the domestic fungi of the ants. Like many ant farmers, termites grow their fungus crops in specialized agricultural chambers on a prepared substratum of dead plant material, and have not (yet) evolved the ability to process living plants as the leaf-cutter ants have.

Unlike ants and termites, the ambrosia beetles do not build specialized agricultural chambers for growing fungus gardens. They bore into tree trunks and excavate extensive galleries of tunnels within the living phloem layer of the tree, and then grow their fungal crops along the walls of their galleries. Besides providing food for the beetles, the fungi also block the chemical pesticides produced by the tree to defend itself from infection—thus, invasion by these beetles can be devastating to forests (Farrell et al. 2001).

Also unlike the ants and termites, agricultural behaviors have independently evolved in seven separate lineages of beetles (table 6.3). Different beetle lineages farm different species of the ophiostomatoid fungi, some species of which they have domesticated (Farrell et al. 2001). Different beetle lineages also use different tree hosts, in that some infest species of conifers, and others attack species of angiosperms.

Animal husbandry in agriculture, where animals are domesticated instead of crops, is also convergent, but rare (table 6.3). Humans domesticated sheep, goats, cattle, and pigs for food some 7,000 years ago (Gupta 2004). In the hexapods, the formicine and dolichoderine ants in particular are known for their domestication of aphids (Aphididae), soft scale insects (Coccidae), tree hoppers (Membracidae), and mealy bugs (Pseudococcidae; Delabie 2001; Stadler and Dixon 2005).

Aphids are herbivores that feed on the phloem sap of plants, which is rich in sugar but poor in nitrogen. To obtain enough nitrogen, aphids have to consume large amounts of sap, and they excrete the excess sugar as honeydew. This honeydew waste product is a food source for ants, who collect it from the aphids. Thus begins the evolution of the trophobiotic relationship between the aphid herbivore and the ant farmer. When tended by the ant farmers, aphids produce much more honeydew, and higher-quality honeydew, than they do when solitary. To offset this metabolic cost, the aphids benefit by being protected from predators by the ant farmers. The ants are so good at protecting their aphid herbivores that farmed aphids can become very numerous and gregarious. The cost of protecting the aphids is therefore offset by the production of aphid "herds," which provide a food source for the ants that is concentrated in a small farmed area—rather than having a diffuse source of food scattered over a wide geographic region. Some aphids have become so domesticated that they depend upon the ant farmer for existence; for example, the aphid *Stomaphis quercus* is cultivated by the ant *Lasius fuliginosus*, and is only found on oak trees farmed by this ant (Stadler and Dixon 2005).

Ants are known to construct protective shelters out of plant debris for some of their farmed insects, both to shelter the insects from the weather and to conceal them from predators. The ant *Formica obscuripes* goes further and shelters its herd of aphids in galleries within its own underground dwelling, leading the herd of aphids out to "pasture" during the day, and herding them back underground for the night (Delabie 2001). Other ants will pick up and transport their farmed insects, particularly soft scale insects, to more favorable plant "pastures" for them to feed. When an ant farmer feeds, it will either drum the aphid's abdomen with its antennae or stroke the aphid's sides to encourage it to excrete honeydew droplets, which the ant collects (Delabie 2001).

At first glance, it would appear that ants have converged on the animal husbandry behavior of the human dairy farmer (table 6.3). The dairy farmer benefits from a herd of cows, which are a concentrated source of milk for food. The cows benefit from being protected from wolves and other predators by the farmer, and farmed cows produce much more milk than wild ones. The farmer leads the herd of cows out of a barn shelter to the open pasture on a daily basis, and later strokes the mammary glands of the cows to encourage them to excrete milk, which the farmer collects. The problem with this analogy is the fact that ants evolved aphid farming in the Early Oligocene, some 30 million years before the existence of humans (Stadler and Dixon 2005). It is the human dairy farmer that has converged on the animal husbandry behavior of ants, not the other way around.

In summary, it took ant agriculturalists 30,000,000 years to go from farming fungus to domesticating fungus: from the Early Eocene paleoattine ants to the Early Miocene neoattine ants (Schultz and Brady 2008). In contrast, human farmers have accomplished the same feat with plants in less than 10,000 years. Yet, as we also saw with the evolution of architectural behaviors, the evolution of stereotypic instinctive behaviors in ants and flexible learned behaviors in humans have both converged on the same result: farming of domesticated crops (table 6.3). Ants and humans also convergently evolved animal husbandry, but in this case the fossil record does not reveal how long it took the ants to accomplish the feat of domesticating aphids.

The complex of agricultural behaviors independently evolved by hexapod arthropods and vertebrate humans is astonishingly convergent. However, as with architecture, using the analytical technique of theoretical morphology to examine the question of convergent agricultural behavior provides a different perspective by focusing on the cultivation process instead: Is there only a limited number of ways to successfully cultivate crops? If so, convergent behavior must necessarily result, for independent lineages of organisms have a limited number of farming options to discover. The convergent usage of chemical herbicides by such disparate organisms as ants and humans results from the fact that such pesticides are necessary for the cultivation of crops. As with architecture, convergent agricultural minds are here the product of functional constraint on evolution.

It is difficult for a single individual to run a really efficient farm. Farming is labor intensive, and is usually the sum product of many individuals who have taken on different tasks. Task partitioning and the division of labor in farming have evolved convergently in both humans and hexapods, but some hexapods have taken it to an extreme not seen in humans—they have become eusocial. Eusociality is an ultimate form of group behavior and, as we shall see in the next section of the chapter, it is also convergent.

Group Behavior

In the first edition of *On the Origin of Species*, Darwin considered convergent minds to present special difficulties for his theory of natural selection. He was troubled by observed cases of identical behavior, or "instincts," in animals within independent evolutionary lineages, and the very worst of these cases was that of the eusocial hexapods: "No doubt many instincts of very difficult explanation could be opposed to the theory of natural selection . . . instincts almost identically the same in animals so remote in the scale of nature, that we cannot account for their similarity by inheritance from a common parent, and must therefore believe that they have been acquired by independent acts of natural selection. . . . I allude to the neuters or sterile females in insect-communities" (Darwin 1859, 235–236). The very heart of the theory of natural selection is the effect of differential reproductive success of individuals with different phenotypes; thus, how can natural selection produce social behaviors in which the majority of individuals do not reproduce at all? Yet eusocial cooperative societies, in which the majority of individuals in those societies have sacrificed their own reproduction, have convergently evolved at least 17 independent times (table 6.4).

In honeybee societies, for example, there is only one fertile female, the queen. The rest of the society consists of numerous female workers, who are the sterile daughters of the queen, and the male drones, who are the fertile sons. Darwin knew nothing of genetics, yet it is in the peculiar haplodiploid genetics of honeybee reproduction that a clue can be seen in the evolution of eusocial behavior. The queen produces both fertilized eggs, which develop into diploid females, and unfertilized eggs, which develop into haploid males. In the classic model of eusocial evolution, Hamilton (1964) pointed out that the haplodiploid genetic system produces sisters that are more closely related to one another than they would be to their own offspring. This is due to the fact that all sisters share an identical set of chromosomes from their father (as the male is haploid), and thus are 75 percent related to one another, whereas they would share only half their chromosomes with a daughter (50 percent

Table 6.4
Convergent evolution of eusocial behavior

Convergent behavior and function: EUSOCIALITY (cooperative social system in which only one female is reproductive, and labor behaviors are complexly divided among the numerous other individuals within the colony, resulting in a highly stable and competitive colonial organization able to sustain high population densities)

Convergent lineages:

1 Regalis sponge-dwelling shrimp (Bilateria: Protostomia: Ecdysozoa: Arthropoda: Mandibulata: Malacostraca: Decapoda: Alpheidae; *Synalpheus regalis*)

2 Chacei sponge-dwelling shrimp (Arthropoda: Mandibulata: Malacostraca: Decapoda: Alpheidae; *Synalpheus chacei*)

3 Small paraneptunus sponge-dwelling shrimp (Arthropoda: Mandibulata: Malacostraca: Decapoda: Alpheidae; *Synalpheus "paraneptunus small"*)

4 Honeybee (Arthropoda: Mandibulata: Hexapoda: Hymenoptera: Apoidea: Apidae: Apini; *Apis mellifera*)

5 Stingless bee (Arthropoda: Mandibulata: Hexapoda: Hymenoptera: Apoidea: Apidae: Meliponini; *Mellipona compressipes*)

6 Halictus sweat bee (Arthropoda: Mandibulata: Hexapoda: Hymenoptera: Apoidea: Halictidae; *Halictus (Halictus) quadricinctus*)

7 Lasioglossum sweat bee (Arthropoda: Mandibulata: Hexapoda: Hymenoptera: Apoidea: Halictidae; *Lasioglossum (Dialictus) figueresi*)

8 Augochlorella sweat bee (Arthropoda: Mandibulata: Hexapoda: Hymenoptera: Apoidea: Halictidae; *Augochlorella pomoniella*)

9 Yellow-jacket wasp (Arthropoda: Mandibulata: Hexapoda: Hymenoptera: Vespoidea: Vespidae: Vespinae; *Vespula maculifrons*)

10 Hover wasp (Arthropoda: Mandibulata: Hexapoda: Hymenoptera: Vespoidea: Vespidae: Stenogastrinae; *Parischnogaster mellyi*)

11 Dracula ant (Arthropoda: Mandibulata: Hexapoda: Hymenoptera: Vespoidea: Formicidae: Ponerinae; *Amblyopone pallipes*)

12 Galling thrip (Arthropoda: Mandibulata: Hexapoda: Thysanoptera: Tubuliferidae: Phlaeothripinae; *Oncothrips tepperi*)

13 Horned-soldier aphid (Arthropoda: Mandibulata: Hexapoda: Hemiptera: Hormaphididae: Cerataphidini; *Pseudoregma sundanica*)

14 Australoplatypus ambrosia beetle (Arthropoda: Mandibulata: Hexapoda: Coleoptera: Curculionidae: Platypodini; *Australoplatypus incompertus*)

15 Damp-wood termite (Arthropoda: Mandibulata: Hexapoda: Isoptera: Termopsidae; *Zootermopsis nevadensis*)

16 Naked mole rat (Bilateria: Deuterostomia: Chordata: Osteichthyes: Sarcopterygii: Reptiliomorpha: Amniota: Synapsida: Therapsida: Mammalia: Eutheria: Euarchontoglires: Rodentia: Hystricognatha: Bathyergidae; *Heterocephalus glaber*)

17 Damaraland mole rat (Mammalia: Eutheria: Euarchontoglires: Rodentia: Hystricognatha: Bathyergidae; *Cryptomys damarensis*)

Note: For data sources, see text.

related; Holmes et al. 2009). Under the predictions of the theory of natural selection, it is more advantageous for sisters to help raise more sisters (produced by their mother, the queen) than it is for them to produce their own offspring. Of the 17 evolutionary lineages that have convergently evolved eusocial behavior (table 6.4), over half are haplodiploid: the bees, in five independent lineages (Cameron and Mardulyn 2001; Brady et al. 2006); the wasps, two independent lineages (Hines et al. 2007); the ants, one lineage (Thorne and Traniello 2003); and the thrips, one lineage (Crespi et al. 1997).

At first glance, the convergent evolution of eusocial behavior might be seen as a function of developmental constraint, in that it is associated with the haplodiploid genetic system. However, eusocial behavior has also convergently evolved in lineages that are diploid (table 6.4), lineages that have no asymmetric degrees of relationships between siblings and offspring of either sex: shrimp, in three lineages (Duffy et al. 2000); aphids, one lineage (Shingleton and Foster 2001); beetles, one lineage (Mueller et al. 2005); termites, one lineage (Johns et al. 2009); and rodents, two lineages (Scantlebury et al. 2006; Holmes et al. 2009).

In contrast to haplodiploid lineages, the convergent evolution of eusocial behavior in diploid lineages has been argued to be the product of functional constraint: "the inability of the haplodiploid hypothesis to explain all eusocial evolution has returned some attention to the ecological benefits of eusocial behavior" (Johns et al. 2009, 17452). In addition, many have argued that eusocial species have a competitive advantage in highly crowded environments (Duffy et al. 2000; Wilson and Hölldobler 2005). Indeed, all of the 17 eusocial lineages are found in organisms that inhabit sheltered, enclosed nesting sites: sponge-dwelling shrimp, plant-gall-dwelling aphids and thrips, wood-boring beetles, hive-dwelling bees and wasps, and subterranean-dwelling ants, termites, and mole rats (table 6.4). Because similarly protected nesting sites may be difficult to find or construct, Holmes et al. (2009) assert, it is less risky to remain at the original site than to seek another, leading to overlapping generations of crowded adults living in a single group, cooperative breeding within the group, and division-of-labor behaviors within the group.

There still exists no universally accepted causal hypothesis for the evolution of all eusocial behavior. Some suggest that the phenomenon is analogous to the convergent self-assembly of molecules (Camazine et al. 2001), while others suggest that it is analogous to the evolution of multicellularity (Boomsma 2009). Wilson and Hölldobler (2005) propose a mix of selective causal factors—an interplay of group selection, kin selection, and individual selection. Others stress ecological factors: the "convergent similarities in behavior . . . suggest that highly eusocial organization has limited permutations. Each tribe has been channeled along a similar behavioral track, responding in similar fashion to similar contingencies" (Cameron and Mardulyn 2001, 209).

In summary, the independent evolution of eusocial behavior may be produced by developmental constraint (intrinsic genetic systems), functional constraint (extrinsic habitat ecologies), or the action of the two constraints in concert. The convergent evolution of hive minds can be spectacularly successful: the number of species of haplodiploid ants and diploid termites constitutes only 2 percent of all known insect species, yet they together constitute more than half of the biomass of all insects on Earth (Wilson and Hölldobler 2005).

Another highly convergent form of group behavior is the collective mind of the herd (table 6.5): "Herding is a form of convergent social behaviour that can be broadly defined as the alignment of the thoughts or behaviours of individuals in a group (herd) through local interaction and without centralized coordination" (Raafat et al. 2009, 420). Everyone in Europe and North America is familiar with the huge flocks of migrating European starlings that flow through our skies like a veritable river of birds. Unlike a river of water acting under the influence of gravity, a bird flock is a river of minds that are acting in concert: "Consider a flock of starlings under attack by a peregrine falcon: The flock contracts, expands, and even splits, continuously changing its density and structure. Yet, no bird remains isolated, and soon the flock reforms as a whole" (Ballerini et al. 2008, 1232). This group behavior is not unique to the birds nor to the terrestrial realm—far out in the oceans, "striking similarities between the internal organization of bird flocks and fish schools" can be seen in the silvery schools of fish responding to the attack of a shark predator (Couzin and Krause 2003, 37).

Couzin (2007, 715) considers the collective minds of a moving herd to function like "an integrated self-organizing array of sensors," in that critical information, like the detection of a source of food or an oncoming predator, "may often be detected by only a relatively small proportion of group members. . . . [B]ehavioural coupling among near neighbours, however, allows a localized change in direction to be amplified, creating a rapidly growing and propagating wave of turning across the group. This positive feedback results from the ability of individuals to influence and be influenced by others, and allows them to experience an 'effective

Table 6.5
Convergent evolution of collective animal behavior

1 Convergent behavior and function: HERDING IN TWO DIMENSIONS (behavioral coupling in a large group of animals to produce an integrated self-organizing array of sensors distributed over a large two-dimensional geographic area to detect food or predators)

Convergent lineages:

1.1 Desert locust (Bilateria: Protostomia: Ecdysozoa: Arthropoda: Mandibulata: Hexapoda: Orthoptera: Acrididae; *Schistocerca gregaria*)

1.2 Leptothorax ant (Arthropoda: Mandibulata: Hexapoda: Hymenoptera: Vespoidea: Formicidae: Myrmicinae; *Leptothorax albipennis*)

1.3 Alamosaur (Bilateria: Deuterostomia: Chordata: Osteichthyes: Sarcopterygii: Reptiliomorpha: Amniota: Sauropsida: Archosauromorpha: Dinosauria: Saurischia: Sauropoda: Titanosauridae; *Alamosaurus sanjuanensis* †Cretaceous)

1.4 Maiasaur (Dinosauria: Ornithischia: Cerapoda: Ornithopoda: Hadrosauridae; *Maiasaura pebblesorum* †Cretaceous)

1.5 African elephant (Amniota: Synapsida: Therapsida: Mammalia: Eutheria: Afrotheria: Proboscidea: Elephantidae; *Loxodonta africana*)

1.6 Blue wildebeest (Mammalia: Eutheria: Laurasiatheria: Cetartiodactyla: Ruminantia: Bovidae; *Connochaetes taurinus*)

1.7 Modern human (Mammalia: Eutheria: Euarchontoglires: Primates: Catarrhini: Hominoidea: Hominidae: Homininae; *Homo sapiens*)

2 Convergent behavior and function: HERDING IN THREE DIMENSIONS (behavioral coupling in a large group of animals to produce an integrated self-organizing array of sensors distributed over a large three-dimensional spatial volume to detect food or predators)

Convergent lineages:

2.1 Antarctic krill (Bilateria: Protostomia: Ecdysozoa: Arthropoda: Mandibulata: Malacostraca: Euphausiacea: Euphausiidae; *Euphausia superba*)

2.2 Honeybee (Arthropoda: Mandibulata: Hexapoda: Hymenoptera: Apoidea: Apidae: Apini; *Apis mellifera*)

2.3 Monarch butterfly (Arthropoda: Mandibulata: Hexapoda: Lepidoptera: Danaidae; *Danaus plexippus*)

2.4 Atlantic herring (Bilateria: Deuterostomia: Chordata: Osteichthyes: Actinopterygii: Clupeiformes: Clupeidae; *Clupea harengus*)

2.5 European starling (Osteichthyes: Sarcopterygii: Reptiliomorpha: Amniota: Sauropsida: Archosauromorpha: Dinosauria: Saurischia: Theropoda: Maniraptora: Aves: Neognathae: Neoaves: Passeriformes: Sturnidae; *Sturnus vulgaris*)

2.6 Bottlenose dolphin (Amniota: Synapsida: Therapsida: Mammalia: Eutheria: Laurasiatheria: Cetartiodactyla: Cetacea: Odontoceti: Delphinidae; *Tursiops truncatus*)

Note: The geological age of extinct species is marked with a †. For data sources, see text.

range' of perception much larger than their actual sensory range." In the European starling (table 6.5), this behavioral coupling is a function of topological distance: each bird interacts with a fixed number of neighboring birds, typically six or seven, instead of with all birds contained in a fixed volume of metric space (Ballerini et al. 2008). This topological interaction allows the size of the flock of birds to expand and contract without altering the behavioral coupling between the birds.

Field studies with the desert locust (table 6.5) have revealed a critical threshold effect that triggers herd behavior. Buhl et al. (2006, 1402) observed that each locust behaves individually until the number of locusts present per square meter of land area reaches the number eight, at which point "a rapid transition occurs from [the] disordered movement of individuals within the group to highly aligned collective motion." Buhl et al. (2006) further argue that this critical density effect is a key triggering factor in all herding behavior, although the threshold number is different for different animal groups.

In table 6.5 I have listed as many disparate convergent lineages that have evolved collective minds, or herd behavior, that I am aware of. I have made no effort to try to determine how many times herd behavior may have convergently arisen within these major lineages; future phylogenetic studies of within-lineage convergence in herd behavior will certainly expand the list given in table 6.5. Within the mammalian clade, I have listed examples of convergent lineages in the three major branches of existent mammals: the afrotherians, laurasiatherians, and euarchontoglires. Some ruminant ungulates form enormous self-organized groups, as seen in the behavior coupling that produces the wavelike front of 100,000 moving individuals in a wildebeest herd in Africa (Couzin and Krause 2003). Elephants form smaller herds, and even we humans exhibit markedly different behavior when in groups than when alone, and unconsciously adopt coordinated behavior when we are moving as pedestrians along the sidewalks of our cities (Raafat et al. 2009). At the opposite extreme, swarming ants and marching locust juveniles can form even larger herds that cover many kilometers in area even though the individual animals themselves are tiny (Buhl et al. 2006; Conradt and Roper 2005).

The independent evolution of herding behavior by the Mesozoic dinosaurs is yet another example of the ecosystem convergence that exists between the dinosaurian and mammalian ecosystems (see table 4.10). In table 6.5 I have listed examples of convergent lineages in the two major branches of the dinosaur clade, the saurischians and the ornithischians.

Preserved fossil trackways demonstrate that sauropods, such as the alamosaur, moved in herds of over 30 individuals. The actual size of the herd was undoubtedly much larger. In the ornithischians, trackways of more than 80 ornithopods in a herd have been found (Martin 2006). In one spectacular example, a herd of maiasaurs was poisoned by gases from a volcanic eruption, producing a mass-kill horizon containing over 10,000 individuals in a bone bed preserved under a layer of volcanic ash in present-day Montana. A range of ages is evident in the herd, from juvenile maiasaurs only 3 meters in length to older adults over 7.5 meters in length (Weishampel and Horner 1990).

Flocking starlings and schooling fish, like the Atlantic herring (table 6.5), have already been mentioned in the introduction to this section of the chapter. Here I simply wish to emphasize that although birds and fish are radically different kinds of animals and live in radically different environments (air versus water), their minds have still converged on the same group behavior. These animals are vertebrates, but the same convergence occurs in the arthropods, as seen in the swarming Antarctic krill in the oceans and the flocking monarch butterflies in our skies (table 6.5). Consensus decision in groups is equally found in large-brained dolphin vertebrates and the tiny-brained honeybee arthropods (Conradt and Roper 2005).

In summary, Sumpter (2006, 5) argues that all herding behavior is driven by the same behavioral principles: "These principles, such as positive feedback, response thresholds and individual integrity, are repeatedly observed in very different animal societies. The future of collective behaviour research lies in . . . asking why they have evolved in so many different and distinct natural systems." Buhl et al. (2006, 1402) note that "models from theoretical physics have predicted that mass-migrating animal groups may share group-level properties, irrespective of the type of animals in the group." Taking these points into consideration, Couzin (2007, 715) concludes that "group behaviour holds clues about the evolution of sociality, and also for the development of novel technological solutions, from autonomous swarms of exploratory robots to flocks of communicating software agents that help each other navigate through complex and unpredictable data environments."

Swarms of robots? Robots are machines, not living organisms. Yet we could construct and program mechanical robots to sense and respond to the presence of, say, six to seven neighboring robots in a topological distance algorithm just like that of European starlings. Would large numbers of such programmed robots spontaneously self-assemble to produce swarm behavior? In addition, from the perspective of theoretical morphology, could it be that there exist only a limited number of ways in which a group of objects can move in a two-dimensional plane, or a three-dimensional volume of space? How much of convergent group behavior in animals is driven by the laws of physics and geometry?

Herding behavior (table 6.5) can serve multiple, disparate functions: predator avoidance, food detection, selective protection of the juvenile members of the herd from predators, successful mass migration from one distant region of the Earth to another without significant loss of group members, and so on. However, there exists one type of group behavior that has a very specific function: pack hunting. Pack hunting is a form of social, coordinated hunting that enables a group of small predators to kill prey animals that are much larger than they are. In table 6.6 I have listed all of the disparate convergent lineages that have evolved pack-hunting behavior that I am aware of. In one instance we know that a

Table 6.6
Convergent evolution of social hunting behavior

1 Convergent behavior and function: PACK HUNTING (social, coordinated hunting by a group of small predators, enabling them to kill prey animals that are much larger than they are)

Convergent lineages:

1.1 New World army ant (Bilateria: Protostomia: Ecdysozoa: Arthropoda: Mandibulata: Hexapoda: Hymenoptera: Vespoidea: Formicidae: Ecitoninae; *Eciton burchelli*)

1.2 Deinonychosaur (Bilateria: Deuterostomia: Chordata: Osteichthyes: Sarcopterygii: Reptiliomorpha: Amniota: Sauropsida: Archosauromorpha: Dinosauria: Saurischia: Theropoda: Maniraptora: Dromaeosauridae; *Deinonychus antirrhopus* †Cretaceous)

1.3 White pelican (Dinosauria: Saurischia: Theropoda: Maniraptora: Aves: Neognathae: Neoaves: Pelicaniformes: Pelicanidae; *Pelecanus erythrorhynchus*)

1.4 African wild dog (Amniota: Synapsida: Therapsida: Mammalia: Eutheria: Laurasiatheria: Carnivora: Caniformia: Canidae; *Lycaon pictus*)

1.5 Wolf (Mammalia: Eutheria: Laurasiatheria: Carnivora: Caniformia: Canidae; *Canis lupus*)

1.6 African lion (Mammalia: Eutheria: Laurasiatheria: Carnivora: Feliformia: Felidae; *Panthera leo*)

1.7 Modern human (Mammalia: Eutheria: Euarchontoglires: Primates: Catarrhini: Hominoidea: Hominidae: Homininae; *Homo sapiens*)

2 Convergent behavior and function: CARRION SCAVENGING (carrion scavenging in social groups for reasons that are not well understood)

Convergent lineages:

2.1 Eurasian black vulture (Dinosauria: Saurischia: Theropoda: Maniraptora: Aves: Neognathae: Neoaves: Falconiformes: Accipitridae; *Aegypius monachus*)

2.2 Turkey vulture (Dinosauria: Saurischia: Theropoda: Maniraptora: Aves: Neognathae: Neoaves: Ciconiiformes: Cathartidae; *Cathartes aura*)

Note: The geological age of extinct species is marked with a †. For data sources, see text.

previously accepted example of the convergent evolution of pack hunting turns out to be not convergent at all: the army ants. It was long thought that pack-hunting behavior in army ants was a convergent trait, and that the New World army ants (Ectoninae) and two lineages of Old World army ants (Aenictinae and Dorylinae) all independently evolved collective minds. This hypothesis has been disproved by the molecular phylogenetic analyses of Brady (2003), which revealed that all the army ants come from a single common ancestor that evolved pack-hunting behavior 105 million years ago, before the Cretaceous breakup of the southern supercontinent Gondwana. Thus, the current existence of army ants in Africa and South American is due to continental drift, and not due to convergent evolution. Army ants hunt in the tens of thousands, and literally tear apart the prey animals that they attack. Apocryphal tales of unstoppable army ant hordes that are capable of entirely defleshing farm animals or even humans abound in Africa.

Pack-hunting behavior has also convergently evolved in the sauropsid dinosaurian and synapsid mammalian lineages of amniote vertebrates, but these animals hunt in much smaller packs. In table 6.6 I have listed one Mesozoic dinosaurian lineage, the dromaeosaurs, for which we have solid fossil evidence of pack-hunting behavior. It is possible that social hunting may have independently evolved in other Mesozoic theropods as well, such as the coelurids (i.e., *Coelophysis baurri*; Martin 2006), and may be revealed in future field work. The dromaeosaur *Deinonychus antirrhopus* hunted in packs of six or more individuals, enabling this relatively small predator (50 to 100 kilograms) to kill large ornithopod prey such as *Tenontosaurus tilleti*, which weighed 1,000 kilograms (Ostrom 1990; Martin 2006). Most modern avian dinosaur predators are solitary hunters, like the hawks and the owls. A few birds have convergently evolved coordinated hunting behavior, but not as an adaptation for hunting large prey animals. The white pelican (table 6.6) is a herding animal, migrating in large numbers, but in addition "[t]hey often capture fish cooperatively, forming a long line, beating their wings and driving the prey into shallow water, where they seize the fish in their large, pouched bills" (Bull and Farrand 1988, 425).

In the mammalian lineage, pack-hunting behavior has independently evolved at least twice in the canids (table 6.6). Ecologically, the modern wolf is the mammalian equivalent of a Cretaceous dromaeosaur (see table 4.10; Ostrom 1990). The smaller African wild dogs are even more vicious than wolves. Nicknamed the "super beast of prey," packs of these canids are seemingly unstoppable when they have chosen a prey animal, and will run down, attack en masse, rip to shreds, and entirely eat a gazelle in ten minutes (Wilson 1980). As such, they approach the defleshing abilities of the army ants.

The top carnivores of the mammalian ecosystem, the felids, are mostly solitary predators. But one lineage, the African lion, has convergently evolved pack-hunting behavior (table 6.6). These cats form social prides, or herds, of numerous females and one to two males. Pack-hunting behavior in the lions is also a gendered behavior, in that the females hunt as a cohesive social unit but the males do not. Last, modern humans are pack hunters but the reverse of the lions, in that the males were traditionally the hunters and the females the gatherers. Pack-hunting behavior enabled our small ancestors (1.5 to 1.8 meters tall, 50 to 100 kilograms in weight) to bring down large prey like the woolly mammoths (*Mammuthus primigenius*) in Europe, which were 3 to 4 meters tall and weighed 8,000 kilograms.

Another odd form of social hunting has convergently evolved in the carrion-eating birds (table 6.6). As discussed in chapter 4, Old World and New World carrion-eating birds have independently evolved many morphological features associated with their feasting on carcasses. Curiously, they also have converged in aspects of their behavior: Eurasian black vultures and North American turkey vultures are gregarious, roosting together in large numbers at night, and often assembling in large groups during the day as well, perched in trees or on top of buildings. Yet Eurasian vultures are related to hawks and eagles, which are usually not gregarious birds, unlike the storks and flamingos that are the close relatives of the turkey vultures. In both Eurasia and North America, these birds use a soaring-in-circular-formation flight pattern in searching for rotting corpses and, when a carcass is discovered, they will often continue to soar in circles until other vultures arrive. Once a group has been assembled, the flock descends to feed.

The function of this convergent behavior is not well understood. Some have proposed that the vultures continue to circle in order to make sure the target animal is indeed dead, or that there are no ground-dwelling predators in the area that might pose a danger to the bird. However, the frequency of this behavior seems to lend more weight to the suggestion that the vultures are actually waiting for other birds to arrive, or even summoning them. They certainly do not need a larger number of vultures to complete the kill, because the prey animal is already dead. Carrion is an unpredictable food resource, and vultures must often wait long periods between feedings. When a carcass is discovered, a vulture will often gorge itself to the point that it has difficulty flying away. It might seem advantageous to the bird to selfishly keep the discovery of a carcass to itself, but at the same time, a large animal carcass is usually too much food for a single bird to consume by itself. Thus, it might be advantageous to share the information with other hungry vultures, in anticipation that they will return the favor in the future when they discover a carrion source, in a case of reciprocal altruism. Whatever the real function of vulture group behavior, it is clear the minds of these birds have converged.

Convergent Mentalities

When confronted with a mirror, or a window acting as a mirror, a territorial bird like a cardinal (*Cardinalis cardinalis*) clearly thinks that it is in the presence of another cardinal, and will attack the image in the mirror as if it were another bird encroaching on its territory. Many animals, when viewing a mirror, realize that another animal of their own kind is present (kittens will often try to play with the kittens they see in the mirror), but they never make the mental leap to the conclusion that the animal they are viewing in the mirror is themselves. They are not self-aware.

Humans are self-aware, and self-awareness was once thought to be a unique human trait. As is the case with so many other mental traits that were once thought to be unique to humans, we now know that self-awareness has independently evolved in other lineages of animals. Within the primates, self-awareness—just like tool construction and usage (table 6.1)—is a synapomorphy for the clade of the Hominoidae, all of the great apes, and not confined to the human species. We did not evolve self-awareness with the evolution of our large brains; we simply inherited it from our ancestors.

Experiments with mirror self-recognition have thus far revealed that elephants, cetaceans, and corvid birds are also consciously aware of their own individual existence (table 6.7): Asian elephants (Prior et al. 2008), bottlenose dolphins (Marino 2002), and magpies (Prior et al. 2008) all have been demonstrated to be aware that the animal they are seeing in the mirror is themselves, and will use the mirror to investigate their own bodies. All of these animals have relatively large brains, but curiously it is difficult to directly compare their brain sizes. The brain of the magpie is tiny compared to the brain of a human, but then the body size of the magpie is also tiny compared to that of the human. If we devise a metric

Table 6.7
Convergent evolution of self-awareness and metacognition

1 Convergent mentality: SELF-AWARENESS (the ability of an animal to be consciously aware of its own existence)

Convergent lineages:

1.1 Magpie (Amniota: Sauropsida: Archosauromorpha: Dinosauria: Saurischia: Theropoda: Maniraptora: Aves: Neognathae: Neoaves: Passeriformes: Corvidae; *Pica pica*)

1.2 Asian elephant (Amniota: Synapsida: Therapsida: Mammalia: Eutheria: Afrotheria: Proboscidea: Elephantidae; *Elephas maximus*)

1.3 Bottlenose dolphin (Mammalia: Eutheria: Laurasiatheria: Cetartiodactyla: Cetacea: Odontoceti: Delphinidae; *Tursiops truncatus*)

1.4 Orangutan (Mammalia: Eutheria: Euarchontoglires: Primates: Catarrhini: Hominoidea: Hominoidae: Pongidae; *Pongo pygmaeus*)

2 Convergent mentality: METACOGNITION (the ability of an animal to monitor or regulate its own cognitive state)

Convergent lineages:

2.1 Bottlenose dolphin (Mammalia: Eutheria: Laurasiatheria: Cetartiodactyla: Cetacea: Odontoceti: Delphinidae; *Tursiops truncatus*)

2.2 Barbary macaque monkey (Mammalia: Eutheria: Euarchontoglires: Primates: Catarrhini: Cercopithecoidea: Cercopithecidae; *Macaca sylvanus*)

2.3 Human (Mammalia: Eutheria: Euarchontoglires: Primates: Catarrhini: Hominoidea: Hominidae; *Homo sapiens*)

Note: For data sources, see text.

measuring the weight of the brain relative to the weight of the body (Prior et al. 2008), the magpie has a brain-body weight index of 31, which is greater than the index of 21 for a human. The bottlenose dolphin has an index of only 9.0, and the Asian elephant an index of 1.6. Clearly the brain-body weight index favors animals of small size over those of large (Prior et al. 2008).

In chapter 5 we considered the effect of size and allometric scaling in producing convergent tissue structures (West et al. 1999). If we plot the logarithm of brain mass versus the logarithm of body mass, most mammals and birds fall along a single linear function of allometric scaling (Gould and Gould 2007). Primates, cetaceans, corvids, and elephants plot above this line, however, indicating that they have larger brains than are strictly necessary for neural control of their body masses. Humans and porpoises have brains that plot the farthest away from the general allometric scaling, but even here problems of brain-size comparison arise. The porpoise lives in a state of weightlessness, of neutral buoyancy in water, and does not need the large muscle masses (which require neural control) that land animals must have in order to exist in the gravitational field of the Earth. Much of their body mass is actually blubber fat, which they use for thermal insulation in the cold waters of the ocean (Gould and Gould 2007). Thus, their large brains may be even more unusual than that of humans.

Although self-aware animals all have unusually large brains (in the allometric sense), their brain structures are radically different: "primate-cetacean cognitive and behavioral convergence is a dramatic example of functional convergence in the face of profound structural or mechanistic divergence" (Marino 2002, 30). Even though the elaboration of the cortical circuitry in primates and cetaceans is very different, they are both members of the clade of the eutherian mammals and share much of their subcortical neuroanatomy (Marino 2002). The corvid birds, however, do not even have a prefrontal cortex! Although these animals are avian dinosaurs, with vastly different brain structures from those of mammals, "intelligence in both corvids and primates has evolved through a process of divergent brain evolution yet convergent mental evolution . . . [showing that] intelligence can evolve in the absence of a prefrontal cortex" (Clayton and Emery 2008, 138–139).

Another mental state associated with high degrees of intelligence is metacognition, the ability of an animal not only to be aware of its own existence, but to think about its own thinking. Humans know when they are uncertain, are aware of that cognitive condition, and are able to think about how to deal with their uncertainty—to seek further information, or to defer a response and to wait (Smith 2009). Thus far in this new area of research, experimentation has revealed that the bottlenose dolphin and the Barbary macaque monkey are aware of their own cognitive states (table 6.7), that capuchin monkeys (*Cebus apella*) show equivocal evidence of metacognition, and that pigeons (*Columba livia*) show no evidence of being aware of their cognitive state (Smith 2009). Smith (2009, 395) notes that the corvid birds, "having proven themselves cognitively sophisticated, would be the usual suspects for testing in this area," as are elephants and great apes other than humans. It is known that both New World and Old World monkeys have convergently evolved tool use (see table 6.1), and the evidence that macaque monkeys have metacognition suggests that these monkeys should be carefully examined for evidence of possible mirror self-recognition as well.

Finally, another mental state associated with convergent intelligence is the awareness of death, as well as the experience of grief and sorrow at the loss of another individual (table 6.8). Both gorillas and dolphins have been observed to mourn the death of their parent or their offspring.

Table 6.8
Convergent evolution of mourning

1 Convergent mentality: FAMILIAL MOURNING (parent mourning the death of an offspring, or offspring mourning the death of a parent)
Convergent lineages:
1.1 Bottlenose dolphin (Mammalia: Eutheria: Laurasiatheria: Cetartiodactyla: Cetacea: Odontoceti: Delphinidae; *Tursiops truncatus*)
1.2 Gorilla (Mammalia: Eutheria: Euarchontoglires: Primates: Catarrhini: Hominoidea: Hominidae: Gorillinae; *Gorilla gorilla*)

2 Convergent mentality: GROUP MOURNING (mourning the death of one's own kind, even if only distantly related)
Convergent lineages:
2.1 African elephant (Mammalia: Eutheria: Afrotheria: Proboscidea: Elephantidae; *Loxodonta africana*)
2.2 Chimpanzee (Mammalia: Eutheria: Euarchontoglires: Primates: Catarrhini: Hominoidea: Hominidae: Homininae: Panini; *Pan troglodytes*)
2.3 Neanderthal human (Mammalia: Eutheria: Euarchontoglires: Primates: Catarrhini: Hominoidea: Hominidae: Homininae: Hominini; *Homo neanderthalensis* †Holocene)
2.4 Modern human (Hominidae: Homininae: Hominini; *Homo sapiens*)

Note: The geological age of extinct species is marked with a †. For data sources, see text.

Both in the wild and in zoos, gorilla and chimpanzee mothers carefully carry around their dead infants for days, caressing them, or placing them on the ground and pacing in circles around them, checking over and over again to see if they have resumed breathing. Dolphin mothers repeatedly lift their infants up in the water for days in apparent attempts to help the calf breathe. The mothers do not eat during this period, and make repeated distress cries.

Other animals are known to extend their mourning behavior to other members of their own kind that are not members of their immediate family (table 6.8). Elephants repeatedly touch the skulls and tusks of long-dead elephants, gently lifting the bones up and down or moving them around. They also repeatedly pace away and then return to the remains of the dead elephant. In a stunning photograph, Monica Szczupider (2009, 12–13) captured the facial expressions and body language of 16 chimpanzees who were mourning the death of an elder female at the Sanaga-Yong Chimpanzee Rescue Center.

Two human species evolved a mourning behavior not seen in our great ape cousins: the burial of our dead. The oldest fossils of anatomically modern humans found to date are 200,000 years old, and are from northeast Africa (modern Ethiopia). The Neanderthal humans have long roots in Europe, going back some 400,000 years. It is thought that the Neanderthal humans evolved from *Homo heidelbergensis* in Europe, sometimes described as "archaic humans," and that *H. heidelbergensis* is derived from *H. erectus*, the first species of human to become geographically widespread outside of Africa (Benton 2005). Modern humans, in contrast, are thought to have evolved from *H. ergaster* back in Africa (Wade 2006).

Thus, modern humans and Neanderthal humans do not share a common species ancestor, but are cousins in the human family tree. Yet both Neanderthal humans and modern humans buried their dead—a mourning behavior that independently evolved in these two separately derived species. We have no evidence that either *H. ergaster*, our ancestor, or *H. heidelbergensis*, the Neanderthal's ancestor, buried their dead. Our two geographically separated species came into contact with one another around 100,000 years ago, in the first attempt of our species to migrate out of Africa through the region that is now Israel. We were stopped by the Neanderthals, who appear to have destroyed all the emigrant modern human populations (Wade 2006). Only 50,000 years ago did modern humans successfully migrate out of Africa, and the last of the Neanderthal human populations went extinct 30,000 years ago (Wade 2006).

In summary, Emery and Clayton (2004, 1903) propose that animals have evolved intelligence not to solve physical problems, but rather to solve social ones: "we argue that complex cognitive abilities evolved multiple times in distantly related species with vastly different brain structures in order to solve similar socioecological problems." That is, rather than evolving as a function of the limited number of ways that exist to solve physical problems with objects, as seen in the convergent evolution of tool-using behavior or architectural behavior, convergent intelligence evolves as a function of the limited number of ways that exist to solve interactive social problems with other individuals. Greater awareness of the behavior and potential motivation of other individuals is hypothesized to lead to the greater awareness of one's own existence as an individual, and of one's own behavior and motivation.

Convergent social systems have evolved in radically different environmental systems on Earth: on land and in the oceans. Whitehead (2008, 114) notes that "terrestrial and oceanic environments provide a tough challenge for convergence. When traits do converge, something remarkable has occurred. Despite the radically different physical environments . . . the social structures and cultures of sperm and killer whales have much in common with those of elephants and humans," and "there are also non-mammalian species, especially birds, that have social, cognitive, cultural, and life-history characteristics in common with the apes, odontocetes [toothed cetaceans], and elephants" (Whitehead 2008, 157). The extension of these observations to alien ecologies and environments is obvious: If the same convergent minds arise in such vastly different ecological settings as the oceans and terrestrial regions of the Earth, what is the probability that they will also arise on extrasolar planets circling alien stars? Will alien minds converge on ours, even though their brain structures may be radically different?

7 Functional and Developmental Constraint in Convergent Evolution

Our results strongly support the hypothesis that the essential elements of organic structure are highly constrained by geometric rules, growth processes, and the properties of materials. This suggests that, given enough time and an extremely large number of evolutionary experiments, the discovery by organisms of "good" designs—those that are viable and that can be constructed with available materials—was inevitable and in principle predictable.
—Thomas and Reif (1993, 342)

Was mich eigentlich interessiert, ist, ob Gott die Welt hätte anders machen können.
—Einstein (quoted in Seelig 1956, 72)

Albert Einstein once mused, "What really interests me is whether God could have made the world in a different way." To Einstein, God represented the laws of nature. He was asking whether the evolution of the universe was so constrained by the initial conditions of the Big Bang, by the observed constants of nature, that only the present universe could have evolved? Or were alternative universes possible, universes that would have evolved along physical pathways not followed by our present universe?

This question can be framed with respect to biological evolution as well. Is the evolution of life so constrained by the geometry of the universe, by the physical constants of nature, that its outcome is predictable? Or are so many alternative evolutionary pathways possible that it would never be possible to predict the trajectory of the evolution of life? We can easily visualize a universe in which every species is morphologically different from every other species, and in which each species has its own unique ecological role, or niche, in nature. That universe does not exist. Instead, we live in a universe where convergence in evolution is rampant at every level, from the external forms of living organisms down to the very molecules from which they are constructed, from their ecological roles in nature to the way in which their minds function.

Since convergent evolution is so ubiquitous in nature, as we have seen in the previous five chapters, the total extent of convergent evolution might best be revealed by studying its opposite: *unique evolution*. That is, rather than compiling lists of convergences, we might compile lists of solitary evolutionary innovations in species that have not been independently discovered by other species in their evolutionary pathways. Vermeij (2006) set out to do just that, and compiled a list of evolutionary innovations said to be unique. He discovered that "purportedly unique innovations either arose from the union and integration of previously independent components or belong to classes of functionally similar innovations" and that "important ecological, functional, and directional aspects of the history of life are replicable and predictable" (Vermeij 2006, 1804).

What are the possibilities for evolution in our universe? Can we even think about considering the total spectrum of what is possible and not possible in biological evolution? The answer is yes, by using the analytical techniques of theoretical morphology, in particular by the construction of theoretical morphospaces (McGhee 2001, 2007). The concept of the theoretical morphospace originated in evolutionary biology (McGhee 1999), but it has subsequently caught the attention of philosophers (Maclaurin 2003), linguists, cultural anthropologists, and neuroscientists (Hauser 2009) who are seeking to explore the spectrum of both possible and impossible languages and cultures. Here we will use the concept to analyze the phenomenon of convergent evolution with respect to the spectrum of existent, nonexistent, and impossible biological form.

Convergent Evolution in Theoretical Morphospace

The analytical techniques of theoretical morphology allow us to take a spatial approach to the concept of convergent evolution. Any given biological form, or *f* in abbreviation, may be described by a set of measurements taken from that form—how tall is it, how wide, how long? Each type of measurement (height, width, length, etc.) can be considered as a dimension of form. The total set of the possible dimensions of form can be used to construct a hyperdimensional morphospace of possible form coordinates (figure 7.1). Each point within this theoretical morphospace represents a specific combination of form measurements that will produce

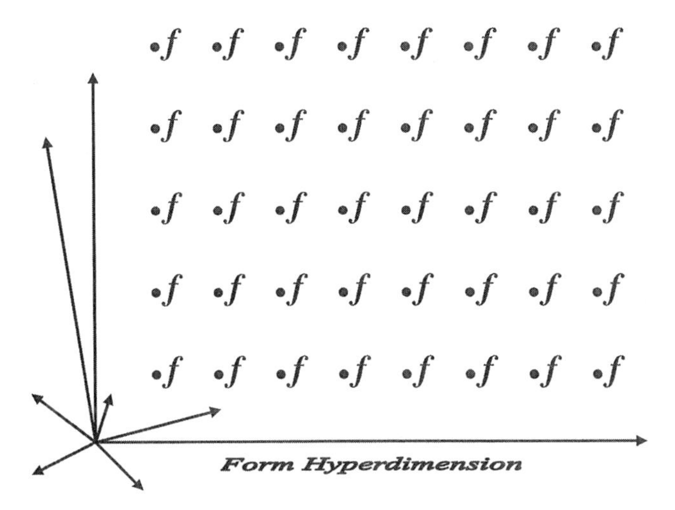

Figure 7.1
A theoretical hyperdimensional space of possible form (modified from McGhee 2007). Each dimension of the space represents a morphological trait that may be measured on a given biological form, *f*. All possible coordinate combinations (points) within the theoretical morphospace represent the set of all possible biological forms. Although only eight dimensions are shown in this schematic diagram, the dimensionality of an actual hyperspace of form will be much larger.

the form coordinate for a hypothetical form *f*. Convergence occurs when forms originally present in different regions of the morphospace evolve in such a way that they move to the same spatial region in the morphospace (figure 7.2).

Returning to figure 7.1, we can begin to consider the effects of evolutionary constraint in producing convergent evolution. First, we know that some types of forms function in nature: these are the myriad forms of life that surround us here on Earth. The opposite of this observation is the concept that there exist forms that do not function in nature, and that if a living organism were to produce one of these forms, it would be lethal. Mutations that produce nonfunctional, lethal forms are well known in biology. We can thus conceptually divide the spectrum of form within the morphospace into spatial regions that contain nonfunctional form and functional form (figure 7.3). The boundary between the spatial regions of form separates nonfunctional from functional forms, and is itself a spatial representation of the concept of functional constraint. That is, if the evolving form is to remain functional, it must remain within the functional region of forms in the morphospace (an aside: these boundaries are not mere idle

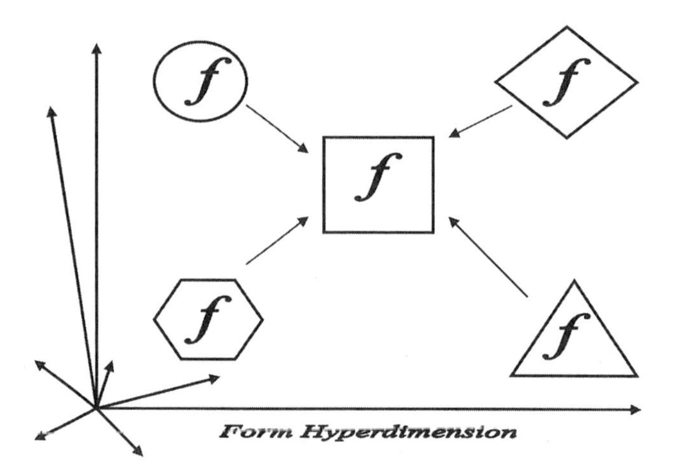

Figure 7.2
Convergent evolution of form within a theoretical morphospace. Organisms in different regions of the morphospace, and thus possessing different initial morphologies, have evolutionary trajectories that take them to the same region within the morphospace, and their morphologies converge.

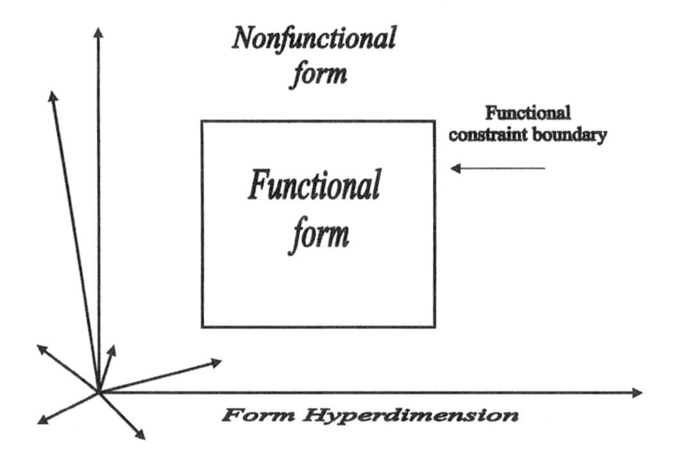

Figure 7.3
Functional constraint in theoretical morphospace. Functional forms are found within the rectangular region in the morphospace, and nonfunctional forms are found outside the rectangular region. The boundaries of the rectangle thus delimit the functional constraint boundary on possible form within the morphospace.

speculation, but can actually be mapped in theoretical morphospaces; see McGhee 2007).

Second, we know that organisms develop biological form from an original cell. We also know that the possible types of form that can be developed from a given cell are limited, that they depend upon the DNA coding within the cell and the interaction of the different molecules and tissue geometries produced as the cell grows. These observations lead to the concept of developmental constraint: the different types of forms that different organisms can develop are limited.

Now let us consider the spatial representation of these two types of evolutionary constraint within the morphospace (figure 7.4). The functional constraint boundary in figure 7.4 is the same as in figure 7.3, but now we have added a developmental constraint boundary as well (the dotted line in figure 7.4). That is, the evolving form must remain within the developmentally possible regions of forms in the morphospace. As can be seen in figure 7.4, the functional constraint boundary within the morphospace does not have to coincide spatially with the developmental constraint boundary.

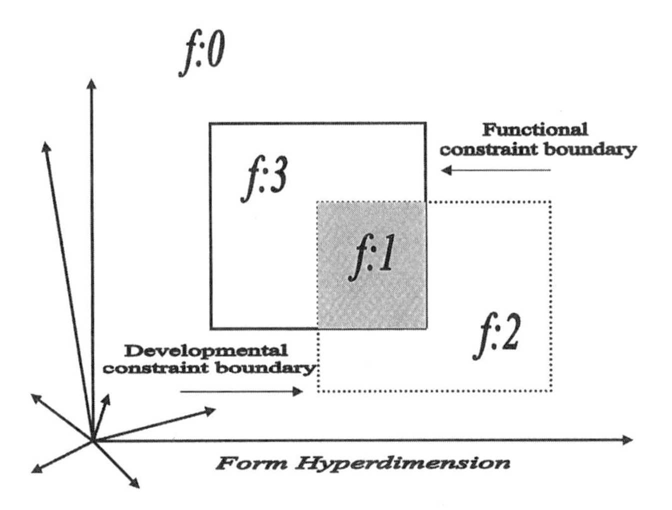

Figure 7.4
Combining functional and developmental constraint in theoretical morphospace. The boundaries of the dotted-line rectangle delimit the developmental constraint boundary on possible form within the morphospace: forms within the dotted-line rectangle are developmentally possible, while forms outside the dotted-line rectangle are developmentally impossible. Forms *f:0* are thus both nonfunctional and developmentally impossible, forms *f:1* (shaded region) are both functional and developmentally possible, forms *f:2* are nonfunctional but developmentally possible, and forms *f:3* are functional but developmentally impossible. See text for discussion.

We now are in the position to consider the spectrum of all possible existent, nonexistent, and impossible biological forms. Let us consider the developmental constraint boundary present in figure 7.4 to spatially portray the spectrum of all biological forms that can be developed by life on Earth. That is, forms within this region can be developed by organisms present on Earth, but forms outside this region cannot. We now see that four different regions of potential form exist within the morphospace, in the form of a Venn diagram:

(1) Biological forms that do not function and that cannot be developed by Earth life, abbreviated *f:0*. These are all of the hypothetical forms shown in figure 7.1 that fall outside the two intersecting rectangles shown in figure 7.4.

(2) Biological forms that do function and that can be developed by Earth life, abbreviated *f:1*. These are all of the hypothetical forms shown in figure 7.1 that fall in the region where the two rectangles shown in figure 7.4 intersect (the shaded region).

(3) Biological forms that could be developed by Earth life, but that do not function and thus are lethal, abbreviated *f:2*. These are all of the hypothetical forms shown in figure 7.1 that are located in figure 7.4 in the region bounded by the dotted-line rectangle but not located in the region where it intersects with the solid-line rectangle (the shaded region).

(4) Biological forms that do function but that cannot be developed by Earth life, abbreviated *f:3*. These are all of the forms shown in figure 7.1 that are located in figure 7.4 in the region bounded by the solid-line rectangle but are not in the region where it intersects with the dotted-line rectangle (the shaded region).

Now let us consider these four regions of possible form from the point of view of actual existent life on Earth. The myriad forms of life that surround us here on Earth clearly are both functional and developable, and we can collect all of these forms into a set of forms {*f:1*}. Every living thing on Earth, from butterflies to bacteria, belong to the form set {*f:1*}. Needless to say, all of the convergent forms of life that we have considered throughout this book are also members of the form set {*f:1*}. The phenomenon of convergent evolution immediately reveals to us that the size of form set {*f:1*} is not infinite, in that life on Earth has been constrained to repeatedly reevolve the same forms within this set over and over. The hypothetical universe in which every species has its own unique functional morphology, is morphologically different from every other species, does not exist.

Biologists over the ages have studied the many different mutant forms of life—two-headed snakes, three-legged frogs, and so on—that are lethal mutations (which are important in that they give us valuable clues about the process of development; see the discussions in Alberch 1989 and Blumberg 2009). Developmental abnormalities like two-headed snakes and frogs with three hind legs instead of two are real and are also nonfunctional, in that they do not survive in the wild (as opposed to in the laboratory). We can collect all such developmental "freaks of nature" (Blumberg 2009) into the set of forms {*f:2*}.

Visualizing the types of forms that are both nonfunctional and that cannot be developed by life on Earth is a bit more difficult, but it can be done mathematically (see McGhee 2007), and we can place all of these hypothetical form coordinates within the morphospace into the form set {*f:0*}. They do not really matter for our discussion here, but they do need to be listed so that we will have considered the complete spectrum of all possible existent, nonexistent, and impossible biological forms. So, thus far we have three sets of form with respect to Earth life: one functional set of forms of Earth life, {*f:1*}; one nonfunctional set of forms of Earth life, {*f:2*}; and one impossible set of forms, {*f:0*}.

The last possible set of forms is crucial to our understanding of the implications of the phenomenon of convergent evolution. These are the possible forms of life that are functional, that work just fine in nature, but that nevertheless cannot be developed by life on Earth. We will place these possible but nonexistent forms on Earth in the final set of forms, {*f:3*}. We have now arrived at the critical question in the analysis of convergent evolution: Does the form set {*f:3*} exist in the universe? We know that it does not exist on Earth, but do living forms exist elsewhere on alien worlds that belong to the form set {*f:3*}?

At first this question may seem to be so abstract as to be of no real importance, but it is of critical importance in our consideration of the question of predictability in evolution (which we will consider in detail in the next chapter). We know that much of the convergent evolution of life on Earth is driven by developmental constraint, yet we do not know the answer to two of the most fundamental questions concerning developmental constraint itself: "How did development originate [on Earth]?" and "How did the developmental repertoire evolve?" (Müller 2007, 944). Only when we have the answers to those two questions can we make predictions about the possible evolution of alien developmental repertoires, and whether or not those developmental repertoires might be similar to those found in Earth life. The existence of the form set {*f:3*} would imply that there exists somewhere in the universe an alien set of functional biological forms that nevertheless cannot be developed by Earth life, and thus these alien life forms would *not be convergent* on any of the forms of life seen on Earth.

Life as We Know It

All life on Earth belongs to the form set {*f:1*}, biological forms that both function and can be developed by Earth life. As we have seen in chapters 2 through 6 in this book, life on Earth is highly convergent. The spectrum of forms in the set {*f:1*} have been reevolved over and over again in the history of life on Earth.

The set of existent forms of life on Earth, {*f:1*}, is a function of the functional constraint boundary and the developmental constraint boundary in the theoretical morphospace (figure 7.4). We believe the functional constraints of physics and geometry to be the same throughout the universe and to be extrinsic with respect to biology (McGhee 2007); therefore, this constraint boundary should also apply to life forms throughout the universe. That is, the spatial position of the functional constraint boundary shown in figure 7.4, relative to the spectrum of hypothetical forms shown in figure 7.1, should remain the same throughout the universe. Alien life forms should belong to the same set of forms {*f:1*} that Earth life forms belong to, given the same physical conditions on that alien planet.

Thus, we could predict that those alien organisms with powered flight will have evolved wings, just as animals on Earth have done (table 2.2), and fast-swimming organisms in alien seas will evolve streamlined, fusiform bodies (table 2.1). Land-dwelling, sessile, photoautotrophic organisms will evolve tree forms, just as plants on Earth have done (table 3.1), and those tree forms will have leaflike structures (table 3.3). The laws of physics and geometry also extend to activities, in that there are a limited number of ways to build structures (table 6.2) and to successfully cultivate crops (table 6.3). The geometries of alien buildings, and the algorithms of farming procedures, should be similar to those found on Earth. If they live in highly social groups, natural selection should drive evolution to produce aliens that are self-aware, just as it has on Earth (table 6.7).

Life as We Do Not Know It

It is the second boundary in theoretical morphospace—the developmental constraint boundary—that undermines our ability to confidently predict that life throughout the universe will converge on the same forms seen here on Earth. Developmental constraints are intrinsic (McGhee 2007), imposed by the laws of biology of specific organisms, in this case the biology of Earth organisms. Will these constraints be the same for alien organisms?

Even in the case of Earth organisms themselves, developmental constraint introduces an element of uncertainty into our ability to precisely predict convergence. We can be confident in predicting that animals that fly will have wings—but how many? Birds, bats, and pterosaurs are developmentally constrained to have only two wings, as their tetrapod phylogenetic legacy provides them with only two forelimbs that can be modified in convergent evolution. This developmental constraint does not apply to insects, however, and so the dragonfly has four wings, not just two. The mythical centaurs have never been evolved by mammals; their tetrapod phylogenetic legacy provides them with only four appendages to modify, not six. The insects, however, are developmentally provided with six appendages, and they have indeed independently evolved centaur forms twice: the mantids and the mantispids (as discussed in chapter 2). Hence, alien organisms, with a totally different developmental repertoire, might evolve forms never even dreamed of in human mythology.

We have already considered one possible example of a potential functional structure that apparently cannot be developed by Earth life: eyes that can see infrared light (chapter 2). Multicellular life with eyes has existed on Earth for over 600 million years, yet never has been able to develop eyes that can see light with long wavelengths, wavelengths in the infrared, even though possession of those eyes would give a significant selective advantage to predators. Both camera-eyed animals and compound-eyed animals have had to evolve alternative organ systems in order to "see" infrared (table 2.5). And we humans have constructed machines that enable us to "see" into the infrared region of the spectrum. Could an alien life form exist that has been able to do so organically? If so, that alien would have a form that belongs to form set {*f:3*}. Or does alien life elsewhere experience the same developmental constraints that forms of life on Earth do? If so, the form set {*f:3*} does not exist—there are no functional forms of life that cannot be developed by life on Earth simply because life on other planets cannot develop these forms either. Several lines of reasoning, however, suggest that this may not true, and that the set of forms {*f:3*} potentially exists in the universe. NASA, the space exploration agency of the United States, is also very interested in the possible existence of non-Earth-type life, as discussed in Peter Ward's book *Life as We Do Not Know It* (2005). In trying to conceive of possible forms of life that are nonexistent on Earth, the scientists at NASA are engaging in classic theoretical morphology.

Let us also engage in a theoretical-morphology thought experiment. Development of life on Earth starts from a single cell and, as far as we currently understand it, depends upon the genetic coding within the cell and the interaction of the different molecules and tissue geometries produced as the cell grows. Fundamentally, we are dealing with molecules and geometries: all Earth life is carbon-based, uses DNA to code for amino acid assembly, and is constructed of amino acids that have left-coiled geometries; furthermore, multicellular life develops complex tissue geometries from simple, spherical, topologically solid cell clusters (Newman 2010). So, let us consider each of these characteristics in turn with regard to the question: Could hypothetical life elsewhere be different? Can we conceive of possible life forms that are nonexistent on Earth?

All life on Earth is carbon-based: non-carbon-based life would surely experience different developmental constraints than those present in Earth life, and therefore could possibly develop functional biological forms that Earth life cannot. The most widely discussed chemical alternative to carbon is silicon-based life. From the periodic table of elements, we see that silicon can form four bonds like carbon (for example, in the formation of monosilane, SiH_4, the analog of methane, CH_4), which is so important for building the carbon-backboned molecules of Earth life. Unfortunately, silicon bonding simply cannot produce the huge macromolecules that carbon can—macromolecules that are very strong and stable at normal Earth temperatures and pressures. Curiously, however, it is thought that silicon-based life might be possible at either very high temperatures or very low temperatures (Ward 2005). At very low temperatures, silicon can form polymers, large macromolecules that are analogs to those formed by carbon. Rather than using water as a solvent, as with carbon-based life, silicon-based life could use liquid methane or nitrogen. As a result, it has been hypothesized that silicon-based life might exist in the methane lakes that exist on Titan, Saturn's planet-like moon that possesses its own nitrogen atmosphere, or in possible liquid nitrogen lakes on Triton, the larger moon of Neptune, and, obviously, in similar worlds elsewhere in the universe (Ward 2005). In particular, McKay and Smith (2005) have shown that is energetically possible for a hypothetical Titan life form to metabolize acetylene (C_2H_2), present on the surface of Titan, using atmospheric hydrogen (H_2) to produce methane (CH_4), analogous to Earth life that metabolizes sugar ($C_6H_{12}O_6$) with atmospheric oxygen (O_2) to produce carbon dioxide (CO_2). They further predict that the presence of such a hypothetical methanogenic life form on Titan should result in chemically anomalous depletions of acetylene on the surface of Titan, and the anomalous depletion of hydrogen in Titan's lower atmosphere. Interestingly, as of this writing, both of these predictions have been observed to occur on Titan by the orbiting Cassini spacecraft (Strobel 2010; Clark et al. 2010), precipitating renewed debate concerning possible life on Titan.

Such hypothetical life forms on cold planets would have developmental pathways radically different from those present in Earth life, and thus could evolve functional morphologies in their cold habitats totally unlike any seen on Earth. If these biological forms exist, they would belong to form set {*f:3*}. Since Earth life cannot exist on Titan or Triton, it is obvious that Earth life cannot develop a biological form that could function in a lake of liquid methane.

Ward (2005) points out that the zone within our solar system containing cold-type environments in which methane and nitrogen are liquids is much wider than the narrow zone in which water is liquid. Only the Earth and Mars exist in this zone today, and Mars is at its very outermost fringe. Consequently, silicon-based life, if it exists, would have a much larger suite of worlds and moons within our own solar system upon which it could exist compared to carbon-based life, and this will be true of other solar systems as well. Could silicon-based life be more abundant in the universe than carbon-based life?

Conway Morris (2003) argues that we should expect to find that life is carbon-based wherever life exists in the universe. According to this argument, functional constraints in molecular evolution will always drive natural selection for the uniquely versatile carbon atom in the formation of polymers, or complex macromolecules, throughout the universe. No other element in the periodic table of elements has carbon's chemical ability to form four bonds that are strong, to form double bonds, to link together to form stable rings and chains, and to form a gas by combining with oxygen (CO_2; silicon readily combines with oxygen, but produces rigid crystalline SiO_2, which presents serious problems if the silicon-based life is to use oxygen in its metabolism). If molecular evolution always converges on carbon in the formation of life, then all life elsewhere will be confined to Earth-like worlds with liquid water. This reasoning also leads Conway Morris (2003) to conclude that life could be quite rare in the universe, as the temperature zone around a star that could contain Earth-like worlds is narrow (Ward 2005).

In summary, if silicon-based life exists in the universe, it could potentially develop forms that belong to form set {*f:3*}, life forms that could potentially not be convergent with any life found on Earth. On the other hand, if all life in the universe is carbon-based and found only on Earth-like worlds, then life elsewhere could be highly convergent on the same life forms found on Earth.

All life on Earth is coded for by base-four DNA, coding with the nucleotides ATCG, which are read in a triplet. Could life exist elsewhere that is not thus coded? Life that uses a different coding system would surely experience different developmental constraints than base-four DNA-coded life, and so could possibly develop functional biological forms that Earth life cannot. The easiest way to envision such life forms would be to keep the same sugar-phosphate chains that are present in DNA, but then to add or subtract the number of nucleotides that code for protein. Laboratory experiments have actually produced synthetic base-six DNA, molecules having six nucleotides rather than four (Ward 2005). Could life have arisen elsewhere that uses base-six DNA coding as part of its developmental system?

Conway Morris (2003) argues that we should expect to find that life uses base-four DNA with triplet codons wherever life exists in the universe. Just as with the element carbon, Conway Morris (2003) points out that computer simulations of alternative coding systems have shown that base-four DNA coding is vastly more efficient than any other hypothetical coding system: less than base-four coding is too limited and restricted, and greater than base-four coding is too cumbersome and error-prone. Freeland and Hurst (1998) have further experimented with base-four DNA with triplet codons by randomizing the codon positions to produce alternative genetic codes. They discovered that, in a sample of over a million alternative codes, only one code was more efficient than the natural code found in life on Earth. While this result can be used to support the view that carbon-based life elsewhere will convergently evolve the same genetic code that was evolved by life on Earth (Conway Morris 2003), from the perspective of theoretical morphology the discovery of a more efficient but nonexistent code is of more interest. Even here, however, the nonexistent code "shows behavior very similar to that of the natural code" (Freeland and Hurst 1998, 247); thus, it is not clear whether life evolved using this alternative code would be that much different from life on Earth.

The amino acid molecules that are used as the building blocks of proteins in Earth life are all left-coiled. Could life exist elsewhere that uses right-coiled amino acids? Experiment shows that chemical selection in protein synthesis produces homochirality, the use of only one type of amino acid geometry. But proteins can be constructed in the laboratory with either all left-coiled or all right-coiled amino acids, and it is not clear why life on Earth has chosen the left-coiled chemical pathway (Castelvecchi 2007). Both right-coiled and left-coiled amino acids are produced by inorganic chemical processes in nature, and so perhaps chance has produced life forms elsewhere that are constructed of proteins made with right-coiled amino acids. Would such life produce biological forms that are simply the chemical mirror-image of Earth life forms? Or would those forms be different from anything seen on Earth, and might they belong to form set {*f:3*}?

Just as the element carbon has unique properties and the DNA coding system found in Earth life is highly efficient, some have argued that molecular evolution preferentially produces left-coiled amino acids. Experiments have suggested (1) that left-coiled amino acids are slightly more stable than right-coiled, and thus could have accumulated more readily in prebiotic environments than right-coiled, and (2) that left-coiled amino acids crystallize slightly faster than right-coiled, and thus could have become more enriched in prebiotic environments than right-coiled (Castelvecchi 2007). Either process would result in prebiotic environments containing more left-coiled amino acids than right-coiled amino acids as potential building blocks for proteins. Still, a chirality-biasing process would not rule out the possibility that life constructed with right-coiled amino acids could exist in the universe; it would simply mean that such life forms would be less abundant in the universe than life forms with left-coiled amino acids.

Modern molecular studies have revealed that a great deal of convergent evolution of life here on Earth is in fact due to developmental constraint: the phenomenon of "deep homology" that we considered in chapter 1. "Studies of deep homology are showing that new structures need not arise from scratch, genetically speaking, but can evolve by deploying regulatory circuits that were first established in early animals. . . . The more that researchers look, the more they will find that the same tools have been used to build a great variety of structures long thought to have independent histories" (Shubin et al. 2009, 822). Would life elsewhere in the universe, even if it is also carbon-based, coded for by base-four DNA, and composed of proteins made of left-coiled amino acids, have evolved the same developmental regulatory systems seen in Earth life? If so, the form set {*f:3*} would not exist, but this at first seems unlikely. The deep-homology regulatory circuits discussed by Shubin et al. (2009) can be traced back to the Cambrian diversification of animal life some 540 to 600 million years ago. Yet life existed on Earth 3.5 billion years ago, and animals are a late addition to the history of life. It took roughly 3 billion years of evolution for life to evolve the deep-homology regulatory circuits present in animal life on Earth—what is the probability that the very same regulatory system would convergently evolve with the convergent evolution of multicellular, heterotrophic life elsewhere? Multicellular, heterotrophic life forms—animals—that have evolved with a different set of deep-homology regulatory circuits could potentially be able to develop forms that cannot be developed by life on Earth, forms that could belong to the set {*f:3*} and that would not be convergent on Earth life.

Because we do not know the answer to the question of how development originated in multicellular life forms on Earth, we also do not know how development might originate in life forms elsewhere in the universe. Theoretical modeling, however, suggests that a significant part of developmental constraint might indeed be a function of physical and geometric constraints (Newman et al. 2006; Müller 2007; Newman 2010), and therefore that developmental systems in alien life might be similar to those found in Earth life. In modeling the evolution of the development of multicellular organisms, Newman et al. (2006) start with four different kinds of physical and chemical patterning mechanisms: diffusion gradients, sedimentation gradients, reaction-diffusion mechanisms, and chemical oscillation mechanisms. Most importantly, these four patterning mechanisms are found in nonliving as well as living chemical systems. Then Newman et al. (2006) add two basic cell properties: differential adhesion and cell polarity. Differential adhesion in pairs of tissues that differ in cohesivity produces mixing behavior like that seen in immiscible liquids, a behavior that is found in nonliving chemical systems and thus not unique to life. Likewise, the acquisition of polarization in the adhesion of cells is a mechanism not unique to life. Newman et al. (2006) next explored the possible interactions of the four patterning mechanisms with the two cell properties in a manner similar to that used in theoretical morphology, producing a spectrum of hypothetical morphological outcomes. In one spectrum, hollow spheres of tissue are produced when diffusion gradients are combined with the cellular properties of differential adhesion and polarity in development, whereas invaginated spheres are produced when sedimentation gradients are applied instead. In the other spectrum, radially periodic tissue structures are produced when reaction-diffusion mechanisms are combined with the cell properties, whereas serially periodic structures are produced when chemical oscillation mechanisms are applied instead. Newman et al. (2006, 289) point out that the spectrum of these hypothetical forms, produced by physical and chemical patterning mechanisms that are not unique to life, are very similar to "the hollow, multilayered and segmented morphotypes seen in the gastrulation stage embryos of modern-day metazoa as well as in Ediacaran fossil deposits of approximately ≈600 Ma." In an extended study, Newman (2010, 285) argues that nine "dynamical patterning modules" (or DPMs) in particular exist within the spectrum of hypothetical developmental forms, and that "the DPMs, in conjunction with cell-type-defining and switching networks, transformed simple, spherical, topologically solid cell clusters into hollow, multilayered, elongated, segmented, folded, and appendage-bearing structures. They thus founded the pathways that evolved into the developmental programs of modern animals."

In conclusion, Newman et al. (2006) and Newman (2010) suggest that the evolution of development on Earth may have been a two-stage process: metazoans originated from multicellular forms and structures first assembled by predominantly physical mechanisms, and then subsequently evolved genetic mechanisms to perpetuate the functionally successful morphologies formed in the first stage. The first step in this process is subject to the laws of physics and geometry, which are assumed to be same throughout the universe. Assembling tissues on alien worlds should also form hollow spheres, invaginated spheres, radially periodic structures, and serially periodic structures in the early evolution of multicellular alien life forms, just as on Earth.

If the early stages of developmental constraint are indeed a function of physical and geometric constraints, then the set {*f:3*} might indeed be empty—multicellular alien life forms may develop (at least initially) in a very similar fashion to those found on Earth. However, the second step in this hypothetical process—the evolution of genetic mechanisms to perpetuate the functionally successful morphologies formed in the first step—returns us to the question of the evolution of the coding mechanism of life that we have considered previously: How likely is it that life elsewhere will evolve the DNA coding system present in Earth life?

On Earth, almost all of the developmental-genetic "tool kit" genes used by the multicellular metazoans (Unikonta: Opisthokonta: Choanozoa: Metazoa; see appendix) are found in the genome of the unicellular choanoflagellate *Monosiga brevicollis* (Unikonta: Opisthokonta: Choanozoa: Choanoflagellata); therefore, the metazoan tool-kit genes predate the evolution of animals themselves (King et al. 2003). Only a few additional tool-kit genes appear in the origin of the Metazoa (the placozoans and sponges), and a few more in the evolution of the simplest Eumetazoa (ctenophores and cnidarians), and then "all the triploblastic metazoan body plans emerged within the space of no more than 20 million years" in the Cambrian explosion of animal evolution (Newman 2010, 284). Since they are determined by the laws of physics and geometry, the multicellular DPMs should be the same for alien life, but could alien life duplicate the feat of the Cambrian explosion in encoding the DPMs into heritable body plans without the preexisting DNA coding system of tool-kit genes? Or does it matter—regardless of the coding system evolved by alien life, will that system not also have to accomplish the exact same result in achieving multicellularity if that alien life is to survive and evolve?

A Periodic Table of Life?

An alternative approach to analyzing the potential existence of the biological form set {*f:3*} elsewhere in the universe would be to try to visualize what these functional forms that are nonexistent on Earth might look like. One of the key features of theoretical morphology is the ability to create both existent and nonexistent form by considering all the possible permutations of form dimension parameters. We have previously seen in chapter 4 that nonexistent ecological roles on Earth could be examined by such a procedure (see table 4.12), resulting in the consideration of the functional possibility of a nonexistent plant that floats in the air by means of a gas-filled bladder or bladders, and in the conclusion that such a plant would have to be carnivorous in addition to being photoautotrophic.

Analogous to the "periodic table of niches" (Pianka 1978) that we considered in chapter 4, it is possible to create a "periodic table of life" in a simple theoretical-morphology thought experiment (McGhee 2008). The periodic table of elements allows chemists to predict not only the behavior of existent elements but the behavior of nonexistent yet possible elements, such as the heavy elements that chemists have subsequently created in nuclear laboratories, elements that do not exist in the natural state. In essence, the rows of the periodic table of elements are based on the complexity of the atomic structure of the elements: elements in the first row have only the electron shell *K*, elements in the second row have the electron shells *K* and *L*, third-row elements have electron shells *K*, *L*, and *M*, and so on. These rows also reflect the evolutionary sequence of the appearance of the elements, with elements in the first row (hydrogen and helium) appearing first in the evolution of the universe, elements in the second row evolving next, and so on. We can use the chemical concepts of elemental complexity and evolutionary sequence in an analogous fashion by arranging the major groups of multicellular life in a similar series of rows of morphological complexity and biological evolutionary sequence (table 7.1).

The columns of the periodic table of elements can be considered to characterize the mobility of the elements in those rows, with highly mobile elements in some columns (elements that chemically combine readily, such as the column containing hydrogen, lithium, sodium, etc.) and low-mobility elements in other columns (elements that are chemically inert, such as the column containing helium, neon, argon, etc.). In an analogous fashion, we can consider the mobility of multicellular forms

Table 7.1
A periodic table of life, based upon locomotory type and evolutionary sequence of origination

	Spectrum of locomotion				
	None	*2D locomotion*		*3D locomotion*	
Sequence of evolution	Sessile	Crawling (legless)	Walking (legs)	Swimming (fusiform body)	Flying (wings)
Plants	Plants	—	—	—	—
Invertebrates	Barnacles	Worms	Arthropods	Squid	Insects
Amphibians	—	Caecilians	Amphibians	Tadpoles	—
Reptiles	—	Snakes	[reptiles]	Ichthyosaurs	Pterosaurs
Dinosaurs	—	—	[dinosaurs]	Penguins	Birds
Mammals	—	—	[mammals]	Porpoises	Bats

Note: Modified from McGhee (2008). See text for discussion.

of life on the basis of locomotory type arranged in a series of columns (table 7.1; I thank George Ellis of Capetown, South Africa, for suggesting to me that I add the column of "no locomotion" to the original thought experiment in McGhee 2008).

Even such a simple attempt to create a periodic table of life immediately reveals major incidences of convergent evolution (table 7.1). Not only is the convergent evolution of fast-swimming fusiform morphologies in tetrapod vertebrates (reptilian ichthyosaurs, mammalian porpoises, and dinosaurian penguins) apparent, but we also see that certain invertebrates animals have also convergently evolved this same fast-swimming morphology (most notably in modern-day squid and cuttlefish cephalopods and their extinct orthoconic and belemnitellid relatives). The major convergences in the evolution of wing structures for powered flight in arthropods (insects), reptiles (pterosaurs), dinosaurs (birds), and mammals (bats) are also apparent in the table.

Two major groups of animals have convergently evolved leg structures for walking: the arthropods and the ancestral amphibians. (The tetrapod reptiles, dinosaurs, and mammals are listed in brackets in the walking column in table 7.1 because their legs are not independent convergences but rather symplesiomorphic structures simply inherited from their amphibian ancestors.) Note, however, that both the amphibians and the reptiles have separately and independently reevolved legless morphologies (amphibian caecilians and reptilian snakes) and morphologically have converged on annelid worms. All of the life forms listed in table 7.1 exist on the Earth and belong to the form set {*f:1*}.

Of major interest are the empty permutations in table 7.1—these are potential candidates for functional life forms that are nonexistent on Earth but could belong to the form set {*f:3*}. Currently, there are no flying frogs on Earth, but we do have gliding frogs that are heading in that evolutionary direction (see table 2.2). No legless, feathered avian-snake forms nor furry mammalian-snake forms exist on Earth, though then again weasels and ferrets—with their elongated bodies and small legs—are headed in this evolutionary direction. These are also life forms that could conceivably be developed by existent frogs, birds, and mammals, and so is their absence on Earth due entirely to functional constraint, and not to developmental constraint?

The major predicted life forms that are nonexistent on Earth revealed in table 7.1 are mobile plants and sessile terrestrial animals. We have previously considered the possibility of flying plants in chapter 4, and concluded that the absence of such forms on Earth was probably due to functional constraint (i.e., inability to deal with the habitat instability produced by weather patterns existent on the Earth). That conclusion might be wrong—could it be instead that Earth plants are unable to develop the gas bladders that would be needed by a flying plant? The closest thing we have in terrestrial ecosystems to the sessile animals that exist in marine ecosystems are the web-spinning spiders. These spiders are mobile, but when feeding, they simply sit sessile in the centers of their webs and wait for prey animals to impact and stick to the web. Ecologically, they are like marine sessile carnivores such as corals. Is the absence of truly sessile terrestrial predators due entirely to functional constraint (i.e., inability to obtain sufficient prey by an immobile carnivore)?

Writers of science fiction have been engaging in theoretical morphology for years, although they are unaware of it. Hypothetical walking plants, the triffids, were created by the science fiction writer John Wyndham in his novel *The Day of the Triffids* (1951). These plants grew normally, using roots for nutrient uptake and leaves for photosynthesis, until they reached maturity, at which time they uprooted and used three specialized root prongs for clumsy walking. The springlike root-prong mechanism used by the triffids for walking is not unlike the springlike trap mechanisms present in Venus fly traps. Interestingly, Wyndham concluded that his mature triffids would have to be carnivorous, just as we concluded in chapter 4 that flying plants would have to use carnivory for nutrient acquisition. Larry Niven (1968a) created hypothetical flying plants, called stage trees, that used rocket propulsion for seed dispersal in his short story "A Relic of the Empire." These plants grew as low shrubs or bushes until their reproductive phase, when they grew tall tree trunks enriched with organic explosives. During the dry season, forest fires would ignite the stage trees, which would lift off like rockets and explode in the skies, scattering their seeds over huge geographic distances. Niven also created sessile mammal-like terrestrial predators, the grogs, in his short story "The Handicapped" (1968). These hypothetical animals evolved a unique adaptation to lure prey animals within striking distance of the sessile predator's long, prehensile tongue (I will not be a spoiler and reveal what the adaptation is). If they existed, triffids, stage trees, and grogs would fill the empty permutations of walking plants, flying plants, and sessile mammals in table 7.1, and they certainly are developmentally impossible for life forms on Earth. Are they candidates for the form set {*f:3*}?

In summary, using the techniques of theoretical morphology, we can visualize possible, but nonexistent, biological forms. But we do not know if the nonexistence of those forms is simply due to functional constraint (that is, that those forms belong to form set {*f:2*}), or if the nonexistence of those forms is due to developmental constraint (that they belong to form set {*f:3*}). The possibility that life elsewhere, evolving with different developmental constraints, might be able to develop functional forms that cannot be developed by life on Earth limits our ability to predict the outcome of evolution with certainty. Only when we fully understand the evolution of the developmental regulatory system present in Earth life will we able to make predictions about possible alien developmental systems.

A strong case has been made in this chapter, however, for the convergent evolution of similar developmental constraints in alien life forms that are carbon-based and that inhabit Earth-type worlds. That is, I suggest that developmental constraints in this restricted context are a function of physical and geometric constraints, just as functional constraints are. If this is true, we should expect radically different developmental constraints to have been evolved only in life forms that are not carbon-based, or that inhabit non-Earth-type planets that are either very hot or very cold, or are gas giants. In our own solar system, Titan is perhaps the best possible candidate for exploration for life forms unlike any seen on Earth.

8 Philosophical Implications of Convergent Evolution

Complexity theory also suggests a new take on an old question, long a staple of science fiction and speculative science: When we do find aliens, or they find us, what will they look like? By revealing many forms of Earth life to be governed by deep geometrical rules of self-organization in nature, complexity suggests a universal geometry of life that should transcend worlds. . . . There is no way to predict precisely what aliens will look like, but the fractal geometry of life gives us reason to believe that when they do finally land on the White House lawn, whatever walks or slithers down the gangplank may look strangely familiar.
—Grinspoon (2003, 272–273)

The laws of physics, it is believed, are the same everywhere in the universe. This is unlikely to be true of biology.
—Crick (1988, 138)

Most scientists tend to ignore philosophers. In evolutionary biology, the scientist Ernst Mayr (1964, xi–xii) traced this tendency back to Charles Darwin himself: "No one resented Darwin's independence of thought more than the philosophers. How could anyone dare to change our concept of the universe and man's position in it without arguing for or against Plato, for or against Descartes, for or against Kant? Darwin had violated all the rules of the game by placing his argument entirely outside the traditional framework of classical philosophical concepts and terminologies. . . . No other work advertised to the world the emancipation of science from philosophy as blatantly as did Darwin's *Origin*." Among other things, philosophers concern themselves with questions about purpose and meaning in life, questions that are unanswerable using the materialistic methodology of science. Because metaphysical questions are unanswerable by science, many scientists consider such questions to be meaningless.

I am a scientist, yet I also find philosophy to be interesting. A scientific commitment to a materialistic methodology does not mean one also has to commit oneself to a materialistic philosophy, although many scientists do. In this chapter I will touch on philosophical questions concerning freedom, purpose, design, destiny, teleology, spirituality, and even God—not usual topics for a scientist. I do so because I am not alone—these questions have been raised by other scientists as well, with reference to a very specific natural phenomenon: convergent evolution.

The difference of opinion between the astrobiologist David Grinspoon and Francis Crick, co-discoverer of the molecular structure of DNA, given in the two epigraphs at the beginning of this chapter, sets the stage for the spectrum of topics to be considered here. In the chapter I will follow the model of discussion used by John Casti in his philosophical-scientific book *Paradigms Lost* (1989), in which arguments for and against a point of view are presented as if in a trial in a courtroom, followed by a ruling from the judge on the bench. I, of course, shall act as judge and, of course, the reader is free to disagree with the bench!

The Argument for Unpredictability: Creative Freedom or Chaotic Randomness?

The best-known evolutionary essayist of the twentieth century, the Harvard paleontologist Stephen Jay Gould, was fond of a thought experiment of his own that he called "replaying life's tape" (Gould 1989). That is, consider the history of the evolution of life on Earth to be similar to a videotape of a popular movie. Imagine what would happen if you could take a copy of the videotape and rewind it to a point early in the movie, erasing everything on the tape that happened after that point, and then could rerun the tape to see what would happen this time on the now blank tape. Would the historical sequence of events in the evolution of life in the rerun of the tape resemble the original? Or would evolution take radically different pathways in the new narrative, producing animal and plant forms totally unlike those of the original? Gould (1989, 51) argues strongly for the second scenario: "Any replay of the tape would lead evolution down a pathway radically different from the road actually taken. . . . The diversity of possible itineraries does demonstrate that eventual results cannot be predicted at the outset. Each step proceeds for cause, but no finale can be specified at the start, and none would ever occur a second time in the same way, because any pathway proceeds through thousands of improbable stages."

In short, in Gould's view, evolution is totally unpredictable: evolution has no predictable direction and no predictable destination. An evolutionary trend in time is solely a chain of contingent historical events, and could not have been deduced from any laws of nature: "Contingency is the affirmation of control by immediate events over destiny. . . . Our own evolution is a joy and a wonder because such a curious chain of events would probably never happen again, but having occurred, makes eminent sense" (Gould 1989, 284–285).

Some evolutionary biologists find Gould's view of evolution as an unpredictable, nonrepeatable, and historically contingent process to be liberating and uplifting. For example, Stuart Kauffman (2008, 5) comments on Gould's view that the path of evolution cannot be deduced from any laws of nature: "a 'natural law' is a compact description beforehand of the regularities of a process. But if we cannot even prestate the possibilities, then no compact descriptions of these processes beforehand can exist. These phenomena, then, appear to be partially beyond natural law itself. This means something astonishingly and powerfully liberating. We live in a universe, biosphere, and human culture that are not only emergent but radically creative. We live in a world whose unfoldings we often cannot prevision, prestate, or predict—a world of explosive creativity on all sides."

Similarly, in writing of Gould's view that evolution could not be repeated, Kauffman (2008, 130) expands on this concept with obvious joy: "The vast nonrepeatability, or nonergodicity, of the universe at all levels of complexity above atoms—molecules, species, human history—leaves room for a creativity in the way the universe unfolds at these levels, a creativity that we cannot predict. . . . Now I want to make my outrageous claim: the evolution of the biosphere is radically nonpredictable and ceaselessly creative."

And last, as I promised at the beginning of this chapter, the concepts of spirituality and of God also arise from this consideration of evolution as an unpredictably creative process: "I believe we can reinvent the sacred. We can invent a global ethic, in a shared space, safe to all of us, with one view of God as the natural creativity in the universe" (Kauffman 2008, xiii).

Other biologists welcome Gould's view of evolution as the ultimate dismissal of the older idea of teleology, that evolution is design-like and progressive, and thus has a direction and ultimately a destination. But Reiss (2009, 174) goes beyond Gould to attack the very idea of adaptive improvement in evolution as being teleological, as in his discussion of Sewall Wright's concept of the adaptive landscape and Darwin's concept of natural selection: "Wright seems never to have realized that while in the context of domestic breeds it may be reasonable to speak of 'improvement' by selection of 'superior' herds, where the standard of value is determined by the goals of the breeder, no such a priori standard of overall value exists in nature. In fact, by focusing on the hypothetical ways in which organisms could evolve toward a future state of improved adaptation, Wright introduced a teleological element into his evolutionary theory, just as Darwin had into his." He further argues, "Darwin introduced a teleological determinism into the heart of his theory. This teleology is expressed in two related conceptions: (1) that evolution is a process going from a less-adapted to a better-adapted state and (2) that natural selection is a deterministic force, or agent, that directs the evolutionary process toward this better-adapted state" (Reiss 2009, 140).

Rather than leading to thoughts of spirituality and God, as in the case of Kauffman (2008), the view that evolution is unpredictable and directionless produces a totally different reaction in Reiss (2009, 356): "In short, when we recognize that organisms exist only by virtue of the fact that they are satisfying their conditions of existence, and that features of organisms likewise exist because they are satisfying *their* conditions for existence, we are freed to consider evolution in a dispassionate manner. Life is not designed, or at least it shows no evidence of design for anything other than continued existence, which needs no designer."

Gould (1989, 290) himself was well aware that his stance on evolution was partially a reaction to "traditionalist" and teleological ideas of evolution: "ultimately, the question of questions boils down to the placement of the boundary between predictability under invariant law and the multifarious possibilities of historical contingency. Traditionalists like [Charles Doolittle] Walcott would place the boundary so low that all major patterns of life's history fall above the line into the realm of predictability (and, for him, direct manifestation of divine intentions). But I envision a boundary sitting so high that almost every interesting event in life's history falls into the realm of contingency."

As one might expect, reaction to Gould's argument from those scientists whose view of evolution he characterized as "traditional" has not been positive; Gould's idea that evolution is unpredictable, contingent, nonrepeating, and directionless has been viewed with horror. Rather than agreeing with Kauffman (2008, 5), who felt Gould's view of evolution is "liberating" and "radically creative," Gould's critics see an abyss of chaotic randomness and meaninglessness: "To say that any of this is in the broadest sense 'adaptive' simply begs the question of how the world came to be ordered in the first place. It is, of course, no accident that the incidental, the chance occurrence, the contingent happenstance, is so influential in our deracinated and nihilistic culture, especially as reflected in the biological sciences that have spent the last century trying to square the circle of a meaningless process, that is, evolution, leading to the appearance of a sentient species that sees meaning all around itself" (Conway Morris 2008, 61).

The Argument for Predictability: Comforting Certainty or Depressing Inevitability?

Surprisingly, some of the strongest statements for evolution as a predictable, law-driven process come from the molecular biologists: "Whether or not there are other sets of lawful organic forms, there is no doubt that the universe of protein folds represents a Platonic universe . . . a universe where abstract rules, like the rules of grammar, define a set of unique immaterial templates which are materialized into a thousand or so natural forms—a world of rational morphology and pre-ordained evolutionary paths. . . . [A]s far as the 1000 protein folds are concerned, we may be sure that they will be present everywhere in the cosmos where there is carbon-based life, utilizing the same 20 protogenic amino acids" (Denton et al. 2002, 340). These authors further claim: "For the lawful nature of the folds provides for the first time evidence that the laws of nature may not only be fine-tuned to generate an environment fit for life (the stage) but may also be fine-tuned to generate the organic forms (the actors) as well, in other words that the cosmos may be even more biocentric than is currently envisaged . . . raising the possibility that all organic forms and indeed the whole pattern of life may finally prove to be the determined end of physics and life a necessary feature of the fundamental order of nature" (338).

Denton et al.'s view of an evolutionary process with "pre-ordained evolutionary paths" and a lawlike behavior such that "the whole pattern of life may finally prove to be the determined end of physics" is the very antithesis of the view of Gould (1999) and Kauffman (2008). Yet it is mirrored by the statements of other molecular biologists: "Darwinian evolution can follow only very few mutational paths to fitter proteins. . . . [W]e conclude that much protein evolution will be similarly constrained. This implies that the protein tape of life may be largely reproducible and even predictable" (Weinreich et al. 2006, 111); and: "Our results show that, despite the dominance of contingency (historical accident) in some recent discussions of evolutionary mechanisms (Gould 1989), purely chemical constraints (that is, the ability of only certain sequences to carry out particular functions) can lead to the repeated evolution of the same macromolecular structures" (Salehi-Ashtiani and Szostak 2001, 84).

On the macroscopic scale of biological form, the paleontologist Simon Conway Morris (2003, 309) argues that most of the potential hyperspace of form is empty, and that the pathways through that hyperspace have been repeatedly traveled in evolution due to evolutionary constraint: "The phenomenon of evolutionary convergence indicates that . . . the number of alternatives is strictly limited, with the interesting implication that the vast bulk of any given 'hyperspace' not only never will be visited during evolutionary exploration but it never can be. These are the howling wildernesses of the maladaptive, the 99.9% recurring of biological space where things don't work, the Empty Quarters of biological non-existence." That is, life is constrained to evolve in only a tiny fraction, 0.1 percent, of the potential form hyperspace, and it is this constraint that produces convergent evolution.

Conway Morris (2003, 309–310) further argues that the available evolutionary pathways in the hyperspace, and the nodes within the hyperspace where functional forms can exist, have been preset from the very beginning of the universe: "It is my suspicion that . . . the nodes of occupation are effectively predetermined from the Big Bang." The idea that evolution must follow predetermined pathways, set by the constants and laws of nature from the very beginning of the universe, leads to the idea that evolution in these predetermined pathways is not only predictable but also inevitable: "The principal aim of this book has been to show that the constraints of evolution and the ubiquity of convergence make the emergence of something like ourselves a near-inevitability" (Conway Morris 2003, 328).

Is evolution predictable, preordained, and inevitable? If so, it should have not only a direction but also an ultimate destination. This was certainly the view of another paleontologist, Pierre Teilhard de Chardin, as described by Aczel (2007, 75–76): "the ideas of evolution became so powerful that they convinced him that everything in the universe, inanimate objects and living systems alike, was in constant flux, ever evolving as decreed by God. The goal was a point where everything would converge to form the body of Christ. This was Teilhard's Omega Point." Teilhard was religious, a Jesuit priest, and it is notable that both Conway Morris's *Life's Solution* (2003) and Barlow's *Let There Be Sight! A Celebration of Convergent Evolution* (2005) have a religious component to their argumentation.

Other scientists have also come to the conclusion that evolution is predictable and lawlike, but from a materialist rather than religious viewpoint, such as the Russian zoologist Leo Berg in his book *Nomogenesis: Or Evolution Determined by Law* (1922). From an admittedly materialist viewpoint, the astrobiologist David Grinspoon (2003, 412) also writes of predictable, goal-directed evolution: "what I mean by spirituality is the religious impulse stripped of religion. . . . I believe the phenomenon of humanity on Earth is a local example of a trend toward higher consciousness and spiritual enlightenment that transpires all over this universe. . . . Cosmic spiritual advancement by Darwinian natural selection!" Kauffman (2008) was able to find spirituality in the view that evolution was unpredictable, contingent, nonrepeating, and directionless, as discussed in the previous section of this chapter. Thus, it is very interesting that spirituality has also been found in the exact opposite view that evolution is predictable, preordained, and inevitable.

It is clear that many who believe that evolution is predictable, preordained, and inevitable find comfort in the lawlike certainty of this point of view. Others react with horror to this point of view, and find it depressingly deterministic, stultifying rather than creative, enslaving rather than liberating. Where is individual freedom, where is creativity, in an evolutionary process that already has a predetermined destination?

Judging the Arguments: Evolutionary Views on Trial

We have now reviewed the arguments for the view that evolution is unpredictable, contingent, nonrepeating, and directionless and those for the view that evolution is predictable, predetermined, and inevitable. What are the implications of the phenomenon of convergent evolution, the subject of this book, with regards to these radically different views of evolution?

First, the view that the evolutionary process is nonrepeating (nonergotic) is demonstrably false. Chapters 2 through 6 of this book have demonstrated that evolution is a highly repeatable process. In fact, repeatability in evolution is rampant at every level of life on Earth, from tiny organic molecules to entire ecosystems of species, and even to the ways in which we think.

Second, the view that evolution is entirely historically contingent, and thus unpredictable (and nonrepeating), is demonstrably false. Convergent evolution—the repeated, independent, evolution of the same trait in multiple phylogenetic lineages at different points in geological time—

reveals to us the limits imposed upon evolution by functional constraints. Once we know what those limits are, we can use them to predict the trajectory of future evolution by predicting that those constraints will continue to operate in the future, just as they have in the past. Given an Earth-like world elsewhere in the universe, we can also predict that the same functional constraints that exist on Earth will be in operation on that world as well. Each step in the evolution of a Jurassic ichthyosaur and a Cenozoic porpoise has its unique contingent aspects, yet the end result was the same: a fusiform body, the same body form that is found in a shark or a swordfish (see table 2.1). If any large, fast-swimming organisms exist in the oceans of Jupiter's moon Europa, swimming under the perpetual ice that covers their world, I predict with confidence that they will have streamlined, fusiform bodies; that is, they will look very similar to a porpoise, an ichthyosaur, a swordfish, or a shark.

Third, the view that evolution is directionless needs more careful examination. Reiss's view that there is no adaptive improvement in evolution is demonstrably false. Reiss (2009, 140) objects to the idea that "evolution is a process going from a less-adapted to a better-adapted state" under the influence of the "deterministic force" of natural selection because he considers that idea to be teleological. Is it teleological that water flows downhill? That it goes from a state of higher potential energy to a state of lower potential energy under the influence of the deterministic force of gravity? Philosophers may argue about whether that phenomenon is teleological or not, but in science it is an empirical observation, an established fact.

It is an empirical observation that adaptive improvement occurs in much of biological evolution. In fact, the adaptive landscape concept of Sewall Wright (which Reiss objects to as teleological) can be used to analyze actual cases of adaptive improvement in biological evolution (McGhee 2007). However, there does exist one model of evolution that corresponds to Reiss's view: the Red Queen hypothesis (Van Valen 1973), named after the Red Queen in Lewis Carroll's *Alice's Adventures in Wonderland* (1865), who told Alice that she had to constantly keep running just in order to stay in the same place. In such a scenario, a species can only evolve fast enough to exactly maintain its same position on a moving adaptive peak; that is, it evolves fast enough to avoid sliding downslope with time (and thus to eventual extinction), yet it cannot evolve fast enough to climb upslope on the peak to higher levels of adaptation (McGhee 2007, 27–28). The species is constantly evolving, but perpetually stuck at the same degree of adaptation. The Red Queen hypothesis is a rather bleak evolutionary possibility, in that no adaptive improvement ever takes place, and the probability of the species going extinct is always a constant regardless of whether the species is young, in existence only a few hundred years, or has been present on the Earth for millions of years.

Reiss (2009, 140) further objects to the idea that natural selection is a deterministic force that directs the evolutionary process toward a better adapted state because that idea appears to be goal directed, or teleological. Is it teleological that water flows in rivers toward the future "goal" of reaching a sea or lake? The flow of water is mindless, and water has no goal in sight, but nevertheless it will reach its lowest possible potential energy state under the deterministic force of gravity, which means that it will wind up in a sea or lake.

Part of Reiss's goal-directed objection stems from the fact that both Darwin and Wright used the analogy of human artificial selection in trying to explain the process of nonhuman natural selection. As quoted earlier, Reiss complains: "Wright seems never to have realized that while in the context of domestic breeds it may be reasonable to speak of 'improvement' by selection of 'superior' herds, where the standard of value is determined by the goals of the breeder, no such a priori standard of overall value exists in nature" (Reiss 2009, 174). That is, Reiss argues that a human breeder may have a standard value or goal that he or she is trying to achieve in artificial selection, but that no such standard value or goal exists in nature. This argument is demonstrably false, because the phenomenon of convergent evolution abundantly demonstrates that the environment does have a priori standards of overall value. The laws of physics impose the functional constraints, the a priori standards, that fast-swimming organisms must be fusiform in shape (see table 2.1) and that flying organisms must have wings (see table 2.2). And so on, for all the myriad examples of convergent evolution that have been considered in chapters 2 through 6 of this book.

However, there is a real difference between the a priori standards of human artificial selection and nonhuman natural selection: the human mind. The a priori standards of nature are mindless functional constraints imposed by the laws of physics and geometry. It is part of Darwin's genius that he realized that the mindless a priori standards of nature would sort out organic variation in a process of mindless natural selection (for a book-length discussion of modern misconceptions regarding teleology and adaptation, see Ruse 2003). Natural selection has a direction only in the sense that it will, in general, operate to move evolving organisms up the slopes of the adaptive landscape to higher states of adaptation (McGhee 2007). It is the empirical observation that the number of these higher states of adaptation, or adaptive peaks, is *limited* that gives evolution its direction. If the number of adaptive peaks were infinite, then each species on Earth would be morphologically different from every other species, and each species would have its own unique ecological role, or niche. Such an Earth does not exist. Instead, repeated evolutionary convergence on similar morphologies, niches, molecules, and even mental states is the norm on Earth.

Fourth: while evolution is not directionless, it has not been proved that the constraint on evolutionary pathways available for life to evolve along was predetermined from the beginning of the universe. It is certainly a possibility, as Einstein himself wondered if there was any freedom left for alternative ways for the physical universe to evolve once the constants and laws of nature were established at the Big Bang, and the same could be true for biological evolution, as discussed in chapter 7. Thus, one could indeed argue that functional constraints, which are themselves a function of the laws of physics and geometry, have been invariant since the beginning of the universe if those same laws of physics and geometry have been.

It is the role of developmental constraint in convergent evolution that leads to uncertainty in the argument for predetermined constraints. Although Stern (2010) has argued for a "predictable genome" and Brakefield (2006, 364) notes that "patterns of parallel morphological evolution given similar ecological opportunities are to be expected and that, given sufficient knowledge of developmental processes, they could be predictable," this possible predictability refers solely to the developmental regulatory processes evolved by Earth life. Would life elsewhere in the universe evolve the same developmental constraints? How much of developmental constraint is lawlike and predictable, a function of the laws of physics and geometry at the level of molecules, cells, and tissues (Newman et al. 2006; Müller 2007; see also chapter 5), and how much is it a function of chance in the evolution of life specific to the Earth, and unpredictable? As we saw in chapter 7, the possibility that life elsewhere, evolving with different developmental constraints, might be able to develop functional forms that cannot be developed by life on Earth limits our ability to predict the outcome of evolution with certainty.

Fifth: it has not been proved that biological evolution is inevitable, that it has a destination. And even if it did, it is not clear how we might determine what that destination might be. Even our models for the physical evolution of the universe have changed radically over the past century: once it was thought the universe was steady-state, and would infinitely persist. Then it was thought that the universe would eventually cease to expand, and then collapse upon itself in a sort of reverse-Big-Bang fireball. Now it is generally thought the universe will continue to expand infinitely and that all the stars will eventually burn out, one by one, in a cold, dark universal death. But, as we currently still do not truly understand dark matter or dark energy, that model of the ultimate destination of the physical evolution of the universe may change in the future as well. As yet, we have no equivalent models for the ultimate destination of biological evolution. If we cannot, at present, confidently predict the ultimate destination of the physical evolution of the universe, how can we hope to predict the ultimate destination of the vastly more complex process of biological evolution?

There are hints that such a destination might exist: the increase in complexity produced by evolution in the past 3.5 billion years of Earth history, and the evolution of self-awareness in the past 5 to 10 million years in multiple independent lineages (see table 6.7). The universe has become aware of itself (Grinspoon 2003). Is such a trend intrinsic to the process of biological evolution? The self-assembling properties of biological evolution, and its constant building upon previous evolutionary steps via the process of natural selection, would lead us to predict that such a process should produce increasing complexity in other life forms as well, on other worlds.

Have life forms on other worlds become self-aware? Has matter awakened, has the universe become aware of itself elsewhere? We have no evidence that this has occurred, although we keep searching the skies for possible electromagnetic or other signals that would point to the existence of extraterrestrial civilizations of self-aware organisms in search of other self-aware organisms. On Earth, the convergent evolution of self-awareness has been linked to the convergent evolution of complex social systems, as discussed in chapter 6. The fact that, on Earth, life in the sea and life on land—life in radically different environments—both have evolved convergent social systems and self-awareness would lead us to predict that self-awareness would evolve on extraterrestrial worlds as well.

Or is self-awareness an intrinsic property of the developmental regulatory systems present in animal life on Earth? Would life elsewhere, evolving under different developmental constraints, have evolved social systems similar to those convergently evolved by Earth life, and hence self-awareness?

The Question of Extraterrestrial Life

The phenomenon of convergent evolution demonstrates that life *repeatedly* evolves in a finite number of preferred directions and, as such, is in principle predictable. Analysis of that phenomenon reveals that it is driven by functional and developmental constraints on evolution. However, we have only one example of that phenomenon in operation: Earth life. Thus, given the existence of an extraterrestrial planet with Earth-like physical conditions, we can predict that the same functional constraints that have shaped the evolution of life on Earth will operate in a similar fashion on that extraterrestrial planet. Life on that planet will look "strangely familiar," as David Grinspoon (2003) argues in the epigraph at the beginning of this chapter.

Francis Crick (1988) may also be right, as well. Convergent evolution of life forms on Earth has also been driven by developmental constraints. Totally different life forms, such as silicon-based life, would undoubtedly evolve with developmental constraints different from those that operate in life on Earth. The "laws of biology" would be different for those life forms, and they might evolve forms that are totally unlike those seen on Earth.

The question of the existence of extraterrestrial life is of crucial importance to the question of the predictability of evolution. Short of intelligent extraterrestrials contacting us and discussing their biochemistry with us, we must search for such life ourselves. Within our own solar system (since, at present, we can only dream of spacecraft that could take us to other star systems), if life is present on Mars or even in the water oceans of Jupiter's moon Europa, it is likely to be convergent on the carbon-based model found on Earth. Only the far-distant planet-like moons Titan and Triton offer physical conditions and chemistries radically different from those found on Earth, and the possibility of discovering life "as we do not know it."

In Conclusion: A Rewrite of Darwin's View of Life

Darwin (1859, 490), in a philosophical mood, closed the first edition of *On the Origin of Species* with the sentence: "There is grandeur in this view of life, with its several powers, having been originally breathed into a few forms or into one; and that, whilst this planet has gone cycling on according to the fixed law of gravity, from so simple a beginning endless forms most beautiful and most wonderful have been, and are being, evolved."

We now have the benefit of an additional 150 years of analysis of biological evolution. The phenomenon of convergent evolution leads us to reconsider Darwin's closing sentence, at least for carbon-based life, and to suggest that it be rewritten (with the modern alterations in the original sentence indicated by italics): "There is grandeur in this view of life *in the universe*, with its several powers *of functional and developmental constraint*, having been originally breathed into a few forms *of life* or into *just the* one *that is carbon based*; and that, whilst this planet *and others have* gone cycling on according to the fixed law of gravity, from so simple a beginning *limited* forms most beautiful and most wonderful have been, and are being, *re*-evolved *throughout the universe*."

Appendix: A Phylogenetic Classification of Life

Table A.1
A phylogenetic classification of life forms discussed in this book

Eubacteria

Archaea

Eukarya

– **Bikonta**

– – Excavobionta

– – Rhizaria

– – Chromoalveolata

– – Green eukaryotes

– – – Glaucophyta

– – – Metabionta

– – – – Rhodobionta

– – – – **Chlorobiota** > see Table A.3

– **Unikonta**

– – Amoebozoa

– – Opisthokonta

– – – Fungi

– – – Choanozoa

– – – – Choanoflagellata

– – – – **Metazoa**

– – – – – Placozoa

– – – – – Demospongiae

– – – – – Hexactinellida

– – – – – Calcarea

– – – – – Eumetazoa

– – – – – – Ctenophora

– – – – – – **Cnidaria**

– – – – – – **Bilateria**

– – – – – – – **Protostomia**

Table A.1
(continued)

– – – – – – – – **Lophotrochozoa**
– – – – – – – – – Lophophorata
– – – – – – – – – – Bryozoa
– – – – – – – – – – Phoronozoa
– – – – – – – – – – – Phoronida
– – – – – – – – – – – Brachiopoda
– – – – – – – – – Eutrochozoa
– – – – – – – – – – Syndermata
– – – – – – – – – – – Rotifera
– – – – – – – – – – – Acanthocephala
– – – – – – – – – – Spiralia
– – – – – – – – – – – Entoprocta
– – – – – – – – – – – Annelida
– – – – – – – – – – – Sipuncula
– – – – – – – – – – – Mollusca
– – – – – – – – – – – Parenchymia
– – – – – – – – – – – – Platyhelminthes
– – – – – – – – – – – – Nemertea
– – – – – – – – Cuticulata
– – – – – – – – – Gastrotricha
– – – – – – – – – **Ecdysozoa**
– – – – – – – – – – **Introverta**
– – – – – – – – – – – Nematozoa
– – – – – – – – – – – – Nematoda
– – – – – – – – – – – – Nematomorpha
– – – – – – – – – – – Cephalorhyncha
– – – – – – – – – – – – Kinorhyncha
– – – – – – – – – – – – Loricifera
– – – – – – – – – – – – Priapulida
– – – – – – – – – – **Panarthropoda**
– – – – – – – – – – – Onychophora
– – – – – – – – – – – Tardigrada
– – – – – – – – – – – **Arthropoda**
– – – – – – – – – – – – Cheliceriformes
– – – – – – – – – – – – Mandibulata
– – – – – – – – Chaetognatha
– – – – – – – **Deuterostomia**
– – – – – – – – **Echinodermata**
– – – – – – – – Pharyngotremata
– – – – – – – – – Hemichordata

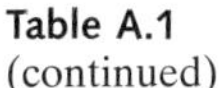

Table A.1
(continued)

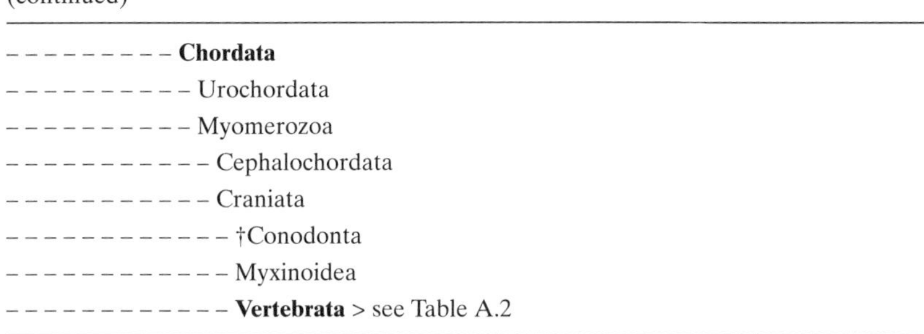

– – – – – – – – – – **Chordata**

– – – – – – – – – – – Urochordata

– – – – – – – – – – – Myomerozoa

– – – – – – – – – – – – Cephalochordata

– – – – – – – – – – – – Craniata

– – – – – – – – – – – – – †Conodonta

– – – – – – – – – – – – – Myxinoidea

– – – – – – – – – – – – – **Vertebrata** > see Table A.2

Note: Major clades used frequently in the text for comparative purposes are in boldface type. Extinct taxa are marked with a †. Classification modified from Lecointre and Le Guyader (2006) and Benton (2005).

Table A.2
A phylogenetic classification of vertebrate taxa discussed in this book

Vertebrata

– Petromyzontiformes

– †Pteraspidomorphi

– †Osteostraci

– Gnathostomata

– – †Placodermi

– – **Chondrichthyes**

– – †Acanthodii

– – **Osteichthyes**

– – – **Actinopterygii**

– – – **Sarcopterygii**

– – – – Dipnoi

– – – – Crossopterygii

– – – – – Actinistia

– – – – Tetrapodomorpha

– – – – – **Tetrapoda**

– – – – – – **Batrachomorpha**

– – – – – – – †Dendrerpetontidae

– – – – – – – †Dissorophidae

– – – – – – – **Lissamphibia**

– – – – – – – – Gymnophiona

– – – – – – – – Batrachia

– – – – – – – – – Anura

– – – – – – – – – Urodela

– – – – – – †Lepospondyli

Table A.2
(continued)

– – – – – – **Reptiliomorpha**

– – – – – – – †Seymouriamorpha

– – – – – – – †Diadectomorpha

– – – – – – – **Amniota**

– – – – – – – – **Sauropsida**

– – – – – – – – – Anapsida

– – – – – – – – – – Testudines

– – – – – – – – – **Diapsida**

– – – – – – – – – – †Ichthyosauria

– – – – – – – – – – **Lepidosauromorpha**

– – – – – – – – – – – †Sauropterygia

– – – – – – – – – – – **Lepidosauria**

– – – – – – – – – – – – Sphenodontia

– – – – – – – – – – – – **Squamata**

– – – – – – – – – – – – – Iguania

– – – – – – – – – – – – – Scleroglossa

– – – – – – – – – – – – – – Amphisbaenia

– – – – – – – – – – – – – – Gekkonta

– – – – – – – – – – – – – – Autarchoglossa

– – – – – – – – – – – – – – – Scincomorpha

– – – – – – – – – – – – – – – Anguimorpha

– – – – – – – – – – – – – – – – †Mosasauridae

– – – – – – – – – – – – – – – – Serpentes

– – – – – – – – – – **Archosauromorpha**

– – – – – – – – – – – **Archosauria**

– – – – – – – – – – – – Crurotarsi

– – – – – – – – – – – – Ornithodira

– – – – – – – – – – – – – †Pterosauria

– – – – – – – – – – – – – **Dinosauria**

– – – – – – – – – – – – – – Saurischia

– – – – – – – – – – – – – – – †Sauropodomorpha

– – – – – – – – – – – – – – – Theropoda

– – – – – – – – – – – – – – – – †Ceratosauria

– – – – – – – – – – – – – – – – Tetanurae

– – – – – – – – – – – – – – – – – †Carnosauria

– – – – – – – – – – – – – – – – – Coelurosauria

– – – – – – – – – – – – – – – – – – Maniraptora

– – – – – – – – – – – – – – – – – – – **Aves**

– Paleognathae

– Neognathae

– Galloanserae

– Neoaves

Table A.2
(continued)

– – – – – – – – – – – – – – †Ornithischia

– – – – – – – – – – – – – – – †Fabrosauridae

– – – – – – – – – – – – – – – †Genasauria

– – – – – – – – – – – – – – – – †Thyreophora

– – – – – – – – – – – – – – – – †Cerapoda

– – – – – – – – – – – – – – – – – †Ornithopoda

– – – – – – – – – – – – – – – – – †Marginocephalia

– – – – – – – – **Synapsida**

– – – – – – – – – †Ophiacodontidae

– – – – – – – – – †Edaphosauridae

– – – – – – – – – †Sphenacodontidae

– – – – – – – – – **Therapsida**

– – – – – – – – – – †Dinocephalia

– – – – – – – – – – †Dicynodontia

– – – – – – – – – – †Gorgonopsia

– – – – – – – – – – Cynodontia

– – – – – – – – – – – **Mammalia**

– – – – – – – – – – – – †Volaticotheria

– – – – – – – – – – – – †Yinotheria

– – – – – – – – – – – – Australosphenida

– – – – – – – – – – – – – †Ausktribosphenida

– – – – – – – – – – – – – **Monotremata**

– – – – – – – – – – – – †Multituberculata

– – – – – – – – – – – – **Theria**

– – – – – – – – – – – – – **Marsupialia**

– – – – – – – – – – – – – **Eutheria**

– – – – – – – – – – – – – – Xenarthra

– – – – – – – – – – – – – – Epitheria

– – – – – – – – – – – – – – – **Afrotheria**

– – – – – – – – – – – – – – – Boreoeutheria

– – – – – – – – – – – – – – – – †Apatemyidae

– – – – – – – – – – – – – – – – **Laurasiatheria**

– – – – – – – – – – – – – – – – – Eulipotyphles

– – – – – – – – – – – – – – – – – Scrotifera

– – – – – – – – – – – – – – – – – – Chiroptera

– – – – – – – – – – – – – – – – – – Fereungulata

– – – – – – – – – – – – – – – – – – – Ferae

– †Creodonta

– Carnivora

– Pholidota

Table A.2
(continued)

-------------------- Perissodactyla
-------------------- †Meridiungulata
-------------------- Cetartiodactyla
----------------- **Euarchontoglires**
------------------ Glires
------------------- Lagomorpha
------------------- Rodentia
------------------ Euarchonta
------------------- Dermoptera
------------------- **Primates**
-------------------- Strepsirrhini
-------------------- Haplorrhini
--------------------- Tarsiiformes
--------------------- Simiiformes
---------------------- **Platyrrhini**
---------------------- **Catarrhini**
----------------------- Cercopithecoidea
----------------------- Hominoidea
------------------------ Hylobatoidae
------------------------ Hominoidae
------------------------- Pongidae
------------------------- Hominidae

Note: Major clades used frequently in the text for comparative purposes are in boldface type. Extinct taxa are marked with a †. Classification modified from Lecointre and Le Guyader (2006) and Benton (2005).

Table A.3
A phylogenetic classification of plant taxa discussed in this book

Chlorobionta
- Ulvophyta
- Prasinophyta
- Streptophyta
-- Chlorokybophyta
-- Klebsormidiophyta
-- Phragmoplastophyta
--- Zygnematophyta
--- Plasmodesmophyta
---- Chaetosphaeridiophyta
---- Charophyta

Table A.3
(continued)

– – – – Parenchymophyta
– – – – – Coleochaetophyta
– – – – – **Embryophyta**
– – – – – – Marchantiophyta
– – – – – – Stomatophyta
– – – – – – – Anthocerophyta
– – – – – – – Hemitracheophyta
– – – – – – – – Bryophyta *sensu stricto*
– – – – – – – – Polysporangiophyta
– – – – – – – – – †Horneophyta
– – – – – – – – – **Tracheophyta**
– – – – – – – – – – **Lycophyta**
– – – – – – – – – – **Euphyllophyta**
– – – – – – – – – – – **Moniliformopses**
– – – – – – – – – – – – †Cladoxylopsida
– – – – – – – – – – – – Equisetophyta
– – – – – – – – – – – – Filicophyta
– – – – – – – – – – – **Lignophyta**
– – – – – – – – – – – – †Aneurophytales
– – – – – – – – – – – – †Archaeopteridales
– – – – – – – – – – – – **Spermatophyta**
– – – – – – – – – – – – – †Lyginopteridales
– – – – – – – – – – – – – †Medullosanales
– – – – – – – – – – – – – †Gigantopteridales
– – – – – – – – – – – – – Ginkgophyta
– – – – – – – – – – – – – Cycadophyta
– – – – – – – – – – – – – Pinophyta
– – – – – – – – – – – – – Gnetophyta
– – – – – – – – – – – – – **Angiospermae**
– – – – – – – – – – – – – – Amborellales
– – – – – – – – – – – – – – Nymphaeales
– – – – – – – – – – – – – – Austrobaileyales
– – – – – – – – – – – – – – **Euangiosperms**
– – – – – – – – – – – – – – – Chloranthales
– – – – – – – – – – – – – – – Ceratophyllales
– – – – – – – – – – – – – – – **Magnoliidae**
– – – – – – – – – – – – – – – – Magnoliales
– – – – – – – – – – – – – – – – Laurales
– – – – – – – – – – – – – – – – Canellales
– – – – – – – – – – – – – – – – Piperales

Table A.3
(continued)

– – – – – – – – – – – – – – – – **Monocotyledons**

– – – – – – – – – – – – – – – – – Acorales

– – – – – – – – – – – – – – – – – Alismatales

– – – – – – – – – – – – – – – – – **Eumonocotyledons**

– – – – – – – – – – – – – – – – – – Miyoshiales

– – – – – – – – – – – – – – – – – – Dioscoreales

– – – – – – – – – – – – – – – – – – Pandanales

– – – – – – – – – – – – – – – – – – Liliales

– – – – – – – – – – – – – – – – – – Asparagales

– – – – – – – – – – – – – – – – – – Commelinidae

– – – – – – – – – – – – – – – – **Eudicots**

– – – – – – – – – – – – – – – – – Ranunculalaes

– – – – – – – – – – – – – – – – – Sabiales

– – – – – – – – – – – – – – – – – Proteales

– – – – – – – – – – – – – – – – – Buxales

– – – – – – – – – – – – – – – – – Trochodendrales

– – – – – – – – – – – – – – – – – **Core eudicots**

– – – – – – – – – – – – – – – – – – Berberidopsidales

– – – – – – – – – – – – – – – – – – Santalales

– – – – – – – – – – – – – – – – – – Dilleniales

– – – – – – – – – – – – – – – – – – Caryophyllales

– – – – – – – – – – – – – – – – – – Saxifragales

– – – – – – – – – – – – – – – – – – Vitales

– – – – – – – – – – – – – – – – – – **Rosidae**

– – – – – – – – – – – – – – – – – – – Crossosomatales

– – – – – – – – – – – – – – – – – – – Geraniales

– – – – – – – – – – – – – – – – – – – Myrtales

– – – – – – – – – – – – – – – – – – – Eurosids I

– – – – – – – – – – – – – – – – – – – Eurosids II

– – – – – – – – – – – – – – – – – – **Asteridae**

– – – – – – – – – – – – – – – – – – – Cornales

– – – – – – – – – – – – – – – – – – – Ericales

– – – – – – – – – – – – – – – – – – – Euasteridae

– Euasterids I

– Euasterids II

Note: Major clades used frequently in the text for comparative purposes are in boldface type. Extinct taxa are marked with a †. Classification modified from Lecointre and Le Guyader (2006), Donoghue (2005), APG II (2003), and Niklas (1997).

References

Aczel, A. D. 2007. *The Jesuit and the skull*. New York: Riverhead Books.

Agosta, W. 1996. *Bombardier beetles and fever trees: A close-up look at chemical warfare and signals in animals and plants*. Reading, Mass.: Helix Books.

Alberch, P. 1989. The logic of monsters: Evidence for internal constraint in development and evolution. *Geobios (Lyon, France): Memoire Special* 12:21–57.

Albert, V. A., S. E. Williams, and M. W. Chase. 1992. Carnivorous plants: Phylogeny and structural evolution. *Science* 257:1491–1495.

APG II. 2003. An update of the Angiosperm Phylogeny Group classification for the orders and families of flowering plants: APG II. *Botanical Journal of the Linnean Society* 141:399–436.

Axe, D. D. 2004. Estimating the prevalence of protein sequences adopting functional enzyme folds. *Journal of Molecular Biology* 341:1295–1315.

Baber, C. 2003. *Cognition and tool use: Forms of engagement in human and animal use of tools*. London: Taylor and Francis.

Bailly, X., S. Vanin, C. Chabasse, K. Mizuguchi, and S. N. Vinogradov. 2008. A phylogenetic profile of hemerythrins, the nonheme diiron binding respiratory proteins. *BMC Evolutionary Biology* 8:244, doi:10.1186/1471-2148-8-244.

Bakker, R. T. 1983. The deer flees, the wolf pursues: Incongruencies in predator-prey coevolution. In *Coevolution*, ed. D. J. Futuyma and M. Slatkin, 350–382. Sunderland, Mass.: Sinauer Associates, Inc.

Bakker, R. T. 1986. *The dinosaur heresies*. New York: William Morrow.

Ballerini, M., N. Cabibbo, R. Candelier, A. Cavagna, E. Cisbani, I. Giardina, V. Lecomte, A. Orlandi, G. Parisi, A. Procaccini, M. Viale, and V. Zdravkovic. 2008. Interaction ruling animal collective behavior depends on topological rather than metric distance: Evidence from a field study. *Proceedings of the National Academy of Sciences of the United States of America* 105 (4): 1232–1237.

Bambach, R. K. 1983. Ecospace utilization and guilds in marine communities through the Phanerozoic. In *Biotic interactions in recent and fossil benthic communities*, ed. M. J. S. Tevesz and P. M. McCall, 719–746. New York: Plenum.

Barlow, C. 2005. Let there be sight! A celebration of convergent evolution. http://www.thegreatstory.org.

Barrett, P. M. 2005. The diet of ostrich dinosaurs (Theropoda: Ornithomimosauris). *Palaeontology* 48 (2): 347–358.

Baskin, D. G., and D. W. Goldring. 1970. Experimental studies on the endocrinology and reproductive biology of the viviparous polychaete annelid, *Nereis limnicola* Johnson. *Biological Bulletin* 139:461–475.

Bastian, J. 1982. Vision and electroreception: Integration of sensory information in the optic tectum of the weakly electric fish *Apteronotus albifrons*. *Journal of Comparative Physiology* 147A:287–297.

Bateman, R. M., and W. A. DiMichele. 1994. Heterospory: The most iterative key innovation in the evolutionary history of the plant kingdom. *Biological Reviews of the Cambridge Philosophical Society* 69:345–417.

Bennett, S. C. 1996. Aerodynamics and thermoregulatory function of the dorsal sail of *Edaphosaurus*. *Paleobiology* 22 (4):496–506.

Benton, M. J. 2005. *Vertebrate palaeontology*. 3rd ed. Oxford: Blackwell Publishing.

Berg, L. S. 1922. *Nomogenesis; or, evolution determined by law*. Cambridge, Mass.: Massachusetts Institute of Technology Press (1969 reprint).

Berkov, A., N. Rodríguez, and P. Centeno. 2008. Convergent evolution in the antennae of a cerambycid beetle, *Onychocerus albitarsis*, and the sting of a scorpion. *Naturwissenschaften* 95 (3):257–261.

Berman, D. S., R. R. Reisz, D. Scott, A. C. Henrici, S. S. Sumida, and T. Martens. 2000. Early Permian bipedal reptile. *Science* 290:968–972.

Bernhardt, P. 2000. Convergent evolution and adaptive radiation of beetle-pollinated angiosperms. *Plant Systematics and Evolution* 222:293–320.

Blackburn, D. G. 1992. Convergent evolution of viviparity, matrotrophy, and specializations for fetal nutrition in reptiles and other vertebrates. *American Zoologist* 32:313–321.

Blackburn, D. G. 1993. Chorioallantoic placentation in squamate reptiles: Structure, function, development, and evolution. *Journal of Experimental Zoology* 266:414–430.

Blackburn, D. G. 2000. Reptilian viviparity: Past research, future directions, and appropriate models. *Comparative Biochemistry and Physiology Part A* 127A:391–409.

Blackburn, D. G., L. J. Vitt, and C. A. Beuchat. 1984. Eutherian-like reproductive specializations in a viviparous reptile. *Proceedings of the National Academy of Sciences of the United States of America* 81:4860–4863.

Block, B. A., J. R. Finnerty, A. F. R. Stewart, and J. Kidd. 1993. Evolution of endothermy in fish: Mapping physiological traits on a molecular phylogeny. *Science* 260:210–214.

Blumberg, M. S. 2009. *Freaks of nature: What anomalies tell us about development and evolution*. Oxford: Oxford University Press.

Boomsma, J. J. 2009. Lifetime monogamy and the evolution of eusociality. *Philosophical Transactions of the Royal Society of London: Series B, Biological Sciences* B364:3191–3207.

Bork, P., C. Sander, and A. Valencia. 1993. Convergent evolution of similar enzymatic function on different protein folds: The hexokinase, ribokinase, and galactokinase families of sugar kinases. *Protein Science* 2:31–40.

Bowmaker, J. K., and D. M. Hunt. 2006. Evolution of vertebrate visual pigments. *Current Biology* 16 (13):R484–R489.

Boyce, C. K., and A. H. Knoll. 2002. Evolution of developmental potential and the multiple independent origins of leaves in Paleozoic vascular plants. *Paleobiology* 28 (1):70–100.

Brady, S. G. 2003. Evolution of the army ant syndrome: The origin and long-term evolutionary stasis of a complex of behavioral and reproductive adaptations. *Proceedings of the National Academy of Sciences of the United States of America* 100 (11):6575–6579.

Brady, S. G., S. Sipes, A. Pearson, and B. N. Danforth. 2006. Recent and simultaneous origins of eusociality in halictid bees. *Philosophical Transactions of the Royal Society of London: Series B, Biological Sciences* B273:1643–1649.

Brakefield, P. M. 2006. Evo-devo and constraints on selection. *Trends in Ecology and Evolution* 21 (7):362–368.

Budelmann, B., and H. Bleckmann. 1988. A lateral line analogue in cephalopods: Water waves generate microphonic potentials in the epidermal head lines of *Sepia* and *Lolliguncula*. *Journal of Comparative Physiology* 164A:1–5.

Buhl, J., D. J. T. Sumpter, I. D. Couzin, J. J. Hale, E. Despland, E. R. Miller, and S. J. Simpson. 2006. From disorder to order in marching locusts. *Science* 312:1402–1406.

Bull, J., and J. Farrand. 1988. *The Audubon Society field guide to North American birds*. New York: Alfred A. Knopf.

Bush, A. M., R. K. Bambach, and G. M. Daley. 2007. Changes in theoretical ecospace utilization in marine fossil assemblages between the mid-Paleozoic and late Cenozoic. *Paleobiology* 33 (1):76–97.

Byrne, M. 2005. Viviparity in the sea star Cryptasterina hysteria (Asterinidae): Conserved and modified features in reproduction and development. *Biological Bulletin* 208:81–91.

Camazine, S., J.-L. Deneubourg, N. R. Franks, J. Sneyd, G. Theraulaz, and E. Bonabeau. 2001. *Self-organization in biological systems*. Princeton: Princeton University Press.

Cameron, K. M., K. J. Wurdack, and R. W. Jobson. 2002. Molecular evidence for the common origin of snap-traps among carnivorous plants. *American Journal of Botany* 89 (9): 1503–1509.

Cameron, S. A., and P. Mardulyn. 2001. Multiple molecular data sets suggest independent origins of highly eusocial behavior in bees (Hymenoptera: Apinae). *Systematic Biology* 50 (2):194–214.

Carrano, M. T., C. M. Janis, and J. J. Sepkoski. 1999. Hadrosaurs as ungulate parallels: Lost lifestyles and deficient data. *Acta Palaeontologica Polonica* 44 (3):237–261.

Carroll, L. [C. W. Dodgson]. 1865. *Alice's adventures in Wonderland*. London: Macmillan.

Carroll, R. L., and W. Lindsay. 1985. Cranial anatomy of the primitive reptile *Procolophon*. *Canadian Journal of Earth Sciences* 22:1571–1587.

Carroll, S. B. 2006. *Endless forms most beautiful: The new science of EvoDevo*. New York: W. W. Norton and Co.

Carter, A. M. 2008. What fossils can tell us about the evolution of viviparity and placentation. *Placenta* 29:930–931.

Cashmore, A. R., J. A. Jarillo, Y.-J. Wu, and D. Liu. 1999. Cryptochromes: Blue light receptors for plants and animals. *Science* 284:760–765.

Castelvecchi, D. 2007. Alien pizza, anyone? Biochemistry may have taken a different turn on other worlds. *Science News* 172:107–109.

Casti, J. L. 1989. *Paradigms lost: Tackling the unanswered mysteries of modern science*. New York: Avon Bard.

Castoe, T. A., A. P. J. de Koning, H.-M. Kim, B. P. Noonan, G. Naylor, Z. J. Jiang, C. L. Parkinson, and D. D. Pollock. 2009. Evidence for an ancient adaptive episode of convergent molecular evolution. *Proceedings of the National Academy of Sciences of the United States of America* 106 (22):8986–8991.

Castoe, T. A., A. P. J. de Koning, and D. D. Pollock. 2010. Adaptive molecular convergence: Molecular evolution versus molecular phylogenetics. *Communicative and Integrative Biology* 3 (1):67–69.

Charnock, S. J., I. E. Brown, J. P. Turkenburg, G. W. Black, and G. J. Davies. 2002. Convergent evolution sheds light on the anti-β-elimination mechanism common to family 1 and 10 polysaccharide lyases. *Proceedings of the National Academy of Sciences of the United States of America* 99 (10):12067–12072.

Chen, L., A. L. DeVries, and C.-H. C. Cheng. 1997. Convergent evolution of antifreeze glycoproteins in Antarctic notothenioid fish and Arctic cod. *Proceedings of the National Academy of Sciences of the United States of America* 94:3817–3822.

Cheng, Y.-N., X.-C. Wu, and Q. Ji. 2004. Triassic marine reptiles gave birth to live young. *Nature* 432:383–386.

Cimino, G., S. De Rosa, S. De Stefano, R. Morrone, and G. Sodano. 1985. The chemical defense of nudibranch molluscs: Structure, biosynthetic origin and defensive properties of terpenoids from the dorid nudibranch *Dendrodoris grandiflora*. *Tetrahedron* 41 (6):1093–1100.

Clack, J. A. 2002. *Gaining ground: The origin and evolution of tetrapods*. Bloomington: Indiana University Press.

Claessens, L. P. A. M., P. M. O'Connor, and D. M. Unwin. 2009. Respiratory evolution facilitated the origin of pterosaur flight and aerial gigantism. *PLoS ONE* 4 (2):e4497. www.plosone.org.

Clark, J. M., L. L. Jacobs, and W. R. Downs. 1989. Mammal-like dentition in a Mesozoic crocodylian. *Science* 244:1064–1066.

Clark, R. N., J. M. Curchin, J. W. Barnes, R. Jauman, L. Sonderblom, D. P. Cruikshank, R. H. Brown, et al. 2010. Detection and mapping hydrocarbon deposits on Titan. *Journal of Geophysical Research-Planets*, doi:10.1029/2009JE003369.

Clayton, N. S., and Emery, N. J. 2008. Canny corvids and political primates: A case for convergent evolution in intelligence. In *The deep structure of biology*, ed. S. Conway Morris, 128–142. West Conshohocken, Pa.: Templeton Foundation Press.

Colosimo, P. F., K. E. Hosemann, S. Balabhadra, G. Villarreal, M. Dickson, J. Grimwood, J. Schmutz, R. M. Myers, D. Schluter, and D. M. Kingsley. 2005. Widespread parallel evolution in sticklebacks by repeated fixation of *Ectodysplasin alleles*. *Science* 307:1928–1933.

Conradt, L., and T. J. Roper. 2005. Consensus decision making in animals. *Trends in Ecology and Evolution* 20 (8):449–456.

Conway Morris, S. 2003. *Life's solution: Inevitable humans in a lonely universe*. Cambridge: Cambridge University Press.

Conway Morris, S. 2008. Evolution and convergence: Some wider considerations. In *The deep structure of biology*, ed. S. Conway Morris, 46–67. West Conshohocken, Pa.: Templeton Foundation Press.

Cooper, A., C. Lalueza-Fox, S. Anderson, A. Rambaut, J. Austin, and R. Ward. 2001. Complete mitochondrial genome sequences of two extinct moas clarify ratite evolution. *Nature* 409:704–707.

Couzin, I. D. 2007. Collective minds. *Nature* 445:715.

Couzin, I. D., and J. Krause. 2003. Self-organization and collective behavior in vertebrates. *Advances in the Study of Behavior* 32:1–75.

Crespi, B. J., D. A. Carmean, and T. W. Chapman. 1997. Ecology and evolution of galling thrips and their allies. *Annual Review of Entomology* 42:51–71.

Crick, F. H. C. 1988. *What mad pursuit: A personal view of scientific discovery*. New York: Basic Books.

Cronk, Q., and I. Ojeda. 2008. Bird-pollinated flowers in an evolutionary and molecular context. *Journal of Experimental Botany* 59 (4):715–727.

Darwin, C. 1859. *On the origin of species by means of natural selection, or the preservation of favoured races in the struggle for life*. London: John Murray.

Darwin, C. 1875. *Insectivorous plants*. London: John Murray.

Daugherty, C. H., G. W. Gibbs, and R. A. Hitchmough. 1993. Mega-island or micro-continent? New Zealand and its fauna. *Trends in Ecology and Evolution* 8 (2):437–442.

Davies, P. L., J. Baardsnes, M. J. Kuiper, and V. K. Walker. 2002. Structure and function of antifreeze proteins. *Philosophical Transactions of the Royal Society of London: Series B, Biological Sciences* B357: 927–935.

Davis, C. C., P. K. Endress, and D. A. Baum. 2008. The evolution of floral gigantism. *Current Opinion in Plant Biology* 11:49–57.

Deeb, S. S., M. J. Wakefield, T. Tada, L. Marotte, S. Yokoyama, and J. A. M. Graves. 2003. The cone visual pigments of an Australian marsupial, the Tammar wallaby (*Macropus eugenii*): Sequence, spectral tuning, and evolution. *Molecular Biology and Evolution* 20 (10):1642–1649.

Delabie, J. H. C. 2001. Trophobiosis between Formicidae and Hemiptera (Sternorrhyncha and Auchenorrhyncha): An overview. *Neotropical Entomology* 30 (4):501–516.

DeMar, R., and J. R. Bolt. 1981. Dentitional organization and function in a Triassic reptile. *Journal of Paleontology* 55:967–984.

Denton, E. J., and J. Gray. 1985. Lateral-line-like antennae of certain of the Penaeidea (Crustacea, Decapoda, Natantia). *Philosophical Transactions of the Royal Society of London: Series B, Biological Sciences* B226:249–261.

Denton, M., and C. Marshall. 2001. Laws of forms revisited. *Nature* 410:417.

Denton, M. J., C. J. Marshall, and M. Legge. 2002. The protein folds as Platonic forms: New support for the pre-Darwinian conception of evolution by natural law. *Journal of Theoretical Biology* 219:325–342.

Dijkgraaf, S. 1967. Biological significance of the lateral line organs. In *Lateral line detectors*, ed. P. H. Cahn, 83–95. Bloomington: Indiana University Press.

Donoghue, M. J. 2005. Key innovations, convergence, and success: Macroevolutionary lessons from plant phylogeny. *Paleobiology* 31 (2 Supplement):77–93.

Donoghue, P. C. J., and M. A. Purnell. 1999. Mammal-like occlusion in conodonts. *Paleobiology* 25 (1):58–74.

Douglas, H. D., III, J. E. Co, T. H. Jones, W. E. Conner, and J. F. Day. 2005. Chemical odorant of colonial seabird repels mosquitos. *Journal of Medical Entomology* 42 (4):647–651.

Duffy, J. E., C. L. Morrison, and R. Ríos. 2000. Multiple origins of eusociality among sponge-dwelling shrimps (*Synalpheus*). *Evolution: International Journal of Organic Evolution* 54 (2):503–516.

Dulvy, N. K., and J. D. Reynolds. 1997. Evolutionary transitions among egg- laying, live-bearing and maternal inputs in sharks and rays. *Philosophical Transactions of the Royal Society of London: Series B, Biological Sciences* B264:1309–1315.

Dunn, R. R., A. D. Gove, T. G. Barraclough, T. J. Givnish, and J. D. Majer. 2007. Convergent evolution of an ant-plant mutualism across plant families, continents, and time. *Evolutionary Ecology Research* 9:1349–1362.

Eisner, T., M. Eisner, and M. Siegler. 2005. *Secret weapons: Defenses of insects, spiders, scorpions, and other many-legged creatures*. Cambridge, Mass.: Belknap Press.

Ellison, A. M., and N. J. Gotelli. 2009. Energetics and the evolution of carnivorous plants—Darwin's "most wonderful plants in the world." *Journal of Experimental Botany* 60 (1):19–42.

Emery, N. J., and N. S. Clayton. 2004. The mentality of crows: Convergent evolution of intelligence in corvids and apes. *Science* 306:1903–1907.

Evans, W. G. 1966. Perception of infrared radiation from forest fires by *Melanophilia acuminata* De Geer (Buprestidae, Coleoptera). *Ecology* 47:1061–1065.

Farrell, B. D., A. S. Sequeira, B. C. O'Meara, B. B. Normark, J. H. Chung, and B. H. Jordal. 2001. The evolution of agriculture in beetles (Curculionidae: Scolytinae and Platypodinae). *Evolution: International Journal of Organic Evolution* 55 (10):2011–2027.

Feduccia, A. 1996. *The origin and evolution of birds*. New Haven, Conn.: Yale University Press.

Finn, J. K., T. Tregenza, and M. D. Norman. 2009. Defensive tool use in a coconut-carrying octopus. *Current Biology* 19 (23):R1070.

Fitzhugh, K. 1989. A systematic revision of the Sabellidae-Caobangiidae-Sabellongidae complex (Annelida: Polychaeta). *Bulletin of the American Museum of Natural History* 192:1–104.

Fitzhugh, K. 1991. Further revisions of the Sabellidae subfamilies and cladistic relationships among the Fabriciinae (Annelida: Polychaeta). *Zoological Journal of the Linnean Society* 102:305–332.

Flemming, A. F., and D. G. Blackburn. 2003. Evolution of placental specializations in viviparous African and South American lizards. *Journal of Experimental Zoology* 299A:33–47.

Fletcher, G. L., C. L. Hew, and P. L. Davies. 2001. Antifreeze proteins of teleost fishes. *Annual Review of Physiology* 63:359–390.

Foster, K. W., and R. D. Smyth. 1980. Light antennas in phototactic algae. *Microbiological Reviews* 44 (4):572–630.

Freeland, S. J., and L. D. Hurst. 1998. The genetic code is one in a million. *Journal of Molecular Evolution* 47:238–248.

Gartrell, B. D. 2000. The nutritional, morphologic, and physiologic bases of nectarivory in Australian birds. *Journal of Avian Medicine and Surgery* 14 (2):85–94.

Gaudin, T. J., and D. G. Branham. 1998. The phylogeny of the Myrmecophagidae (Mammalia, Xenarthra, Vermilingua) and the relationship of *Eurotamandua* to the Vermilingua. *Journal of Mammalian Evolution* 5 (3):237–265.

Gensel, P. G., and A. E. Kasper. 2005. A new species of the Devonian lycopod genus, *Leclercqia*, from the Emsian of New Brunswick, Canada. *Review of Palaeobotany and Palynology* 137:105–123.

Givnish, T. J., J. C. Pires, S. W. Graham, M. A. McPherson, L. M. Prince, T. B. Patterson, H. S. Rai, et al. 2005. Repeated evolution of net venation and fleshy fruits among monocots in shaded habitats confirms a priori predictions: Evidence from an *ndhF* phylogeny. *Philosophical Transactions of the Royal Society of London: Series B, Biological Sciences* B272:1481–1490.

Gould, E. 1965. Evidence for echolocation in the Tenrecidae of Madagascar. *Proceedings of the American Philosophical Society* 109 (6):352–360.

Gould, J. L., and C. G. Gould. 2007. *Animal architects: Building and the evolution of intelligence*. New York: Basic Books.

Gould, S. J. 1989. *Wonderful life: The Burgess Shale and the nature of history*. New York: W. W. Norton.

Grant, K. A., and V. Grant. 1968. *Hummingbirds and their flowers*. New York: Columbia University Press.

Gregory, J. E., A. Iggo, A. K. McIntyre, and U. Proske. 1987. Electroreceptors in the platypus. *Nature* 326:386–387.

Grinspoon, D. 2003. *Lonely planets: The natural philosophy of alien life*. New York: HarperCollins Publishers Inc.

Gupta, A. K. 2004. Origin of agriculture and domestication of plants and animals linked to early Holocene climate amelioration. *Current Science* 87 (1):54–59.

Hagan, H. R. 1951. *Embryology of the viviparous insects*. New York: Ronald Press.

Hamilton, W. D. 1964. The genetical evolution of social behavior I, II. *Journal of Theoretical Biology* 7:1–52.

Hansell, M. 2005. *Animal architecture*. Oxford: Oxford University Press.

Hansell, M., and G. D. Ruxton. 2008. Setting tool use within the context of animal construction behaviour. *Trends in Ecology and Evolution* 23 (2):73–78.

Hansen, D. M., and M. Galetti. 2009. The forgotten megafauna. *Science* 324:42–43.

Hardison, R. C. 1996. A brief history of hemoglobins: Plant, animal, protist, and bacteria. *Proceedings of the National Academy of Sciences of the United States of America* 93:5675–5679.

Harshman, J., E. L. Braun, M. J. Braun, C. J. Huddleston, R. C. K. Bowie, J. L. Chojnowski, S. J. Hackett, et al. 2008. Phylogenetic evidence for multiple losses of flight in ratite birds.

Proceedings of the National Academy of Sciences of the United States of America 105:13462–13467.

Hauser, M. D. 2009. The possibility of impossible cultures. *Nature* 460:190–196.

Heinrich, B. 1993. *The hot-blooded insects: Strategies and mechanisms of thermoregulation*. Cambridge, Mass.: Harvard University Press.

Heinrich, B. 1996. *The thermal warriors: Strategies of insect survival*. Cambridge, Mass.: Harvard University Press.

Hines, H. M., J. H. Hunt, T. K. O'Connor, J. J. Gillespie, and S. A. Cameron. 2007. Multigene phylogeny reveals eusociality evolved twice in vespid wasps. *Proceedings of the National Academy of Sciences of the United States of America* 104 (9):3295–3299.

Holmes, M. M., B. D. Goldman, S. L. Goldman, M. L. Seney, and N. G. Forger. 2009. Neuroendocrinology and sexual differentiation in eusocial mammals. *Frontiers in Neuroendocrinology* 30:519–533.

Hopkins, C. D. 2008. Commentary: Evolution of electric organs. *Journal of Physiology, Paris* 102:162–163.

Hosler, J. 2003. *The Sandwalk adventures: An adventure in evolution told in five chapters*. Columbus, Ohio: Active Synapse.

Hoy, R. R., and D. Robert. 1996. Tympanal hearing in insects. *Annual Review of Entomology* 41:433–450.

Hunt, D. M., K. S. Dulai, J. A. Cowing, C. Julliot, J. D. Mollon, J. K. Bowmaker, W.-H. Li, and D. Hewett-Emmett. 1998. Molecular evolution of trichromacy in primates. *Vision Research* 38:3299–3306.

Hunter, J. P., and J. Jernvall. 1995. The hypocone as a key innovation in mammalian evolution. *Proceedings of the National Academy of Sciences of the United States of America* 92:10718–10722.

Janis, C. M. 1990. The correlation between diet and dental wear in herbivorous mammals, and its relationship to the determination of diets in extinct species. In *Evolutionary paleobiology of behavior and coevolution*, ed. A. J. Boucot, 241–259. Amsterdam: Elsevier Press.

Jernvall, J. 2000. Linking development with generation of novelty in mammalian teeth. *Proceedings of the National Academy of Sciences of the United States of America* 97:2641–2645.

Ji, Q., Z.-X. Luo, C.-X. Yuan, J. R. Wible, J.-P. Zhang, and J. A. Georgi. 2002. The earliest known eutherian mammal. *Nature* 416:816–822.

Johns, P. M., K. J. Howard, N. L. Breisch, A. Rivera, and B. L. Thorne. 2009. Nonrelatives inherit colony resources in a primitive termite. *Proceedings of the National Academy of Sciences of the United States of America* 106 (41):17452–17456.

Jost, M. C., D. M. Hillis, Y. Lu, J. W. Kyle, H. A. Fozzard, and H. H. Zakon. 2008. Toxin-resistant sodium channels: Parallel adaptive evolution across a complete gene family. *Molecular Biology and Evolution* 25 (6):1016–1024.

Kauffman, S. A. 2008. *Reinventing the sacred: A new view of science, reason and religion*. New York: Basic Books.

Kellogg, E. A. 1999. Phylogenetic aspects of the evolution of C4 photosynthesis. In *C_4 plant biology*, ed. R. F. Sage and R. K. Monson, 411–444. San Diego: Academic Press.

Kelly, D. A. 2002. The functional morphology of penile erection: Tissue designs for increasing and maintaining stiffness. *Integrative and Comparative Biology* 42:216–221.

Kemp, A. 1977. The pattern of tooth plate formation in the Australian lungfish, *Neoceratodus forsteri* Krefft. *Biological Journal of the Linnean Society* 60:223–258.

Kielan-Jaworowska, Z. 1979. Pelvic structure and nature of reproduction in Multituberculata. *Nature* 277:402–403.

King, A. S. 1981. Phallus. In *Form and function in birds*. Vol. 2, ed. A. S. King and J. McLelland, 107–147. London: Academic Press.

King, N., C. T. Hittinger, and S. B. Carroll. 2003. Evolution of key cell signaling and adhesion protein families predates animal origins. *Science* 301:361–363.

Kivic, P. A., and P. L. Walne. 1983. Algal photosensory apparatus probably represent multiple parallel evolutions. *Bio Systems* 16:31–38.

Klok, C. J., R. D. Mercer, and S. L. Chown. 2002. Discontinuous gas-exchange in centipedes and its convergent evolution in tracheated arthropods. *Journal of Experimental Biology* 205:1019–1029.

Koenigswald, W. v., K. D. Rose, L. Grande, and R. D. Martin. 2005. First apatemyid skeleton from the Lower Eocene Fossil Butte Member, Wyoming (USA), compared to the European apatemyid from Messel, Germany. *Palaeontographica Abteilung* A272:149–169.

Koenigswald, W. v., and H.-P. Schierning. 1987. The ecological niche of an extinct group of mammals, the early Tertiary apatemyids. *Nature* 326:595–597.

Kornegay, J. R., J. W. Schilling, and A. C. Wilson. 1994. Molecular adaptation of a leaf-eating bird: Stomach lysozyme of the Hoatzin. *Molecular Biology and Evolution* 11 (6):921–928.

Kuhn-Schnyder, E., and H. Rieber. 1986. *Handbook of paleozoology*. Baltimore: Johns Hopkins University Press.

Kunz, T. H., M. S. Fujita, A. P. Brooke, and G. R. McCracken. 1994. Convergence in tent architecture and tent-making behavior among Neotropical and Paleotropical bats. *Journal of Mammalian Evolution* 2:57–58.

Kuroki, T. 1967. Theoretical analysis of the role of the lateral line in directional hearing. In *Lateral line detectors*, ed. P. H. Cahn, 217–237. Bloomington: Indiana University Press.

Labandeira, C. C., J. Kvaček, and M. B. Mostovski. 2007. Pollination drops, pollen, and insect pollination of Mesozoic gymnosperms. *Taxon* 56:663–695.

Laughlin, S., A. D. Blest, and S. Stowe. 1980. The sensitivity of receptors in the posterior median eye of the nocturnal spider, *Dinopis*. *Journal of Comparative Physiology* 141:53–65.

Lawn, R. M., K. Schwartz, and L. Patthy. 1997. Convergent evolution of apolipoprotein(a) in primates and hedgehog. *Proceedings of the National Academy of Sciences of the United States of America* 94:11992–11997.

Lecointre, G., and H. Le Guyader. 2006. *The tree of life: A phylogenetic classification*. Cambridge, Mass.: Belknap Press of Harvard University Press.

Lee, M. S. Y. 1998. Convergent evolution and character correlation in burrowing reptiles: Towards a resolution of squamate relationships. *Biological Journal of the Linnean Society* 65:369–453.

Lee, M. S. Y., and R. Shine. 1998. Reptilian viviparity and Dollo's law. *Evolution: International Journal of Organic Evolution* 52 (5):1441–1450.

Loewer, P. 1995. *Seeds: The definitive guide to growing, history, and lore*. Portland, Ore.: Timber Press.

Logsdon, J. M., and W. R. Doolittle. 1997. Origin of antifreeze protein genes: A cool tale in molecular evolution. *Proceedings of the National Academy of Sciences of the United States of America* 94:3485–3487.

Long, J. A., K. Trinajstic, G. C. Young, and T. Senden. 2008. Live birth in the Devonian period. *Nature* 453:650–653.

Longrich, N. R., and P. J. Currie. 2008. *Albertonykus borealis*, a new alvarezsaur (Dinosauria: Theropoda) from the Early Maastrichtian of Alberta, Canada: Implications for the systematics and ecology of the Alvarezsauridae. *Cretaceous Research* 30 (1):239–252.

Luo, Z.-X., R. L. Cifelli, and Z. Kielan-Jaworowska. 2001. Dual origin of tribosphenic mammals. *Nature* 409:53–57.

Luo, Z.-X., Q. Ji, and C.-X. Yuan. 2007. Convergent dental adaptations in pseudo-tribosphenic and tribosphenic mammals. *Nature* 450:93–97.

Maclaurin, J. 2003. The good, the bad and the impossible. *Biology and Philosophy* 18:463–476.

Maclaurin, J., and K. Sterelny. 2008. *What is biodiversity?* Chicago: University of Chicago Press.

Malmström, T., and R. H. H. Kröger. 2006. Pupil shapes and lens optics in the eyes of terrestrial vertebrates. *Journal of Experimental Biology* 209:18–25.

Manger, P. R., R. Collins, and J. D. Pettigrew. 1997. Histological observations on presumed electroreceptors and mechanoreceptors in the beak skin of the long-beaked echidna, *Zaglossus bruijnii*. *Philosophical Transactions of the Royal Society of London: Series B, Biological Sciences* B264:165–172.

Marino, L. 2002. Convergence of complex cognitive abilities in cetaceans and primates. *Brain, Behavior and Evolution* 59:21–32.

Martin, A. J. 2006. *Introduction to the study of dinosaurs*. Oxford: Blackwell.

Martin, V. J. 2002. Photoreceptors of cnidarians. *Canadian Journal of Zoology* 80: 1703–1722.

Martinez, R. N., C. L. May, and C. A. Forster. 1996. A new carnivorous cynodont from the Ischigualasto Formation (Late Triassic, Argentina), with comments on eucynodont phylogeny. *Journal of Vertebrate Paleontology* 16:271–284.

Maryańska, T., H. Osmólska, and M. Wolsan. 2002. Avialan status for Oviraptorosauria. *Acta Palaeontologica Polonica* 47:97–116.

Masonjones, H. D., and S. M. Lewis. 1996. Courtship behavior in the dwarf seahorse, *Hippocampus zosterae*. *Copeia* (3):634–640.

Mayr, E. 1964. Introduction. In *A facsimile of the first edition of* On the origin of species, ed. E. Mayr, vii–xxvii. Cambridge, Mass.: Harvard University Press.

Mayr, G., and D. S. Peters. 2006. Response to comment on "A well-preserved Archaeopteryx specimen with theropod features." *Science* 313: 1238c, doi: 10.1126/science.1130964.

Mayr, G., B. Pohl, and D. S. Peters. 2005. A well-preserved Archaeopteryx specimen with theropod features. *Science* 310: 1483–1486.

McGhee, G. R. 1999. *Theoretical morphology: The concept and its applications*. New York: Columbia University Press.

McGhee, G. R. 2001. Exploring the spectrum of existent, nonexistent and impossible biological form. *Trends in Ecology and Evolution* 16:172–173.

McGhee, G. R. 2007. *The geometry of evolution: Adaptive landscapes and theoretical morphospaces*. Cambridge: Cambridge University Press.

McGhee, G. R. 2008. Convergent evolution: A periodic table of life? In *The deep structure of biology*, ed. S. Conway Morris, 17–31. West Conshohocken, Pa.: Templeton Foundation Press.

McGhee, G. R., P. M. Sheehan, D. J. Bottjer, and M. L. Droser. 2004. Ecological ranking of Phanerozoic biodiversity crises: Ecological and taxonomic severities are decoupled. *Palaeogeography, Palaeoclimatology, Palaeoecology* 211:289–297.

McIntosh, J. S. 1997. Sauropoda. In *Encyclopedia of dinosaurs*, ed. P. J. Currie and K. Padian, 654–658. San Diego: Academic Press.

McKay, C. P., and H. D. Smith. 2005. Possibilities for methanogenic life in liquid methane on the surface of Titan. *Icarus* 178:274–276.

McKinney, F. K. 2007. *The Northern Adriatic ecosystem: Deep time in a shallow sea*. New York: Columbia University Press.

McKinney, F. K., S. J. Hageman, and A. Jaklin. 2007. Crossing the ecological divide: Paleozoic to modern marine ecosystem in the Adriatic Sea. *Sedimentary Record* 5 (2):4–8.

McPherron, S. P., Z. Alemseged, C. W. Marean, J. G. Wynn, D. Reed, D. Geraads, R. Bobe, and H. A. Béarat. 2010. Evidence for stone-tool-assisted consumption of animal tissues before 3.39 million years ago at Dikika, Ethiopa. *Nature* 466:857–860.

Meyers, J. J., and A. Herrel. 2005. Prey capture kinematics of ant-eating lizards. *Journal of Experimental Biology* 208:113–127.

Milewski, A. V., and W. J. Bond. 1982. Convergence of myrmecochory in Mediterranean Australia and South Africa. In *Ant-plant interactions in Australia*, ed. R. C. Buckley, 89–98. The Hague: Dr. W. Junk Publishers.

Moller, P. 1995. *Electric fishes: History and behavior*. London: Chapman and Hall.

Mooney, H. A. 1977. *Convergent evolution in Chile and California: Mediterranean climate ecosystems*. Stroudsburg, Pa.: Dowden, Hutchinson and Ross, Inc.

Moore, J., and P. Willmer. 1997. Convergent evolution in invertebrates. *Biological Reviews of the Cambridge Philosophical Society* 72:1–60.

Morell, V. 2002. Placentas may nourish complexity studies. *Science* 298:945.

Mueller, U. G., N. M. Gerardo, D. K. Aanen, D. L. Six, and T. R. Schultz. 2005. The evolution of agriculture in insects. *Annual Review of Ecology Evolution and Systematics* 36:563–595.

Müller, G. B. 2007. Evo-devo: Extending the evolutionary synthesis. *Nature Reviews: Genetics* 8:943–949.

Mummenhoff, K., A. Polster, A. Mühlhausen, and G. Theißen. 2008. *Lepidium* as a model system for studying the evolution of fruit development in Brassicaceae. *Journal of Experimental Biology online*, doi:10.1093/jxb/ern304.

Mundy, N. I. 2005. A window on the genetics of evolution: *MC1R* and plumage colouration in birds. *Philosophical Transactions of the Royal Society of London: Series B, Biological Sciences* B272:1633–1640.

Naylor, G. J. P., T. M. Collins, and W. M. Brown. 1995. Hydrophobicity and phylogeny. *Nature* 373:565–566.

Nevo, E. 1999. *Mosaic evolution of subterranean mammals: Regression, progression, and global convergence*. Oxford: Oxford University Press.

Newman, S. A. 2010. Dynamical patterning modules. In *Evolution: The extended synthesis*, ed. M. Pigliucci and G. B. Müller, 281–306. Cambridge, Mass.: Massachusetts Institute of Technology Press.

Newman, S. A., G. Forgacs, and G. B. Müller. 2006. Before programs: The physical origination of multicellular forms. *International Journal of Developmental Biology* 50:289–299.

Niklas, K. 1997. *The evolutionary biology of plants*. Chicago: University of Chicago Press.

Niven, L. 1968. The handicapped. In *Neutron star*, 209–235. New York: Ballantine Books.

Niven, L. 1968a. A relic of the empire. In *Neutron star*, 29–49. New York: Ballantine Books.

Norberg, R. Å. 1977. Occurrence and independent evolution of bilateral ear asymmetry in owls and implications on owl taxonomy. *Philosophical Transactions of the Royal Society of London: Series B, Biological Sciences* B280:375–408.

Oakley, T. H., and C. W. Cunningham. 2002. Molecular phylogenetic evidence for the independent evolutionary origin of an arthropod compound eye. *Proceedings of the National Academy of Sciences of the United States of America* 99:1426–1430.

O'Connor, P. M., and L. P. A. M. Claessens. 2005. Basic avian pulmonary design and flow-through ventilation in non-avian theropod dinosaurs. *Nature* 436:253–256.

O'Connor, P. M., J. J. W. Sertich, N. J. Stevens, E. M. Roberts, M. D. Gottfried, T. L. Hieronymus, Z. A. Jinnah, R. Ridgely, S. E. Ngasala, and J. Temba. 2010. The evolution of mammal-like crocodyliforms in the Cretaceous Period of Gondwana. *Nature* 466: 748–751.

Ödeen, A., and O. Håstad. 2003. Complex distribution of avian color vision systems revealed by sequencing the SWS1 opsin from total DNA. *Molecular Biology and Evolution* 20 (6):855–861.

O'Reilly, J. C., D. A. Ritter, and D. R. Carrier. 1997. Hydrostatic locomotion in a limbless tetrapod. *Nature* 386:269–272.

Ostrom, J. H. 1990. Dromaeosauridae. In *The Dinosauria*, ed. D. B. Weishampel, P. Dodson, and H. Osmólska, 269–279. Berkeley: University of California Press.

Peet, R. K. 1978. Ecosystem convergence. *American Naturalist* 112:441–459.

Pianka, E. R. 1978. *Evolutionary ecology*. 2nd ed. New York: Harper and Row.

Pianka, E. R., and W. S. Parker. 1975. Ecology of horned lizards: A review with special reference to *Phrynosoma platyrhinos*. *Copeia* 1975:141–162.

Piatigorsky, J. 2007. *Gene sharing and evolution*. Cambridge, Mass.: Harvard University Press.

Poelwijk, F. J., D. J. Kiviet, D. M. Weinreich, and S. J. Tans. 2007. Empirical fitness landscapes reveal accessible evolutionary paths. *Nature* 445:383–386.

Popov, I. Y. 2002. "Periodical systems" in biology (a historical issue). *Verhandlungen zur Geschichte und Theorie der Biologie* 9:55–68.

Prior, H., A. Schwarz, and O. Güntürkün. 2008. Mirror-induced behavior in the magpie (*Pica pica*): Evidence of self-recognition. *PLoS Biology* 6 (8):1642–1650.

Pritchard, G., M. H. McKee, E. M. Pike, G. J. Scrimgeour, and J. Zloty. 1993. Did the first insects live in water or in air? *Biological Journal of the Linnean Society* 49:31–44.

Proske, U., and E. Gregory. 2003. Electrolocation in the platypus: Some speculations. *Comparative Biochemistry and Physiology: Part A, Molecular and Integrative Physiology* 136 (4):821–825.

Protas, M. E., C. Hersey, D. Kochanek, Y. Zhou, H. Wilkens, W. R. Jeffery, L. I. Zon, R. Borowsky, and C. J. Tabin. 2005. Genetic analysis of cavefish reveals molecular convergence in the evolution of albinism. *Nature Genetics* 38(1): 107–111.

Prothero, D. R., and R. M. Schoch. 2002. *Horns, tusks, and flippers: The evolution of hoofed mammals*. Baltimore: Johns Hopkins University Press.

Prud'homme, B., and S. B. Carroll. 2006. Monkey see, monkey do. *Nature Genetics* 38 (7):740–741.

Purnell, M. A. 1995. Large eyes and vision in conodonts. *Lethaia* 28:187–188.

Raafat, R. M., N. Chater, and C. Frith. 2009. Herding in humans. *Trends in Cognitive Sciences* 13 (10):420–428.

Rasmussen, D. T., E. L. Simons, F. Hertel, and A. Judd. 2001. Hindlimb of a giant terrestrial bird from the Upper Eocene, Fayum, Egypt. *Palaeontology* 44 (2):325–337.

Reardon, D., and G. K. Farber. 1995. The structure and evolution of α/β barrel proteins. *Journal of the Federation of American Societies for Experimental Biology* 9:497–503.

Reep, R. L., C. D. Marshall, and M. L. Stoll. 2002. Tactile hairs on the postcranial body in Florida manatees: A mammalian lateral line? *Brain, Behavior and Evolution* 59:141–154.

Reiss, J. O. 2009. *Not by design: Retiring Darwin's watchmaker*. Berkeley: University of California Press.

Rey, M., S. Ohno, J. A. Pintor-Toro, A. Llobell, and T. Benitez. 1998. Unexpected homology between inducible cell wall protein QID74 of filamentous fungi and BR3 salivary protein of the insect *Chironomus*. *Proceedings of the National Academy of Sciences of the United States of America* 95:6212–6216.

Reznick, D. N., M. Mateos, and M. S. Springer. 2002. Independent origins and rapid evolution of the placenta in the fish genus *Poeciliopsis*. *Science* 298:1018–1020.

Rich, T. H., J. A. Hopson, A. M. Musser, T. F. Flannery, and P. Vickers-Rich. 2005. Independent origins of middle ear bones in monotremes and therians. *Science* 307:910–914.

Robert, D., J. Amoroso, and R. R. Hoy. 1992. The evolutionary convergence of hearing in a parasitoid fly and its cricket host. *Science* 258:1135–1137.

Robert, D., and U. Willi. 2000. The histological architecture of the auditory organs in the parasitoid fly *Ormia ochracea*. *Cell and Tissue Research* 301:447–457.

Robinson, P. L. 1956. An unusual saurposid dentition. *Zoological Journal of the Linnean Society* 43:283–293.

Robson, P., G. M. Wright, J. H. Youson, and F. W. Keeley. 2000. The structure and organization of lamprin genes: Multiple-copy genes with alternative splicing and convergent evolution with insect structural proteins. *Molecular Biology and Evolution* 17 (11):1739–1752.

Rodríguez-Trelles, F., R. Tarrio, and F. J. Ayala. 2003. Convergent neofunctionalization by positive Darwinian selection after ancient recurrent duplications of the *xanthine dehydrogenase* gene. *Proceedings of the National Academy of Sciences of the United States of America* 100 (23):13413–13417.

Roff, D. A. 1994. The evolution of flightlessness: Is history important? *Evolutionary Ecology* 8:639–657.

Rokas, A., and S. B. Carroll. 2008. Frequent and widespread parallel evolution of protein sequences. *Molecular Biology and Evolution* 25 (9):1943–1953.

Roots, C. 2006. *Flightless birds*. Westport, Conn.: Greenwood Press.

Rose, R. D., and R. J. Emry. 2005. Extraordinary fossorial adaptations in the Oligocene palaeanodonts *Epoicotherium* and *Xenocranium* (Mammalia). *Journal of Morphology* 175:33–56.

Rowe, T. 1993. Phylogenetic systematics and the early history of mammals. In *Mammal phylogeny: Mesozoic differentiaton, multituberculates, monotremes, early therians, and marsupials*, ed. F. S. Szalay, M. J. Novacek, and M. C. McKenna, 129–145. Berlin: Springer-Verlag.

Ruse, M. 2003. *Darwin and design: Does evolution have a purpose?* Cambridge, Mass.: Harvard University Press.

Sage, R. F. 1999. Why C4 photosynthesis? In *C_4 plant biology*, ed. R. F. Sage and R. K. Monson, 3–16. San Diego: Academic Press.

Sage, R. F. 2001a. C4 plants. In *Encyclopedia of biodiversity*. Vol. 1, ed. S. A. Levin, 575–598. San Diego: Academic Press.

Sage, R. F. 2001b. Environmental and evolutionary preconditions for the origin and diversification of the C_4 photosynthetic syndrome. *Plant Biology* 3:202–213.

Salehi-Ashtiani, K., and J. W. Szostak. 2001. *In vitro* evolution suggests multiple origins of the hammerhead ribozyme. *Nature* 414:82–84.

Salvini-Plawen, L. von. 2008. Photoreception and the polyphyletic evolution of photoreceptors (with special reference to Mollusca). *American Malacological Bulletin* 26 (1/2):83–100.

Salvini-Plawen, L. von, and E. Mayr. 1977. On the evolution of photoreceptors and eyes. *Evolutionary Biology* 10:207–263.

Scantlebury, M., J. R. Speakman, M. K. Oosthuizen, T. J. Roper, and N. C. Bennett. 2006. Energetics reveal physiologically distinct castes in a eusocial mammal. *Nature* 440:795–797.

Schierwater, B., M. Eitel, W. Jakob, H.-J. Osigus, H. Hadrys, S. L. Dellaporta, S.-O. Kolokotronis, and R. DeSalle. 2009. Concatenated analysis sheds light on early metazoan evolution and fuels a modern "urmetazoan" hypothesis. *PLoS Biology* 7 (1):36–44.

Schmitz, A., M. Gebhardt, and H. Schmitz. 2008. Microfluidic photomechanic infrared receptors in a pyrophilous flat bug. *Naturwissenschaften* 95:455–460.

Schneider, E. L., and S. Carlquist. 1995. Vessel origins in Nymphaeaceae: *Euryale* and *Victoria*. *Botanical Journal of the Linnean Society* 119:185–193.

Schultz, T. R., and S. G. Brady. 2008. Major evolutionary transitions in ant agriculture. *Proceedings of the National Academy of Sciences of the United States of America* 105 (14):5435–5440.

Seelig, C. 1956. *Helle Zeit—Dunkle Zeit. In Memoriam Albert Einstein*. Zurich: Europa Verlag.

Sereno, P. 1997. Psittacosauridae. In *Encyclopedia of dinosaurs*, ed. P. J. Currie and K. Padian, 611–613. San Diego: Academic Press.

Sereno, P., Z. Xijin, and T. Lin. 2009. A new psittacosaur from Inner Mongolia and the parrot-like structure and function of the psittacosaur skull. *Philosophical Transactions of the Royal Society of London: Series B, Biological Sciences*, doi:10.1098/rspb.2009.0691.

Shen, Z., and M. Jacobs-Lorena. 1999. Evolution of chitin-binding proteins in invertebrates. *Journal of Molecular Evolution* 48: 341–347.

Shi, Y., and S. Yokoyama. 2003. Molecular analysis of the evolutionary significance of ultraviolet vision in vertebrates. *Proceedings of the National Academy of Sciences of the United States of America* 100 (14):8308–8313.

Shingleton, A. W., and W. A. Foster. 2001. Behaviour, morphology and the division of labour in two soldier-producing aphids. *Animal Behaviour* 62:671–679.

Shubin, N., C. Tabin, and S. Carroll. 2009. Deep homology and the origins of evolutionary novelty. *Nature* 457:818–823.

Sinha, N. R., and E. A. Kellogg. 1996. Parallelism and diversity in multiple origins of C_4 photosynthesis in the grass family. *American Journal of Botany* 83 (11):1458–1470.

Smith, J. D. 2009. The study of animal metacognition. *Trends in Cognitive Sciences* 13 (9):389–396.

Spinney, L. 2007. Back to their roots. *New Scientist* 194 (2608):48–51.

Stadler, B., and A. F. G. Dixon. 2005. Ecology and evolution of aphid-ant interactions. *Annual Review of Ecology Evolution and Systematics* 36:345–372.

Steadman, D. W. 2006. *Extinction and biogeography of tropical Pacific birds*. Chicago: University of Chicago Press.

Stein, W. E., F. Mannolini, L. V. Hernick, E. Landing, and C. M. Berry. 2007. Giant cladoxylopsid trees resolve the enigma of the Earth's earliest forest stumps at Gilboa. *Nature* 446:904–907.

Stensiö, E. 1963. The brain and the cranial nerves in fossil lower craniate vertebrates. *Skrifter utgitt av det Norske Videnskaps-Akademi i Oslo. I. Matematisk-Naturvidenskapelig Klasse* 13:1–120.

Stern, D. L. 2010. *Evolution, development, and the predictable genome*. Greenwood Village, Colo.: Roberts and Company.

Stern, D. L., and V. Orgogozo. 2008. The loci of evolution: How predictable is genetic evolution? *Evolution: International Journal of Organic Evolution* 62 (9):2155–2177.

Stern, D. L., and V. Orgogozo. 2009. Is genetic evolution predictable? *Nature* 323: 746–751.

Stewart, A. 2009. *Wicked plants*. Chapel Hill: Algonquin Books.

Stewart, C. B., J. W. Schilling, and A. C. Wilson. 1987. Adaptive evolution in the stomach lysozymes of foregut fermenters. *Nature* 330:401–404.

Strobel, D. F. 2010. Molecular hydrogen in Titan's atmosphere: Implications of the measured tropospheric and thermospheric mole fractions. *Icarus*, doi:10.1016/j.icarus.2010.03.003.

Sumpter, D. J. T. 2006. The principles of collective animal behaviour. *Philosophical Transactions of the Royal Society of London: Series B, Biological Sciences* B361:5–22.

Szczupider, M. 2009. Visions of Earth: Cameroon Sanaga-Yong chimpanzee rescue center. *National Geographic* 216 (5):12–13.

Terada, T., O. Nureki, R. Ishitani, A. Ambrogelly, M. Ibba, D. Söll, and S. Yokoyama. 2002. Functional convergence of two lysyl-tRNA synthetases with unrelated topologies. *Nature Structural Biology* 9 (4):257–262.

Thien, L. B., P. Bernhardt, M. S. Devall, Z.-D. Chen, Y.-B. Luo, J.-J. Fan, L.-C. Yuan, and J. H. Williams. 2009. Pollination biology of basal angiosperms (ANITA grade). *American Journal of Botany* 96 (1):166–182.

Thomas, R. D. K., and W.-E. Reif. 1993. The skeleton space: A finite set of organic designs. *Evolution: International Journal of Organic Evolution* 47:341–360.

Thompson, D'A. W. 1942. *On growth and form.* Cambridge: Cambridge University Press.

Thompson, M. B., and B. K. Speake. 2006. A review of the evolution of viviparity in lizards: Structure, function and physiology of the placenta. *Journal of Comparative Physiology* 176B:179–189.

Thorne, B. L., and J. F. A. Traniello. 2003. Comparative social biology of basal taxa of ants and termites. *Annual Review of Entomology* 48:283–306.

Tomescu, A. M. F. 2008. Megaphylls, microphylls and the evolution of leaf development. *Trends in Plant Science* 14 (1):5–12.

Tripp, B. C., K. Smith, and J. G. Ferry. 2001. Carbonic anhydrase: New insights for an ancient enzyme. *Journal of Biological Chemistry* 276 (52):48615–48618.

Turner, A. 1997. *The big cats and their fossil relatives*. New York: Columbia University Press.

van Holde, K. E., K. I. Miller, and H. Decker. 2001. Hemocyanins and invertebrate evolution. *Journal of Biological Chemistry* 276 (19):15563–15566.

van Schaik, C. P., E. A. Fox, and A. F. Sitompui. 1996. Manufacture and use of tools in wild Sumatran orangutans. *Naturwissenschaften* 83 (4):186–188.

Van Valen, L. 1973. A new evolutionary theory. *Evolutionary Theory* 1:1–30.

Varricchio, D. J. 1997. Troodontidae. In *Encyclopedia of dinosaurs*, ed. P. J. Currie and K. Padian, 749–754. San Diego: Academic Press.

Vermeij, G. J. 2006. Historical contingency and the purported uniqueness of evolutionary innovations. *Proceedings of the National Academy of Sciences of the United States of America* 103 (6):1804–1809.

von Frisch, K. 1974. *Animal architecture*. New York: Harcourt Brace Jovanovich.

Wade, N. 2006. *Before the dawn: Recovering the lost history of our ancestors*. New York: Penguin Press.

Wake, M. H. 1992. Evolutionary scenarios, homology and convergence of structural specializations for vertebrate viviparity. *American Zoologist* 32:256–263.

Wake, M. H. 1993. Evolution of oviductal gestation in amphibians. *Journal of Experimental Zoology* 266:394–413.

Wald, G., and S. Raypart. 1977. Vision in annelid worms. *Science* 196:1434–1439.

Ward, P. D. 2005. *Life as we do not know it*. New York: Viking Penguin.

Watson, T. 1993. Why some fishes are hotheads. *Science* 260:160–161.

Weinreich, D. M., N. F. Delaney, M. A. DePristo, and D. L. Hartl. 2006. Darwinian evolution can follow only very few mutational paths to fitter proteins. *Science* 312:111–114.

Weishampel, D. B., and J. R. Horner. 1990. Hadrosauridae. In *The Dinosauria*, ed. D. B. Weishampel, P. Dodson, and H. Osmólska, 534–561. Berkeley: University of California Press.

Went, F. W. 1971. Parallel evolution. *Taxon* 20:197–226.

West, G. B., J. H. Brown, and B. J. Enquist. 1999. The fourth dimension of life: Fractal geometry and allometric scaling of organisms. *Science* 284:1677–1679.

Whitehead, H. 2008. Social and cultural evolution in the ocean: Convergences and contrasts with terrestrial systems. In *The deep structure of biology*, ed. S. Conway Morris, 143–160. West Conshohocken, Pa.: Templeton Foundation Press.

Wicher, K. B., and E. Fries. 2007. Convergent evolution of human and bovine haptoglobin: Partial duplication of the genes. *Journal of Molecular Evolution* 65:373–379.

Williams, D. S., and P. McIntyre. 1980. The principal eyes of a jumping spider have a telephoto component. *Nature* 288:578–580.

Wilson, E. O. 1980. *Sociobiology: The abridged edition*. Cambridge, Mass.: Harvard University Press.

Wilson, E. O., and B. Hölldobler. 2005. Eusociality: Origin and consequences. *Proceedings of the National Academy of Sciences of the United States of America* 102 (38): 13367–13371.

Wourms, J. P., and J. Lombardi. 1992. Reflections on the evolution of piscine viviparity. *American Zoologist* 32:276–293.

Wu, X.-C., H.-D. Sues, and A. Sun. 1995. A plant-eating crocodyliform reptile from the Cretaceous of China. *Nature* 376:678–680.

Wyndham, J. 1951. *The day of the triffids*. New York: Doubleday and Company.

Yanoviak, S. P., R. Dudley, and M. Kaspart. 2005. Directed aerial descent in canopy ants. *Nature* 433:624–626.

Yeager, M., and A. L. Hughes. 1999. Evolution of the mammalian MHC: Natural selection, recombination, and convergent evolution. *Immunological Reviews* 167:45–58.

Yee, M. S. Y. 1999. Molecular phylogenies become functional. *Trends in Ecology and Evolution* 14 (5):177–178.

Yokoyama, R., and S. Yokoyama. 1990. Convergent evolution of the red- and green-like visual pigment genes in fish, *Astyanax fasciatus*, and human. *Proceedings of the National Academy of Sciences of the United States of America* 87:9315–9318.

Zakon, H. H., and G. A. Unguez. 1999. Development and regeneration of the electric organ. *Journal of Experimental Biology* 202:1427–1434.

Zhang, J. 2003. Paleomolecular biology unravels the evolutionary mystery of vertebrate UV vision. *Proceedings of the National Academy of Sciences of the United States of America* 100 (14):8045–8047.

Zhang, J. 2006. Parallel adaptive origins of digestive RNases in Asian and African leaf monkeys. *Nature Genetics* 38 (7):819–823.

Zhang, J., and S. Kumar. 1997. Detection of convergent and parallel evolution at the amino acid sequence level. *Molecular Biology and Evolution* 14 (5):527–536.

Zusi, L., and D. Bridge. 1981. On the slit pupil of the Black Skimmer (*Rhynchops niger*). *Journal of Field Ornithology* 52 (4):338–340.

Index of Common Names

Index of Species

Index of Topics